AF805187

Building Sustainable Futures

Mohammad Dastbaz • Ian Strange
Stephen Selkowitz
Editors

Building Sustainable Futures

Design and the Built Environment

Springer

Editors
Mohammad Dastbaz
Faculty of Arts, Enviro & Tech
Leeds Beckett University
Leeds
United Kingdom

Ian Strange
Centre for Urban Dev & Environmental
Mgt
Leeds Beckett University
Leeds
United Kingdom

Stephen Selkowitz
Building Technology and Urban Systems
Lawrence Berkeley National Laboratory
Berkeley
California
USA

ISBN 978-3-319-19347-2 ISBN 978-3-319-19348-9 (eBook)
DOI 10.1007/978-3-319-19348-9

Library of Congress Control Number: 2015947255

Springer Cham Heidelberg New York Dordrecht London
© Springer International Publishing Switzerland 2016
This work is subject to copyright. All rights are reserved by the Publisher, whether the whole or part of the material is concerned, specifically the rights of translation, reprinting, reuse of illustrations, recitation, broadcasting, reproduction on microfilms or in any other physical way, and transmission or information storage and retrieval, electronic adaptation, computer software, or by similar or dissimilar methodology now known or hereafter developed.
The use of general descriptive names, registered names, trademarks, service marks, etc. in this publication does not imply, even in the absence of a specific statement, that such names are exempt from the relevant protective laws and regulations and therefore free for general use.
The publisher, the authors and the editors are safe to assume that the advice and information in this book are believed to be true and accurate at the date of publication. Neither the publisher nor the authors or the editors give a warranty, express or implied, with respect to the material contained herein or for any errors or omissions that may have been made.

Printed on acid-free paper

Springer International Publishing AG Switzerland is part of Springer Science+Business Media (www.springer.com)

Acknowledgments

We would like to thank all the contributing authors for their tireless efforts in providing us with their valuable research that has made this book possible. We also would like to thank Tiffany Gasbarrini and the Springer team for working closely with us and providing advice and support when needed.

Special thanks to Carole Smith for her constant and valuable administrative support, for being patient, sending numerous e-mails, as deadlines loomed, and for gathering all the information and material required getting the book to the production stage.

We would like to thank all the colleagues at Leeds Sustainability Institute (LSI) at Leeds Beckett University, who provided us with their support and research material and helped shape the structure of the book.

Mohammad Dastbaz
Ian Strange
Stephen Selkowitz

Contents

Contributors

Dr. Quintin Bradley School of Built Environment and Engineering, Leeds Beckett University, Leeds, UK

Matthew Brooke-Peat Leeds Sustainability Institute, Leeds Beckett University, Leeds, UK

Dr. Azad Camyab Pearlstone Energy Limited, London, UK

Prof. Mohammad Dastbaz Leeds Beckett University, Leeds, UK

Dr. Aitor Erkoreka ENEDI Research Group, University of the Basque Country (UPV/EHU), ETSI Bilbao, Alameda Urquijo, Spain

David Farmer Leeds Sustainability Institute, Leeds Beckett University, Leeds, UK

Martin Fletcher Leeds Sustainability Institute, Leeds Beckett University, Leeds, UK

Dr. Fiona Fylan Leeds Sustainability Institute, Leeds Beckett University, Leeds, UK

Prof. Parisa Ghodous Computer Science Department (Département Informatique), University Claude Bernard Lyon 1, Lyon, France

Dr. David Glew Leeds Sustainability Institute, Leeds Beckett University, Leeds, UK

Prof. Christopher Gorse Leeds Sustainability Institute, Leeds Beckett University, Leeds, UK

Dr. David Haigh School of Built Environment and Engineering, Leeds Beckett University, Leeds, UK

Dr. David Johnston Leeds Sustainability Institute, Leeds Beckett University, Leeds, UK

Prof. Farzad Khosrowshahi School of Built Environment and Engineering, Leeds Beckett University, Leeds, UK

Dr. Silvija Krajter Ostoić Leeds Sustainability Institute, Leeds Beckett University, Leeds, UK

Dr. Jamileh Manoochehri Leicester School of Architecture, De Montfort University, The Gateway, Leicester, UK

Dominic Miles Shenton Leeds Sustainability Institute, Leeds Beckett University, Leeds, UK

Guy Nevill Max Fordham LLP, London, UK

James Parker Leeds Sustainability Institute, Leeds Beckett University, Leeds, UK

Dr. Andrew Platten Faculty of Arts, Environment and Technology, Leeds Beckett University, Leeds, UK

Prof. Alan Simson Leeds Sustainability Institute, Leeds Beckett University, Leeds, UK

Dr. Lindsay Smales School of Built Environment and Engineering, Leeds Beckett University, Leeds, LS2, UK

Melanie Smith Leeds Sustainability Institute, Leeds Beckett University, Leeds, UK

Dr. Albert T. P. So City University of Hong Kong, Hong Kong, China

Dr. Anne Stafford Leeds Sustainability Institute, Leeds Beckett University, Leeds, UK

Prof. Ian Strange School of Built Environment and Engineering, Leeds Beckett University, Leeds, UK

Dr. Maria Theodorou Leeds School of Architecture, Leeds Beckett University, LEEDS, UK

Felix Thomas Leeds Sustainability Institute, Leeds Beckett University, Leeds, UK

Prof. David Turnbull Leeds School of Architecture, Leeds Beckett University, LEEDS, UK

Jim W. Ure ABS consulting (UK) Limited, London, UK

Pam Warhurst CBE, School of Built Environment and Engineering, Leeds Beckett University, Leeds, LS2, UK

Conrad T. C. Wong Yau Lee Holdings Limited, Hong Kong, China

Jo Wright FCB Studios, Bath, UK

Part I
Sustainable Construction and Measurement

Chapter 1
Building Sustainable Futures: An Ever Changing Policy Agenda

Mohammad Dastbaz and Ian Strange

1.1 Introduction

The rapid growth of population in the twentieth century and its continuation in the twenty-first century (pushing the world population to over 7 billion), together with ever-increasing demands on our planet's dwindling natural resources, has created a crisis of enormous magnitude that can no longer be denied. Numerous global initiatives led by the United Nations (UN) and other international and national agencies, aimed at the growing impact of environmental damage on every aspect of our lives, have created a sense of urgency to act (and to act now) before it is too late.

The UN "Climate Summit" in September 2014, where 120 world leaders attended, started its deliberations with the assumption that "climate change" is not a problem for tomorrow but is here today, and is the one which is disrupting our lives and every aspect of our development. Sustainable use of our natural resources, sustainable production, use of low carbon and renewable energy sources, sustainable transport and a planned approach to the controlled urbanisation were among some of the key challenges discussed at the summit.

Perhaps the most worrying and disappointing aspect of the 2014 summit was to note that despite early initiatives and declarations signed by world leaders, little progress had been made to halt the tide of further large-scale environmental damage to our planet in advanced industrial as well as developing countries. It is also worth mentioning that it was almost exactly a decade ago that the International Conference on Population and Development (ICPD), held in Cairo, Egypt in 1994, placed humans at the centre of development and stated that:

M. Dastbaz (✉)
Faculty of Arts, Enviro & Tech, Leeds Beckett University,
Broadcasting Place, LS2 9EN Leeds, UK
e-mail: m.dastbaz@leedsbeckett.ac.uk

I. Strange
School of Built Environment and Engineering, Leeds Beckett University,
Northern Terrace, LS2 8AG Leeds, UK
e-mail: i.strange@leedsbeckett.ac.uk

© Springer International Publishing Switzerland 2016

M. Dastbaz et al. (eds.), *Building Sustainable Futures*,
DOI 10.1007/978-3-319-19348-9_1

> Sustainable development as a means to ensure human well-being, equitably shared by all people today and in the future, requires that the interrelationships between population, resources, the environment and development should be fully recognized, appropriately managed and brought into harmonious, dynamic balance. To achieve sustainable development and a higher quality of life for all people, States should reduce and eliminate unsustainable patterns of production and consumption and promote appropriate policies, including population related policies, in order to meet the needs of current generations without compromising the ability of future generations to meet their own needs.
> ICPD Programme of Action, Chap. II, Principle 6 (UN 1994)[1].

1.2 Historical Context

A large portion of the existing literature seems to suggest that the idea of "sustainable development" first surfaced in the early part of the twentieth century and was put on various national and international agendas following the "Earth Day" held on 22nd April in 1970, where millions gathered (organised by G. Nelson in the USA), and then the first international conference on "the human environment" held in Stockholm, Sweden in 1972.

A more comprehensive exploration of historical events suggests that as early as the fourteenth century, processes were being put in place to safeguard our environment. Among these, one can name the English Parliament Sanitary Act, passed in 1388, which prohibited the throwing of filth and garbage into waters and rivers. However, it is in the nineteenth century that the majority of our contemporary sustainability challenges have their origin. The dawn of the Industrial Revolution and the growth of machine-based industries changed the face of our planet for good. The replacement of the farming/cottage-type production industry with large factories changed our world, and signalled the beginnings of monumental change, innovation and immense scientific discoveries. The Industrial Revolution also fundamentally changed Earth's ecology and humans' relationship with their environment. One of the most immediate and drastic repercussions of the Industrial Revolution was the explosive growth of the world's population.[2] According to Eric McLamb, with the dawn of the Industrial Revolution in the mid-1700s, the world's population grew by about 57 % to 700 million, would reach 1 billion in 1800, and within another 100 years it would finally grow to around 1.6 billion.

A 100 years later, the human population would finally surpass the 6 billion mark.[3] This phenomenal growth in population put enormous pressure on the planet, forcing it to cope with a continuously expanding deficit of resources.

It is clear from Fig. 1.1 that between the 1950s and 2000, the planet's population doubled in size, and it is also important to state that at this stage for the first time in our history, there are more people living in the cities and urban areas than outside the cities.

---

[1] UN Publication. (2012)

[2] Dastbaz et al. 2015

[3] McLamb 2011

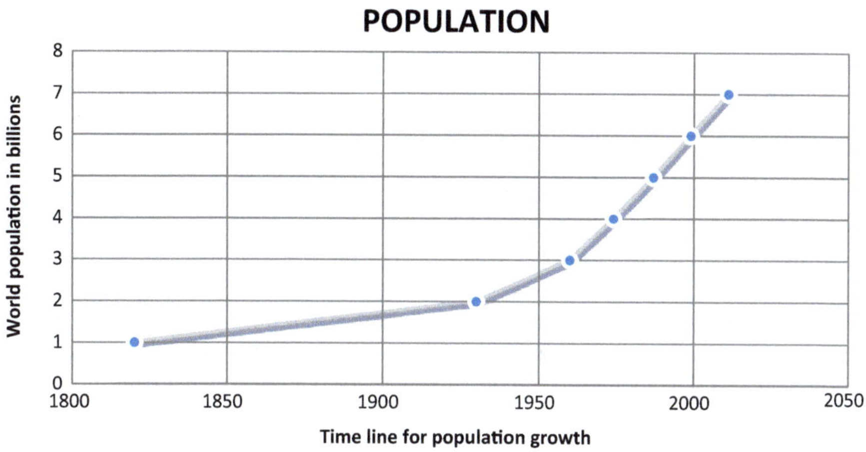

Fig. 1.1 Population growth timeline

In a report by Stuart Hart, published in the Harvard Review in 1997[4], it is stated that the total "environmental burden" (*EB*) emerging from our impact on the planet is the function of three key factors. These are: population (*P*), affluence (*A*—proxy for consumption) and technology (*T*—proxy for how wealth is generated). According to this report, the total *EB* can then be expressed as:

$$EB = P \times A \times T$$

1.3 Sustainable Development

In 1983, the UN set up the "World Commission on Environment and Development" (WCED) headed by Harlem Brundtland, the Prime Minister of Norway. The aim of this commission (that came to be known as the "Brundtland Commission") was to focus on looking into environmental and developmental problems and to suggest possible solutions to the UN and its members. In 1987, the Brundtland Commission published its first major report titled: "Our Common Future", which had a major impact on the "sustainable development" debate and significantly influenced the Earth Summit in Brazil in 1992 and the third UN Conference on Environment and Development in South Africa in 2002.

"Our Common Future"—Chap. 2[5], "Towards Sustainable Development", provides a comprehensive and insightful debate covering the key aspects of sustainable development and the key challenges facing human society. The report stated:

[4] Hart 1997

[5] Bruntland Commission. (1987)

> The satisfaction of human needs and aspirations in the major objective of development. The jobs—are not being met, and beyond their basic needs these people have legitimate aspirations for an improved quality of life. A world in which poverty and inequity are endemic will always be prone to ecological and other crises. Sustainable development requires meeting the basic needs of all and extending to all the opportunity to satisfy their aspirations for a better life…

Following the Brundtland Commission report, the "Earth Summit" in Rio in 1992 was viewed as a turning point in the debate about sustainable development. For the first time since the end of World War II, over 100 heads of state and government together with delegations from over 170 countries attended the "Earth Summit" to chart the course of human sustainable development into the next century. As part of this summit, world leaders discussed and signed an international convention on "environment and development" and an "agenda" for the twenty-first century (which was later called Agenda 21). The secretary general of the United Nations Conference on Environment and Development (UNCED), Maurice Strong, summarised Agenda 21 as, a "program of action for a sustainable future for the human family and a first step towards ensuring that the world will become a more just, secure and prosperous habitat for all humanity".[6]

1.4 Current Agenda

The "Millennium Summit" of the United Nations in 2000 established a number of "Millennium Development Goals" (MDGs) which highlighted areas of concern and the agenda for developing the sustainability debate across the world. Eight MDGs were identified which are listed below:

- Eradicate poverty and hunger (MDG1)
- Achieve universal primary education (MDG2)
- Gender equality and empowering women (MDG3)
- Reduce child mortality (MDG4)
- Improve maternal health (MDG5)
- Combat HIV/AIDS, malaria and other diseases (MDG6)
- Environmental sustainability (MDG7)
- Global partnership and development (MDG8)

The main issue of focus for this book of course is MDG7, which is "environmental sustainability" with particular attention on sustainable design and the built environment.

One of the challenges identified in the literature dealing with "sustainable design and built environment" is the different viewpoints and approaches between industry, business and environmental campaigners and researchers and academia, and how to bridge the gap between the differences and, more importantly, how to

[6] For more information see http://yosemite.epa.gov/r10/oi.nsf/Sustainability/History.

tackle the issues facing our environment. In a report published by Price Waterhouse Cooper titled: "The Sustainability Agenda: The Industry Perspective", the authors point out that:

> Many in business have long viewed sustainability—efforts to avert climate change, for example, or to improve education in underserved neighbourhoods—a matter of corporate philanthropy, with no relevance to their corporations' core strategies. The costs of such activities were seen as detracting from profitability and accounted for on a public relations line under marketing; their scope was limited but they were promoted with considerable fanfare.[7]

The report goes on to suggest that there is a clear shift in business attitudes, with responsiveness and sensitivity to environmental requirements gaining ascendency and becoming the dominant discourse amongst large corporations. Sceptics would point out that, to date, most large-scale environmental damage is caused and continues to be caused by big corporations (one could name: the Exxon Valdez oil spill that occurred in Prince William Sound, Alaska in 1987; or the BP Oil Spill, where more than 200 million gallons of crude oil was pumped into the Gulf of Mexico for a total of 87 days in 2010) or by reckless states policies across the world. In our view, this remains one of the greatest challenges facing our society and sustainable development. One would hope that the growing worldwide awareness and public opinion (supported by substantial research-informed evidence and suggested alternatives) would pave the way for a different course of development.

The substantial and growing body of literature dealing with sustainable development and the built environment points to a number of critical factors that need serious consideration. These are:

- Population growth
- Urbanisation and poverty
- Pollution and the challenge of developing renewable and sustainable energy
- Resources—(land and water availability and use)

The chapters in this book have addressed some of these key challenges. In particular, they have illustrated the variety of interpretations and views on what constitutes sustainable development between industry, business, government, environmental campaigners and community groups, and the complex nature of navigating through these interpretations. As mentioned already, the literature on sustainability and sustainable development is a large and ever-growing one. At one level, it is a literature that is steeped in representations of how we globally construct collective ways to understand sustainability, producing the knowledge and belief systems that underpin global response to global threats and challenges[8]. This is also about how, for example the UN works to establish geopolitical commitments to realising sustainable development, or how European policy and sustainable development indicators are embedded in the political and policy practices of the European Union, or how na-

[7] See: www.pwc.com/co/es/responsabilidad-corporativa/assets/sust-agenda.pdf published 2008.

[8] Bruntland Commission. (1987)

tion states implement sustainable development policy across internal departmental and policy sectors.[9]

Despite this extensive literature, with its numerous policy documents and statements at different scales, at international, national and sub-national levels, the voluminous amounts of words and text speaks to continuing shortfalls and weaknesses in how we put into practice the principles of sustainable development, as well as the superficiality of the ways we seek to integrate economic, social and environmental agendas over time. In terms of economic impacts, such as, for example fluctuations in growth and/or recessionary trends, persistence of cycles of economic boom and bust and continuing levels of high unemployment or variations or imbalances between economic activities among regions,[10] it is clear is that in neoliberalist economies there are disparities and inequalities across all the dimensions of sustainability, which are likely to continue over the long run. Similarly, the societal impact of this is difficult to address, with much of the literature pointing to a continuing problem in addressing poverty and/or creating stronger prospects for social inclusion.[11] Moreover, the ways in which we are currently addressing environmental sustainability provide further evidence of the challenges facing us, and that significant commitment and substantial change is needed to avoid future economic, social and environmental catastrophes[12].

It is clear that tackling sustainable development at whatever level is to be a long haul. The transitory and dynamic nature of the dimensions of sustainability almost without question ensures that both the practice and process of becoming more sustainable (let alone achieving some level of sustainable development) is a slow, complex and incremental task. The United Nations World Economic and Social Survey[13] report is instructive here. It reminds us that despite progress in reaching world poverty reduction where there were 700 million fewer people living in conditions of extreme poverty in 2010 compared with 1990, there is still much to be done to achieve the UN MDGs. The world continues to be faced with major challenges across the economic, social and environmental dimensions of sustainability. Despite the actions to address some of these issues (for example the implementation of Agenda 21 and the UN Conference on Sustainable Development 1993) major issues of global poverty, inequality and unsustainable consumption and production patterns persist. The 2013 UN World Economic and Social Survey makes it clear that a "business as usual" approach will not work as we continue to face challenges around, which are as follows:

- Climate change and the need to develop more integrated approaches to the management of natural resources and ecosystems;
- Hunger, malnourishment and food security issues in many countries (both developing and developed);

[9] European Commission (2010); Department for Environment (2011); United Nations (2012)

[10] Aiginger et al.(2011)

[11] MacInnes et al. (2014)

[12] United Nations. (2011)

[13] United Nations World Economic and Social Survey. (2013)

- Continuing and rapid urbanisation (particularly in the developing countries);
- Income inequality and the threat of social conflict;
- Increasing energy demands that are likely to remain unmet for millions of people (unless there are improvements in access to energy services;
- Instability in global financial markets and the threat of financial crisis leading to difficulties in financing policies and actions for greater sustainable development.

As outlined earlier in this chapter, these challenges have been informed and influenced by broader trends that have had a major global impact, for example shifting demographic characteristics and trends across the world, continuing environmental deterioration, resource depletion and significant advances in new technology. As the work of the UN highlights, more integrated understanding of these trends and how they link together to have varying economic, social and environmental impacts on particular places is necessary for more effective policy responses to be formulated.[14]

1.5 Buildings, Technology and Design

Chapters in Sect. 1.1 of this book are dedicated to demonstrating that the key to achieving any degree of success in making sustainable buildings is how we use technology, and the ways technology is used innovatively in the construction process. Such technological innovation is key to the transformative responses we need in relation to how we tackle, for example environmental pollution and energy consumption. As the UN Word Economic and Social Survey states:

> Changes in consumption patterns can drive the creation of new technologies necessary for sustainability and their adoption and diffusion at the desired pace. Success in bringing about these changes will require substantial reorganization of the economy and society and changes in lifestyles. Economic and financial incentives for the creation and adoption of new technologies will be needed which may include innovative policy reforms.[15]

At a global level, changes in the way we use energy will be a key feature of the continuing drive towards environmental sustainability. Emissions trends point towards only limited success in reducing greenhouse gases by 2050, even with greater use of renewable energy sources and more efficient use of energy. Similarly, while some of our cities may be less polluted now than they otherwise would have been if no actions had been taken, global targets for reductions in CO_2 will be difficult to meet. The examples in this book show that there are a number of ways to reduce the energy consumption of buildings in their construction and in the use and occupancy of a building. A key area of challenge facing us is the energy performance of buildings. The chapters by Gorse et al. demonstrate that what makes a building sustainable is difficult to define and is highly dependent on the approaches to sustainable construction adopted at the outset of a project. In Chap. 2, Gorse et al. highlight the key attributes of what makes a building sustainable, but more importantly shows

[14] United Nations. (2012)

[15] UN United Nations World Economic and Social Survey. (2013)

how sustainable goals need to be determined at an early stage in the development of a project to give the best possible opportunity for sustainability goals and targets to be met. In its explanation of the surveying and measuring of the thermal properties of buildings, Chap. 3 by Gorse et al. points to the significance of how the energy performance of buildings varies and that a building's differing patterns of energy consumption is a complex process and necessitating closer measurement and assessment if buildings are to become more sustainable. By monitoring the building performance in the field as it were, their results point to key insights that will inform future iterations of building design in a move towards greater energy efficiency. Indeed, many of the chapters in this section are illustrative of the following:

> …sustainable development pathways share common features. First, the sooner the implementation of policies starts, the greater the technological flexibility and the less costly the actions required. Second, policies increasing efficiency in the delivery of energy services can go a long way. Perhaps the most important insight provided by scenario analyses is that the world can go a long way towards controlling emissions, if it invests decisively in energy efficiency.[16]

Such a transition to better uses of energy and to a low carbon future requires policy environments that enable change and that creates *policy space and coherence; international financing; international cooperation; and enabling international institutions, establishing rules and norms.*[17]

The technology options outlined throughout the chapters in Sect. 1.1 show that there are potential paths to reductions in energy consumption that the construction sector can offer. However, what they also highlight is that such paths require boldness, ambition, cooperation between stakeholders, financial investment and, crucially, changes in the behaviour of those who build and use the spaces in which we live and work. In particular, the chapter by Wong, So and Platten reminds us that the sustainability agenda does not solely relate to fixing the energy demands and performance of buildings. Other concerns such as water conservation and the use of green infrastructure in the construction of the building are also key to a building's sustainability credentials. Equally, their analysis is a useful reminder that sustainable construction objectives are not simply a result of following building codes or new design methodologies. What is equally important is how projects are led and managed (by client and developer) if more sustainable outcomes are to flow through the construction process. In their discussion of a key development in Hong Kong, it is clear that developer's objectives to maximise the building's green credentials was a key reason for the use of building information modelling (BIM) from the beginning of the construction process. Coupled with features such as modular off-site production of standard components, time was saved along with the elimination of waste in the building process. The example used by Wong, So and Platten also shows that a development in a densely populated and complex setting such as Hong Kong can produce buildings constructed to strong environmental standards. What this suggests is that the future moves to the zero carbon building could be

[16] United Nations World Economic and Social Survey. (2013)

[17] United Nations World Economic and Social Survey. (2013)

focussed around developing the performance of the building fabric and the use of more passive technologies.

Section 1.2 of the book is dedicated to a number of world class case studies and exemplars that demonstrate that for cities and urban areas, cooperation between different layers and levels of interest is necessary for effective approaches to sustainability to be developed. The examples used in the chapters' are suggestive of how more democratic and plural approaches to policy offer the best opportunities for stronger and more effective policy and sustainable development outcomes. Integration and coordination across a range of areas such as land use, infrastructure development, food security and the provision of housing (which the chapters highlight) are crucial for greater urban sustainability requiring the cooperation and co-working of public, private and voluntary sectors in the effective management of urban development. Equally, the chapters make it clear that a singular approach to responding to the local impacts of global sustainability challenges does not work, and that more nuanced approaches that take account of the priorities, policies and local contexts of towns and cities is what is more likely to produce locally effective responses to global problems. As the 2013 United Nations World Economic and Social Survey reminds us: ... *measures of sustainable development progress also need to be tailored to the particular challenges and opportunities identified and prioritized by the cities' main stakeholders.*[18]

The chapter by Bradley and Haigh focusses on the problematic nature of the representation of what constitutes sustainable communities providing a clear demonstration of how the notion of "community" in the sustainability debate has shifted to accommodate particular practices of sustainability. They provide a powerful argument that the idea of the sustainable community has been used as a way to further exacerbate inequality and social division rather than address and balance economic, social and environmental objectives. Yet it also demonstrates how the idea of the sustainable community can be seen as a form of localised politics infused with redistributive justice, but one offering an ambiguous and contradictory model for a reconfiguration of the welfare state. This might be seen in the movement toward neighbourhood planning that they describe in their chapter, a movement which could offer a way to work at a more local/community and people-oriented level. Rather than simply seeing localism as yet another policy programme that privileges planners and planning and that pushes a political preference for growth and denies a space for environmental and social agendas, it is possible to see this as another opportunity to challenge planning's assumptions and orthodoxies (as was tried in the 1970s, but which was overrun in the long wave of neo-liberalism from the 1980s onwards). Even if we do not agree with the ideological drive of the variant of localism outlined by the sustainable communities agenda, or if we see the turn to neighbourhood planning considered in Bradley and Haigh's chapter as just another example of an ill thought out policy, the idea of a more locally accountable planning practice is a seductive and attractive one, which it seems, will only assume greater importance and significance—and not just in the UK.

[18] United Nations World Economic and Social Survey. (2013)

1.6 Focus on Design

A focus on design is key to making places better and to improving place sustainability. The chapter by Smales and Warhurst illustrates this through its focus on how key design principles can be used to help local communities develop a way to think (and rethink) what constitutes sustainability for their local areas and neighbourhoods. While the examples used in the chapter are local, the implications of how to create and deliver better places through (such examples as the Incredible Edible Todmorden (IET) project) have much wider applicability. The bottom up approach of the IET project (in its attempt to tackle local food security issues) has much to tell us about local action through local leadership and how local action can work to address key sustainability issues such as climate change and developing greater inclusivity. Such action reminds us that while we are all subject to broader patterns of change which often feel (and indeed are) beyond our individual control, there is still space for individual and community action to offer alternative responses to dealing with that change, and in so doing, help to create the context within which to help shape the places and spaces that we inhabit and use on a daily basis.

At a scale that moves beyond the local, the discussion by Simson and Ostoic shows us that urban change is a constant feature of contemporary life, and moreover that the pace of urban change is only likely to quicken as we move through the twenty-first century. Adjusting to this feature of our lives (or rather readjusting to continuous urban change) will be crucial to the creation and maintenance of sustainable urban environments. Here, Simson and Ostoic make a strong case that does not just call for urban densification as a way to create more multifunctional and liveable and open space, as all too often this approach has not been as successful as had been hoped. Equally, their analysis draws our attention to the sometimes limited impact of urban design in *not* creating more sustainable spaces (compare for example with the arguments of Smales and Warhurst in this volume). Perhaps their most important point is that the role of urban design and the creation of green infrastructure in cities is at a key moment, or turning point, where hard choices have to be made over whether our current approaches to the physical greening of cities is to be merely cosmetic or is to forge a new alliance between built form, grey, green and blue infrastructure that will establish a hybrid, more creative, approach to sustainable urbanism.

1.7 A Final Note

In a message to the 2012 Conference of the Association of European Schools of Planning (AESOP) the geographer and social theorist David Harvey issued a warning to planners (and by implication others in the built environment professions) that we were entering the "the end game of urbanisation" which would present those charged with making urban spaces better with three key challenges requiring a radical rethink about what type of urban environments we might want to live in. The first of these was the need to tackle global inequality and poverty. The second was the threat of environmental degradation and environmental change making planners have to rethink what urbanisation is about and focussing their attention on the qualities of land, air and water—encouraging them to think about this in imaginative ways. And, the third challenge was the need to rethink the growth complex of capitalism, urging us to move towards growth that is both environmentally beneficial as well as being socially just. All of this of course requires new ways of urban living, combining new ways of production and consumption. In small, but significant ways, the issues discussed by the authors in this book have in many ways responded to that call, and more importantly offered both socially informed and technically literate responses to the global and local challenge of working to make the place and spaces we inhabit more sustainable.

References

Aiginger K, Barenthaker-Sieber S, Vogel J (2011) Competitiveness under new perspectives. Working Paper No. 44. Welfare Wealth Work for Europe, European Commission

Bruntland Commission (1987) Our common future. World commission on environment and development report. http://www.un-documents.net/ocf-02.htm#I. Accessed 24 Feb 2015

Dastbaz M, Pattinson C, Aghkar B (2015) Green information and communication technologies—a sustainable approach. Morgan Koffman, Burlington

Department for Environment (2011) Mainstreaming sustainable development: the government's vision and what this means in practice. Department for Environment, Food and Rural Affairs

European Commission (2010) Communication from the Commission, Europe 2020, A strategy for smart, sustainable and inclusive growth, European Commission, Brussels. http://ec.euro,pa.eu/europe2020/index_en.htm. Accessed 24 Feb 2010

Hart SL (1997) Beyond greening: strategies for a sustainable world. Harv Bus Rev 75:66–76

MacInnes T, Aldridge H, Bushe S, Tinson A, Born TB (2014) Monitoring poverty and social exclusion 2014. Joseph Rowntree Foundation, New Policy Institute, pp 6–8

McLamb E (2011) Ecological impact of industrial revolution. http://www.ecology.com/2011/09/18/ecological-impact-industrial-revolution/. Accessed 27 Feb 2015

United Nations Population Fund (2012) Population matters for sustainable development. http://www.unfpa.org/public/cache/offonce/home/publications

UN United Nations World Economic and Social Survey (2013) DESA. p viii

United Nations (2011) Cities and climate change: global report on human settlements

United Nations (2012) Report on the United Nations Conference on Sustainable Development, United Nations, New York. http://www.uncsd2012.org/content/documents/814UNCSD%20REPORT%20final%20revs.pdf. Accessed 27 Feb 2015

United Nations World Economic and Social Survey (2013) DESA 2013. p vi

United Nations World Economic and Social Survey (2013) DESA 2013. P xi

United Nations World Economic and Social Survey (2013) DESA. p xiv

United Nations World Economic and Social Survey (2013) DESA. p xv

Chapter 2
Surveying and Measuring the Thermal Properties of Buildings

Christopher Gorse, Melanie Smith, David Glew, Felix Thomas, Dominic Miles Shenton and David Farmer

The energy performance of buildings and the ability to accurately predict energy demand is of global importance. As the relative cost and environmental impact of harnessing energy increases so does our need for energy efficiency. Designing, constructing and retrofitting buildings to be more energy efficient requires a thorough understanding of the way each building behaves and responds to its climatic variations. Although the measurement of a building's energy consumption is straightforward, understanding why consumption differs from that expected requires a detailed and systematic building performance analysis. The way a building is assembled and retrofitted affects performance, and thus each aspect of a building's makeup should be measured or monitored to understand its behaviour. When attempting to understand the performance of a building, it is important to consider each element, the components used and the way that they interface to perform as a whole. The measurement of building components in the laboratory is relatively well documented but the testing and measuring of buildings once constructed in the field is an emerging science. This chapter presents the methods used to survey, measure and monitor building performance in the field and how the work is being used to inform the next generation of energy-efficient buildings (Fig. 2.1).

2.1 Energy Performance Directive: The Need for Measurement and Control

The built environment is a major energy user. Across the EU, buildings account for approximately 40% of total energy used, currently consuming more energy than all forms of transportation combined. As power sources change and become more

C. Gorse (✉) · M. Smith · D. Glew · F. Thomas · D. M. Shenton · D. Farmer
Leeds Sustainability Institute, Leeds Beckett University, Broadcasting Place, Leeds LS2 9EN, UK
e-mail: c.gorse@leedsbeckett.ac.uk

© Springer International Publishing Switzerland 2016
M. Dastbaz et al. (eds.), *Building Sustainable Futures*,
DOI 10.1007/978-3-319-19348-9_2

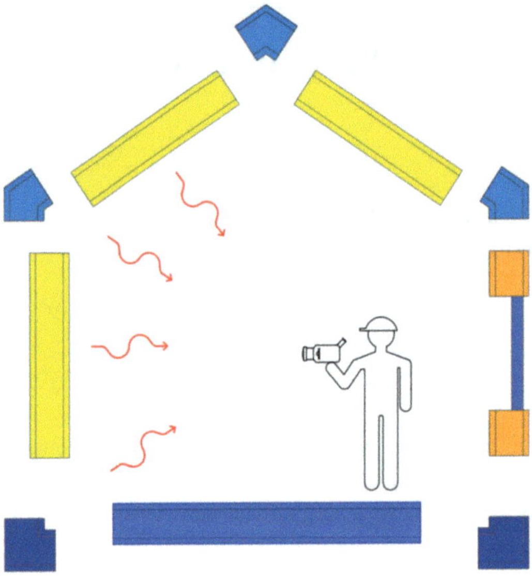

Fig. 2.1 Surveying to understand thermal properties

expensive, this statistic is increasingly unacceptable and the need to reduce and control energy demand is crucial.

The majority of the energy spent in buildings is used for space heating and conditioning, totalling 70 % of energy consumed. To tackle the problem, the European Union has enacted regulations such as the EPBD (2010) and the EED (2012) to promote energy efficiency in the built environment and achieve 20 % of its primary energy saving target by 2020.

One of the challenges to deliver the energy-saving targets is to ensure that the energy efficiency actions taken in the built environment affect real-world energy consumption. Unfortunately, recent studies (Bell et al. 2010; Bordass et al. 2001; Carbon Trust 2011; Galvin 2014; Gorse et al. 2011, 2012a; Johnston et al. 2014) indicate that large 'gaps' exist between the theoretical and actual performance of buildings, regardless of the level of energy-efficient measures adopted. Furthermore, the studies of retrofit projects have demonstrated a similar deviation between that of the intended and as-built performance (Miles-Shenton et al. 2011; Stafford et al. 2012a, 2014).

2.2 Factors Contributing to Performance Variance

Variables that affect building energy performance arise from laboratory predictions, simulations and designs that do not account for real-world environmental conditions. The building exposure, user operation, surrounding buildings and vegetation will all contribute to differences in real-world performance. In addition, variances in workmanship and build quality often go undetected during commissioning processes and add to the discrepancy. Once built, a lack of user understanding and inadequate energy feedback to the user or facility manager result in substandard operation and maintenance. All of these factors result in actual energy performance and operation being significantly different from predictions. Optimal performance of buildings will only be achieved where information is fed back to the user and all of those involved in design, construction and operation.

In order to ensure real performance, it is essential that the energy-efficient technologies used in buildings do more than simply satisfy regulations based on design, theory and laboratory testing. Genuine, measurable efficiency improvements in real-world environments must be achieved.

Unfortunately, many current claims of thermal efficiency for buildings and building products are based on 'nominal' assumptions, theoretical models and extrapolations from laboratory test data; the result of which often do not correspond to real-world performance. The in-situ data currently available regarding real-world performance of energy-efficient building technologies are sparse and limited to only a few research institutes. Yet, these data universally suggest that there is a significant discrepancy between the predicted and the measured real-world performance.

Without real-world measured data, it is impossible to guarantee energy performance claims and to design and implement effective programmes and practices that can result in the intended impact on the energy performance of buildings. The first step towards ensuring and validating the effectiveness of efficient building programmes, therefore, is to measure, identify, understand and address the gap (Fig. 2.2).

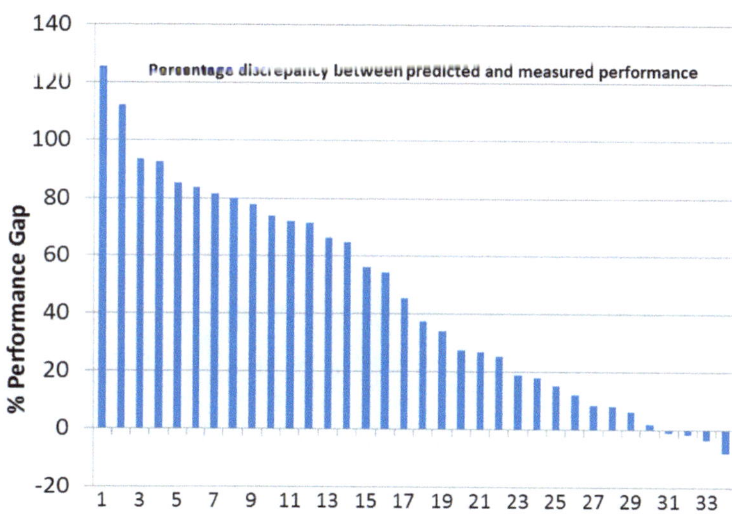

Fig. 2.2 Percentage discrepancy between measured and predicted heat loss in buildings tested by the LSI. Most buildings showing a greater percentage of heat loss than designed. (Stafford et al. 2012b; Gorse et al. 2012b)

The work published in the UK revealed very wide variations in expected fabric performance for both new and retrofit properties (see Fig. 2.2). This variance was found regardless of whether the buildings were of standard or built to enhance their thermal capabilities. With regard to some retrofit properties, buildings were found to exceed their expected performance, and this was due to a failure to obtain reliable existing building data. With the existing buildings, where the materials and fabric are unknown, assumptions are made with regard to the materials used and their thermal properties. Where the information is inadequate, variation in real performance can be better or worse, but only because inadequate information is fed into the initial assumptions and thermal models.

A more recent evaluation of the new buildings studied also found that in the majority of cases, the difference between the measured and the predicted fabric thermal performance was considerable, with the measured fabric performance of the dwellings, being on average, 50% greater than that predicted (Fig. 2.3). The results are obtained from whole building heat loss tests, conducted in unoccupied dwellings. Despite this, the results obtained from three of the dwellings (those on the far right of Fig. 2.3) suggest that it is possible to construct dwellings that perform very close to that predicted. All three of these dwellings were Passivhaus certified, designed to have a very low predicted heat loss. Although the variance in all the buildings is of concern, the results from the nearly zero carbon buildings are encouraging. With the development of tests and monitoring methods, advances in building surveys that are able to demonstrate actual building performance and feedback loops through smart monitoring systems, there is no reason why the positive results cannot become the norm.

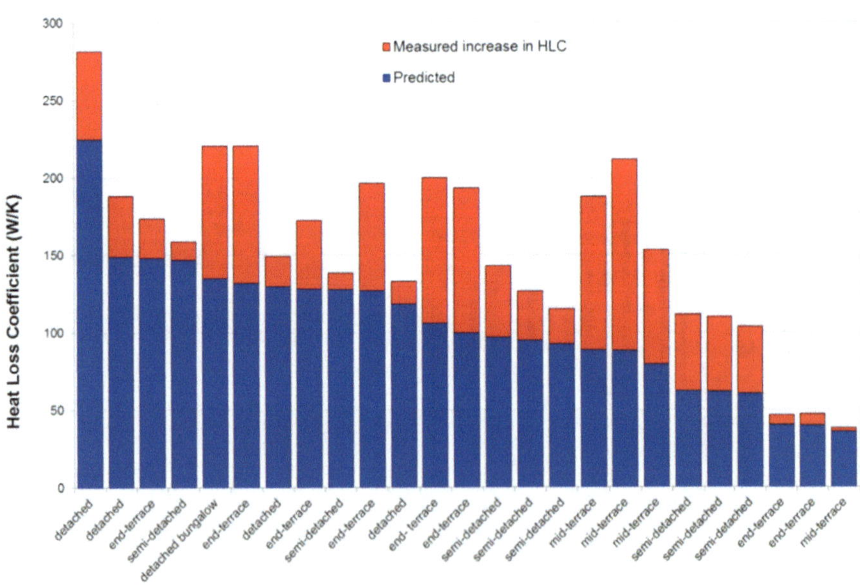

Fig. 2.3 Predicted versus measured heat loss, new and retrofit buildings in the UK tested by the LSI. (Johnston et al. 2014)

The solution to the problem requires more than the mere recognition of its existence. To be useful, such recognition must lead to an effective data collection, analysis, study and modelling, which in turn can influence design and implementation, ultimately leading to the reduction of the performance gap.

The energy performance of any building depends on an interactive system consisting of four main elements: fabric, services, behaviour and external environmental conditions. All of the fixed and fitted components can be built to specification and controlled. If not performing correctly, they can be changed, modified or updated. The fixed fabric and services should be relatively easy to control. The occupant behaviour cannot be controlled but can be influenced by the information that they receive on the energy efficiency and buildings energy use, by financial incentives and by regulation. As energy demand increases, there is a greater need to ensure that buildings are used effectively and energy use is controlled. There are many incentives in place to encourage more effective use of buildings. Over time, and as building performance knowledge increases, the financial incentives, regulatory requirements and frameworks to encourage people to change and become energy efficient will increase. Figure 2.4 shows the main components of the building and the ability to control or influence their use.

The building *fabric* consists of the thermal envelope that contains the internal environment. *Services* include operations that heat, ventilate and cool the environment (HVAC systems), as well as controlling smart energy systems. The *behaviour* component reflects the actions of the building user to maintain comfort and health through the use of temperature control, ventilation, solar control shading and lighting.

The gap between theoretical and real-world energy performance for any building can result from a multitude of reasons that impact on the three above-mentioned

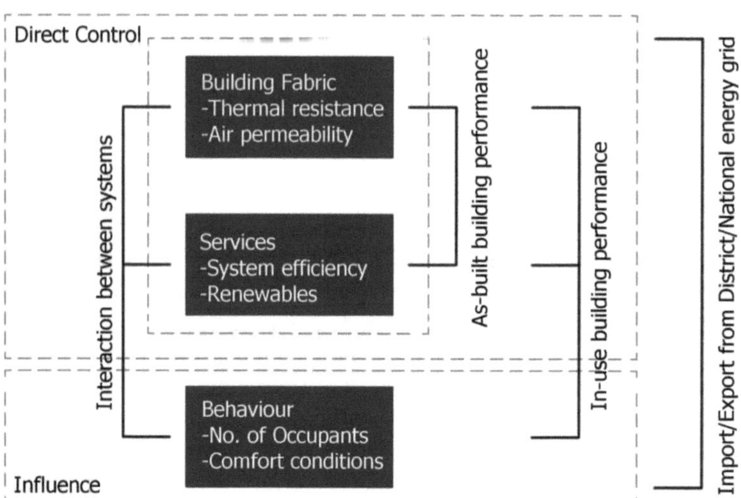

Fig. 2.4 Key building elements affecting energy efficiency and performance. (Thomas and Gorse 2013)

elements. Part of the gap originates during the design and construction phase of the building due to incorrect or inadequate construction and installation practices, poor building design and/or inaccurate theoretical models that fail to consider many influential factors. Thermal bridges, moisture and air flow, weather conditions, user behaviour, control strategies, sun protection systems, temporary heat protection measures and buffer zones are all too often overlooked.

Unfortunately, buildings do need maintaining, but energy efficiency can also reduce during the lifetime of the building, due to errors occurring in the building system, poor software management and faults, user operation and users unaware of wasted energy.

Considering this complexity, the collection and analysis of real-world building energy performance data requires measurement and monitoring that can be applied consistently regardless of building location, usage or environment. Only by monitoring and measuring building energy performance together with gathering and analysing energy data, is it possible to use this performance data to its fullest effect. To create models that can generate accurate and reasonable expectations of real-world energy performance, energy data need to be captured and fed back throughout a building's lifetime.

With the existing buildings, it is extremely important to understand the building, before any retrofit upgrades are applied. Building surveys are an inherent part of assessing the condition, characteristics and thermal properties.

2.3 Building Properties: Variations Construction

Over the years, the methods of building have changed. When examining new or the existing buildings, energy analysts and surveyors must ensure that the properties of a building can be recognised and defined, although this is not always an easy task. With new buildings, the process is relatively simple. New buildings are designed, specified and built to specification or at least this is the preferred assumption. In some cases, designs are not buildable and contractors make ad hoc decisions which prevent delays and allow building to continue, while in other cases products are switched to bring in a more cost-effective solution or switched because the specified product is in short supply or unobtainable. When products are switched, the properties often vary and performance will differ from that specified. The way a building is assembled can create differences, the quality of construction will vary depending on the skill and knowledge of the builder, and the quality control processes in place. Regardless of all of these factors, new buildings are supported by design, and structural and thermal specifications. The information from the specifications is fed into a Standard Assessment Process (SAP) or Reduced Data RdSAP (DECC 2013a). The end result of all of these factors feeding into the assessment process is an energy rating in the form of an energy performance certificate as shown in Fig. 2.5. As we have previously seen, for example as shown at Fig. 2.3 for new builds assessments, there can be a big difference between expected performance and that achieved.

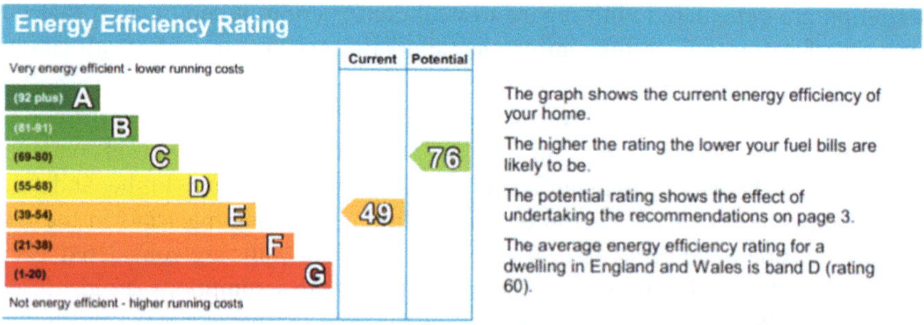

Top actions you can take to save money and make your home more efficient

Recommended measures	Indicative cost	Typical savings over 3 years	Available with Green Deal
1 Increase loft insulation to 270 mm	£100 - £350	£141	✓
2 Cavity wall insulation	£500 - £1,500	£537	✓
3 Draught proofing	£80 - £120	£78	✓

Fig. 2.5 Extract from an energy performance certificate. (DECC 2013b)

2.4 Common Factors that Lead to Differences in New Build Performance

The research reported has shown there there are common factors which can lead to differences in the performance of new-build properties from that envisaged. For example:

- Missing or incomplete design information leading to ad hoc design on site
- Last-minute changes in specification
- Product switching
- Construction process clashes, products fitted out of sequence
- Poor quality
- Workmanship, skill level and understanding
- Incorrect installation, alignment, fixing and sealing
- Malicious practice—failure to undertake installation, damage to property
- Lack of commissioning and quality checks

2.5 Variation in Existing Buildings

Existing buildings are much more difficult to assess, and hence, the reason that a reduced level of data is needed to complete the required SAP form. It might be assumed that it is relatively easy to characterise an existing building and to define how it is constructed, but if the changes in build styles, practice and regional variation are considered with regard to wall construction, it can be seen that the process of defining even the simple characteristics and features of a wall can be difficult.

Traditionally, buildings in the UK have been built using solid wall construction made of cob, wattle and daub, stone, brick or other materials and in various regional styles, using whatever local materials were easiest to source. While timber and steel frame construction are also common, the traditional building materials in the UK are still considered to be stone or brick. If just masonry walls are considered, each wall type will have different characteristics; some of the assumptions are identified and listed in Figs. 2.6 and 2.7. Thus, to understand the properties of the building, it is important that a full survey is undertaken and the characteristic behaviour 'in-situ' determined.

The characteristics of the building assemblies need to be understood, e.g., many masonry walls have gaps or cavities in them and this will cause them to behave differently. Some cavities are intentional and others are not. Initially cavities were introduced to create a break in the wall, reducing the passage of water driven by rain (precipitation) and other moisture sources. The first cavities were no more than a few centimetres wide (10–30 mm). The separation between the two skins was narrow. In some cases, such walls constructed with narrow cavity are often termed solid walls and are not typically referred to as cavity walls at all.

The introduction of cavities and separation of the masonry skins significantly reduced the passage of moisture. Small air gaps also offer a thermal break, reducing the risk of condensation on the internal surface. Unfortunately, as we now require internal temperatures to be higher, traditional solid walls with cavities do not offer the thermal efficiency demanded. Where unfilled cavities once provided the required standard of construction, the free flowing air within the cavity can reduce the thermal performance, allow thermal bypass and increase the potential for cold spots and localised condensation.

The properties of the materials that make up the cavity walls affect the way moisture and air moves through the building fabric. Early clay bricks were formed using locally selected dense clays and were relatively impervious. Due to the low porosity, the cavity did not need to be wide to prevent the passage of moisture. However, in exposed situations where stone and brick were more porous, the width of the cavity had to be wide to prevent the wind-blown rain penetration across the cavity. In some wide cavities, loose rubble or stone was introduced, to act as fill and introduce some lateral stability and mass. The stone cavity fill was loose and moisture that penetrated through the outside leaf was restricted and less likely to cross through as the rubble did not properly connect. Penetrating moisture dripped down the external leaf percolating and evaporating away at the foot of the wall.

As access to fossil fuels became increasingly difficult to source, and pressure to reduce related carbon emission was added, it became important to improve the thermal performance of buildings and reduce the level of energy required to heat them. Cavity walls provide an convenient place to improve thermal performance and place insulation materials. Initially thin layers of insulation were used in new-build, although these have steadily built up, both in thickness and improved thermal resistance (Fig. 2.7).

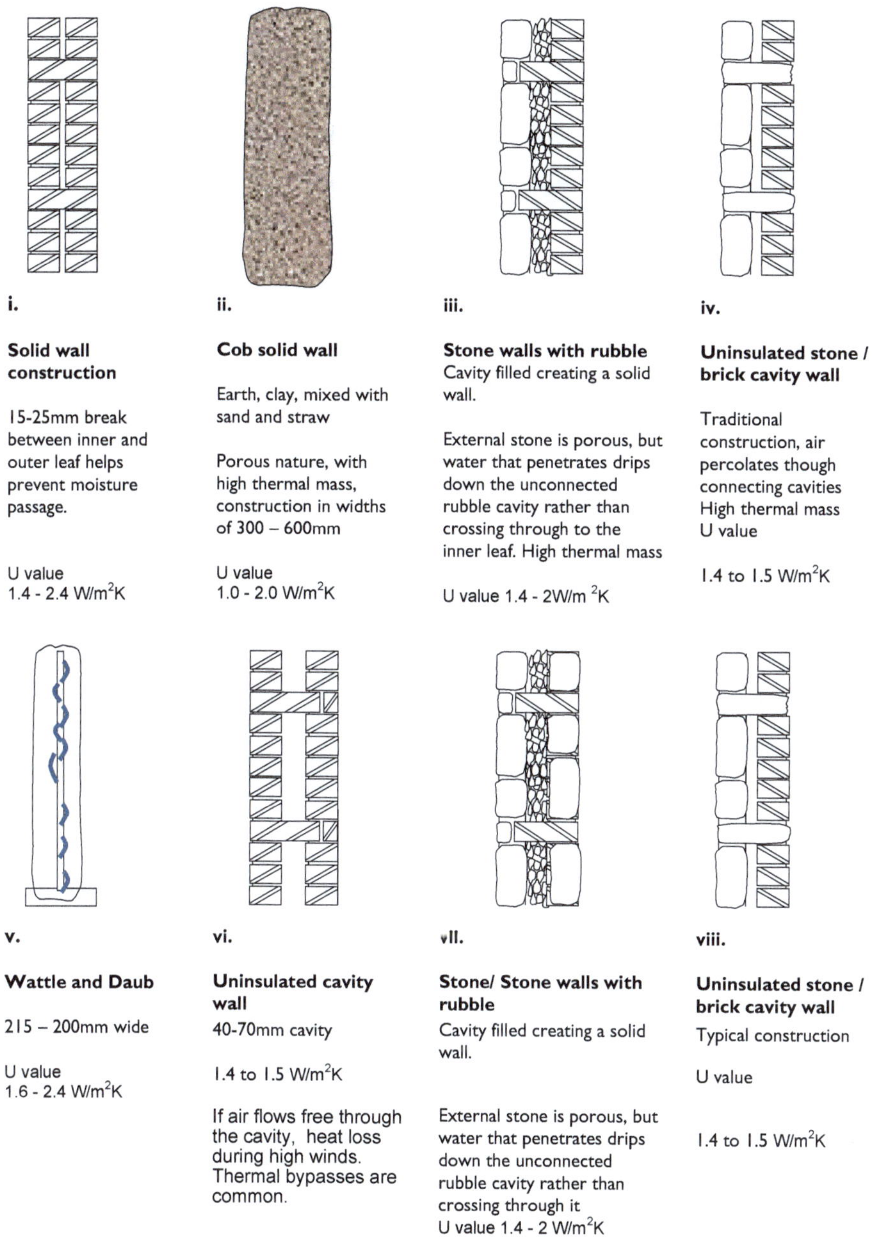

Fig. 2.6 Some traditional variations in wall construction. (Thomas and Gorse 2013a)

From the short description of how walls have changed, it can be seen that it is important to undertake good building surveys to, as far as is practicable, determine the building characteristics and properties of the building elements.

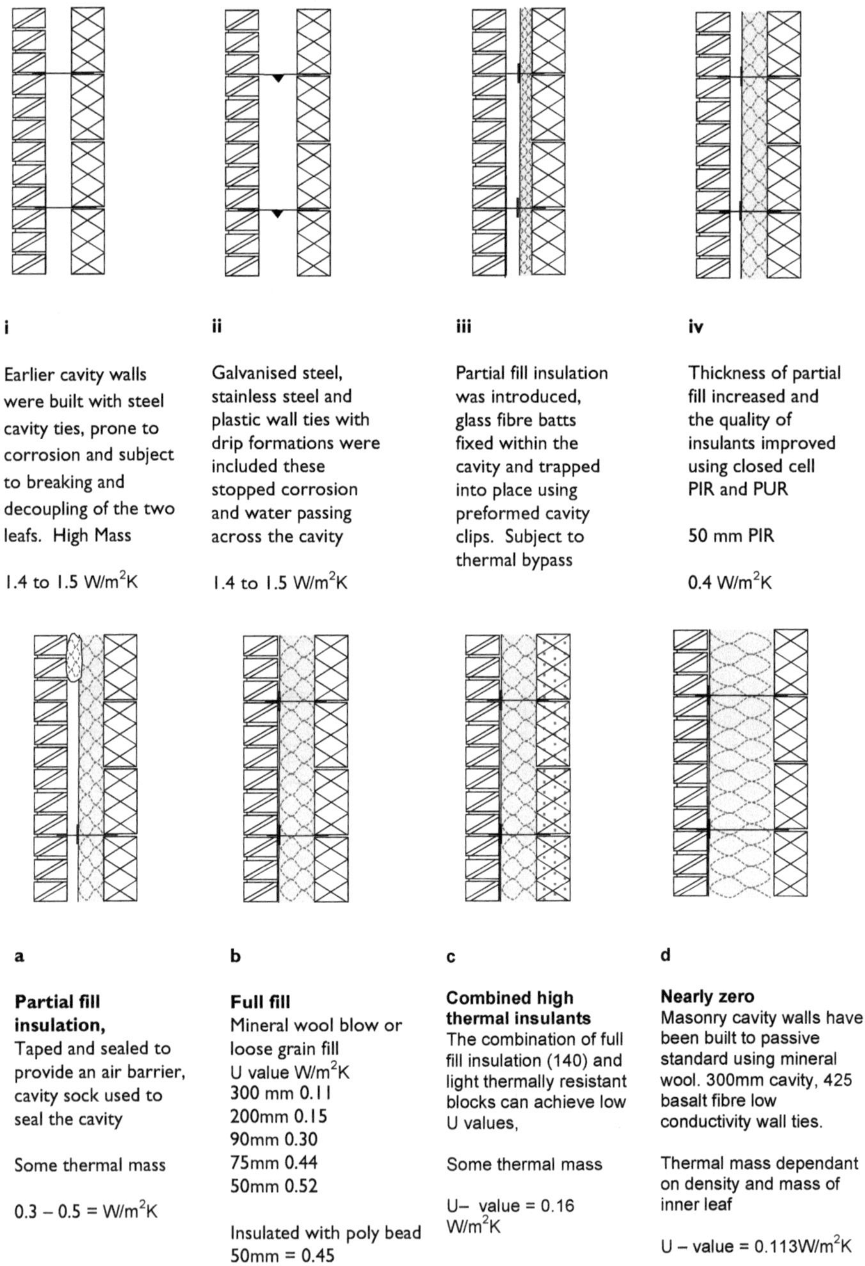

Fig. 2.7 Some traditional variations in cavity wall construction and thermal insulation. (Thomas and Gorse 2013b)

2.6 Building Surveys

New properties are required to be built to regulations and standards which are in place, pertinent to that property type and country. Existing properties are not usually required to be upgraded to these performance standards, but where improved performance is required to existing buildings, then it is more usual that they adhere to a proportionally 'improved' standard. Even when circumstances dictate that an enhanced energy performance is required, there are often parts or elements where a compromise or much more relaxed approach needs to be taken (e.g., English Heritage 2011). This may depend on the building's construction period or style. In the UK before the year 1919, most buildings were often built using 'traditional' construction, usually with solid walls. After this period, most buildings have included some form of cavity and damp-proofing to their external walls. However, after 1945, some more modern buildings have solid walls, e.g., 'no-fines' solid walls built between the 1950s and 1970s.

When upgrade is being considered or implemented, for a reliable assessment of a building's potential performance (i.e, where its assumed and modelled performance corresponds closely with its actual performance) we need to understand the building's construction and its performance both elementally and holistically. One aspect of the performance gap for buildings when an existing property is being thermally refurbished is a miscalculation or misunderstanding of its performance prior to any interventions (e.g. Historic Scotland 2011). This is important because to determine any improvements, the initial condition must first be known. If the initial performance is better than expected and the performance following invention is not as good as expected, then gains and pay-back periods would be unfavourably affected. A further consideration is that the existing building's actual performance may be such that a particular possible intervention may not be cost-effective at a given time.

In addition to any anticipated carbon and monetary savings, the effect of any intervention on the fabric of an element (i.e., walls, floors, roofs) should be considered. Changing the way an element or a building functions in relation to air and moisture movement can have adverse effects such as mould growth internally or interstitially, deterioration of masonry through freeze–thaw action, or decay or corrosion of materials in the fabric (Melville and Gordon 1973).

Prior to works being carried out on a property, some level of inspection is carried out to assess the condition and size of the property. For a single element in a property or for the whole building, this can vary between four basic levels: a desktop assumption of type and condition from estate records; a brief visual assessment solely for design purposes; a short condition survey using standard surveying equipment; or a full structural building survey with initial performance testing taking some hours or even days (RICS 2013). Property surveys are carried out for a variety of reasons, but can usually be summarised as aiming to inform the client what the material and condition is, what, if anything, is wrong and why it is wrong, what damage or failing has occurred or is expected to occur, how serious this is, what is needed to put it right, how much this is likely to cost, when the remedial work should be carried out, who is responsible for rectification works and what further action is to be taken by the client (RICS 2010). The level of survey undertaken is dependent on the client's requirements for the depth of answers to these issues, and the level of resources (in time and money) that are available for finding the answers.

The RICS produce standard pro forma which are used for their stated three levels of residential building survey for property purchase, currently known as RICS residential Condition Survey, RICS Home Buyers report, and RICS Building Survey. Samples of these can be found from the RICS Web site or through search engines. These are produced for the specific services required by clients and usually for the purchase of the property, but published pro forma currently are not in standard use for thermal performance investigation. Figure 2.5 provides an example of a surveying pro forma used for thermal performance investigations.

Knowledge of existing property performance is growing but there are many unknowns. If the designer of any intervention, be they architect, engineer or contractor, is not aware of the effect of the intervention, this in itself can contribute to the performance gap. The current knowledge of the existing performance and the likely performance of the property following intervention, whether this is adverse or advantageous, should always be included in the design considerations.

A second aspect of this is that an initial high level of knowledge and understanding of the existing construction is usually assumed. One of the prerequisite assumptions of the basic model for research or design is that the property construction and condition is known to the extent of the fourth level of survey briefly described above, i.e., the full structural building survey with initial performance testing, whereas in practice, especially where the client has a large portfolio, a desktop assumption or brief visual inspection only may have been made. Although exercises such as desktop assumption or a brief sample visual inspection are useful for determining average commercial conditions for improvement of the properties, each individual property is likely to have its own idiosyncrasies and anomalies to the extent that no individual property will actually perform as expected from the averaged model.

A protocol for an initial survey assessment of a property's construction, condition and performance should therefore be considered. It may not be cost-effective for in-depth full building surveys and testing to be conducted on each property, but the clients and designers should at least be aware of this and that occasional surprises can therefore be expected.

When considering interventions for improving thermal performance, there are particular pointers for surveys of any level. These would include the following:

- External wall construction: materials, thickness, presence and size of cavities, infilling, internal lining(s), homogeny, condition
- External coatings, materials, finish, condition
- Floor levels in relation to ground level, damp proof course level, and in relation to the existing and proposed external coating levels
- Positions of gulley's, inspection chambers, rodding points, drains
- Dampness: rising, penetration, resulting from services/splashing/faults, interstitial, condensation
- Eaves construction: overhangs, soffit boards, rainwater goods, soil vents, roof void ventilation, room in roof space, collar roofs

- Window positions and proposed positions in relation to any proposed wall insulation
- Existing vents for human ventilation, appliance combustion air, open fire drawing air
- Existing insulation to roof voids: material, condition, thickness, evenness, adequacy

An example of pro forma is shown at Fig. 2.8.

When conducting the survey, if possible, the proposed intervention(s) should be known and considered (STBA 2012). This would enable the surveyor to alert the designer to possible issues with the proposals. There are a number of typical aspects which raise concern. For example, with external wall insulation, the insulation would normally be stopped 150 mm above ground level to avoid sustained wetting and bridging of the damp proof course from rain splash, etc. which may be reasonable but not if the floor level is below this and a thermal bridge introduced. Thermal bridges around floor level exacerbate the effects of stratification, resulting in reduced occupier comfort. Thermal bridges are also often introduced when the client or designer decides not to extend roof eaves due to the complication of external services and instead stops the external wall insulation short at the eaves. In this instance, the thermal bridge can result in condensation and mould growth inside the property which is damaging to the occupant. If in addition interstitial condensation arises, further damage to the fabric is likely (Oreszczyn et al. 2005). For these reasons the survey should identify and highlight, to designers, any possible challenges.

Internal wall insulation can also present problems. Removal and replacement of internal architectural features obviously need to be considered, such as coving, picture rails, architraves. A physical and cost challenge is the treatment for suspended floors/wall junctions to avoid thermal bridges. Determining where the new thermal envelope starts and finishes can be highlighted by the survey. For example, if there is a cellar, the spandrel wall and stair soffit to the first floor adjacent to the stairs leading to the cellar (a typical arrangement) along with the cellar door, will need insulating if the cellar is to be outside the thermal element.

Another consideration for internal insulation is the presence of dampness (STBA 2014). Although it is generally agreed that the existing buildings can perform better than assumed in the past, a damp building generally performs less well. Dampness in timber and masonry can also lead to more ideal conditions for fungal and insect attack in the former and erosion in the latter (Viitanen et al. 2010). Timber can be found in the roof timbers, and in the external walls as door and window joinery, floor joist ends, and also in older properties as lintels and beams and simply as filling pieces in place of brick or stone in external walls. Internal insulation leads to brickwork and timber outside the thermal envelope, being cooler than originally designed. If the thermal envelope is not airtight, moisture-laden air moving through the construction may reach dew-point inside the construction, i.e., interstitially, and condense. Additionally, if the thermal envelope is airtight, drying out of the envelope (made damp through normal weather conditions, or through faults in the construction) is likely to be delayed or reduced (Künzel and Kiessl 1996). Both of these typical scenarios will lead to timber and masonry remaining moist with

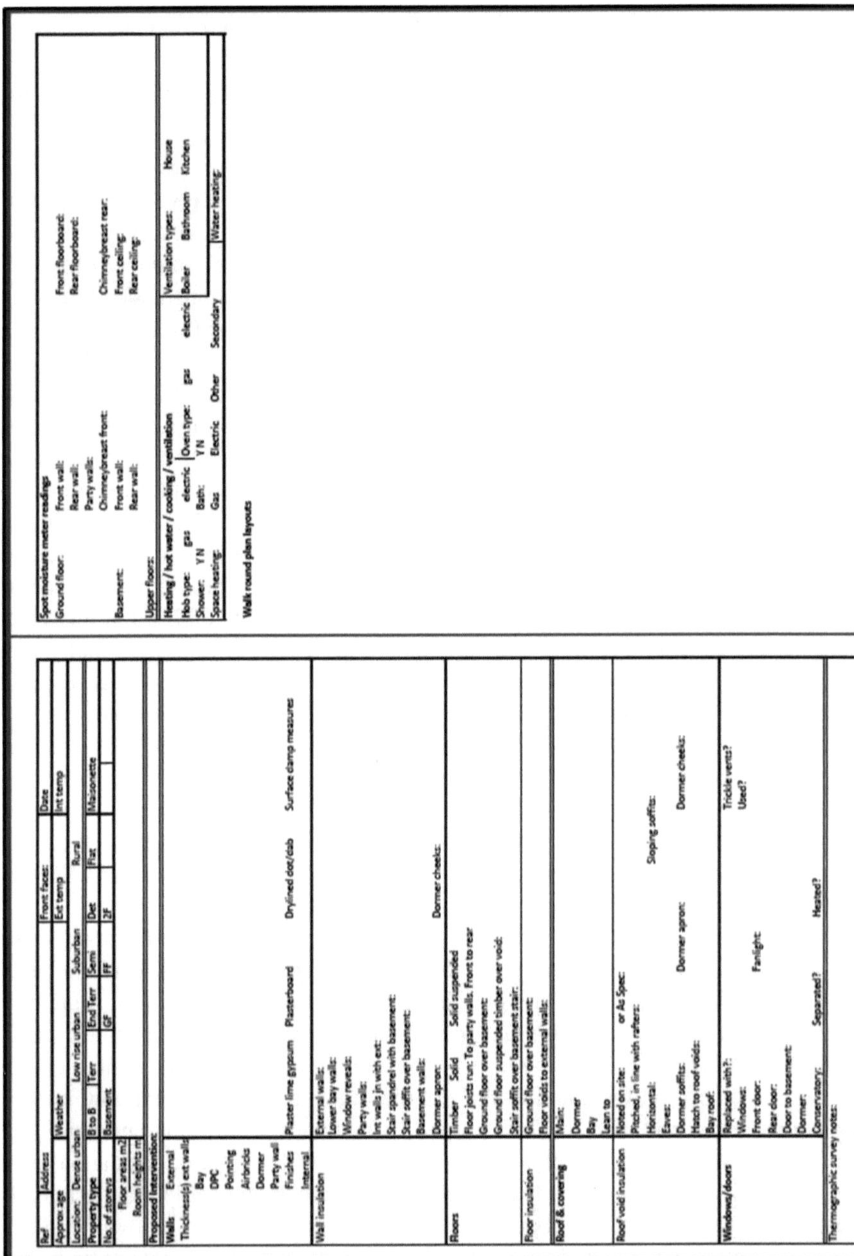

Fig. 2.8 Example of survey pro forma for thermal investigation. Intended for use with detailed notes, moisture survey, photographs and thermography survey

ensuing problems (May 2013). The building survey should therefore raise awareness of the existing damp issues in the property occurring before any works/designs begin.

Research into the long-term effects of increasing the thermal performance of the existing buildings, especially those with solid walls, is ongoing. The importance of carrying out building surveys for properties undergoing this type of intervention to determine pre-works, construction and performance, and post-work, performance and prognosis, is therefore an important aspect of ensuring sustainable buildings.

2.7 Surveying and Thermography for Building Forensics

Thermal images can provide much useful information about the building. If a thermal survey is undertaken by a competent person and with induced air pressure or depressurisation, it can provide indications of construction assemblies, materials used, cold spots, thermal bridges, air leakages and bypasses, highlight moisture and difference densities of materials. However, such information is normally supported with inspections and other field tests that can quickly confirm visual thermal data (Fig. 2.9). There is benefit in examining the thermal images taken when there is little pressure difference and when the pressure differentials increase. The thermal images can help to identify the passage of heat through air permeable buildings due to wind.

2.8 Infra-Red Thermography

Infra-red thermography, as it relates to buildings, is the use of thermal imaging equipment to visualise the surface temperature of building elements. Thermography allows the temperature of large areas to be read quickly and easily. Thermal images can be interpreted and analysed to draw conclusions about the thermal characteristics of a building's fabric, potentially locating defects.

Fig. 2.9 Thermal image being used to assess variations in temperature following a retrofit upgrade

2.9 What is Thermal Imaging?

Thermal imaging equipment (commonly referred to as thermal cameras) is used to capture thermal images, which represent the surface temperatures of objects within the field of view. This is done by focusing infra-red radiation onto an infra-red detector (much like the image sensor found inside a digital camera, but with IR radiation rather than visible spectrum radiation); the signals generated are then processed to generate a thermal image (FLIR 2011). Thermal cameras interpret the intensity of infra-red radiation received and output an image representing the apparent surface temperatures across the cameras field of view (Pearson 2011).

Infra-red radiation is a band of the electro-magnetic spectrum, with wave lengths longer than visible light, ranging from 0.7 to 1000 μm (Pearson 2011); this radiation is invisible to the human eye, but its effects can be felt as heat. All objects with a temperature above absolute zero (−273 °C or 0 K) emit infra-red radiation, and the intensity and wavelength of the emitted radiation vary with the surface temperature of the object.

The surface emissivity affects the intensity of infra-red radiation given off. Emissivity ranges from 0 representing an ideal, non-emissive object to 1, representing an ideally emissive 'black body' object. Objects with emissivity below 1 will appear to be cooler than they actually are, and this can be corrected with on most thermal cameras and in thermal image processing software.

2.10 Using Infra-red Thermography

Infra-red thermography is very useful in building inspection, as thermography is a non-contact, non-destructive method of building inspection it can be carried out quickly, and with relatively little inconvenience to building occupants. Using a thermal camera to observe the fabric of a building allows surface temperatures to be observed quickly and easily. By observing the distribution of surface temperatures in a building's fabric, anomalies and defects can be located, which would not be otherwise discoverable without long-term or destructive investigation of the building's fabric, defects such as:

- Missing, misplaced or damaged insulation
- Location of air infiltration paths
- Thermal bridges
- Damp areas
- Heating system faults.

(Hart 1991) (Fig. 2.10)

Thermal image shown in Fig. 2.10 reveals that the section of roof between wall and ceiling is not insulated, as shown by the colder areas between the rafters. This is typical of missing insulation in timber-framed elements (Fig. 2.11).

Fig. 2.10 Thermal image showing the heat transfer through the uninsulated eaves

Fig. 2.11 Thermal image showing heat loss through air vents Leeds Sustainability Institute 2014

Figure 2.11 shows a view from the outside of a building. Warm patches indicate areas where heat is escaping from the internal space, in this case through an air vent and the edges of the window frames (Fig. 2.12).

These images show an example of air leakage. Dark ray patterns are visible around a sealed air vent. This is typical of the penetration of cold air into a building. As the

Fig. 2.12 Thermal image showing air leakage (Leeds Sustainability Institute 2014)

air infiltrates the building, it cools the surfaces along its path, causing the distinctive cool region. Dark pattern is palette specific. This image was taken under normal conditions. For air leakage detection, a pressure difference of at least 10 Pa (pascals) and 10 degrees K is advised (Hart 1991).

2.11 Practicality

Though Infra-red thermography has a number of benefits, it is not without its drawbacks. Undertaking of thermal survey requires a high level of skill in order to correctly interpret findings (Usamentiaga 2014). It is important to be aware of misleading results such as:

- Thermal reflections, Infra-red radiation bouncing off a reflective surface, a reflection will change position when the camera is moved.
- Low emissivity materials are particularly prone to reflections as the lower a material's emissivity the higher its reflectance.
- Apparent surface temperature differences that occur without a temperature gradient can often be due to differing surface emissivities, roughness, angle of incidence and/or variations in surface moisture.
- Hot objects in frame can cause thermal cameras that are set to automatically range to change contrast and range, which may cause some details to be obscured making on-site interpretation difficult. This can be overcome by manually adjusting settings to give the desired image.

Thermography is also reliant on a number of environmental conditions in order to give reliable results, where conditions do not match with requirements and a survey may be difficult to carry out:

- Stable temperatures: external temperature should not vary by more than 10 °C for 24 h before the test. External and internal temperatures should not vary by more than 5 and 2 °C respectively during the period of the investigation. Variable temperatures may lead to misinterpretation due to variation in heat capacity loss across the target area.
- A temperature difference of at least 5 °C is required between internal and external conditions during the test.
- The faces of the building under investigation should not be exposed to direct sun light for 12 h before or during the test.

(BS EN 13187:1999)

Where these conditions are not met, it should be noted in any report resulting from the investigation.

The use of infra-red thermography allows a great deal of information concerning the thermal performance of a buildings fabric to be collected in a relatively short period of time. This information can then be interpreted to find possible defects that would otherwise have been difficult to detect. Thermography does not determine defects with absolute certainty, but it does give a strong indication that can guide further, targeted investigation.

References

Bell M, Wingfield J, Miles-Shenton D, Seavers J (2010) Elm Tree and Stamford Brook technical report: low carbon housing: lessons from Elm Tree Mews. Joseph Rowntree Foundation, November 2010. http://www.jrf.org.uk/publications/low-carbon-housing-elm-tree-mews. Accessed June 2011

Bordass B, Cohen R, Standeven M, Leaman A (2001) Assessing building performance in use 3: energy performance of Probe buildings. Build Res Inform 29(2):114–128

BSI (1999) BS EN 13187:1999 Thermal performance of buildings—Qualitative detection of thermal irregularities in building envelopes—infrared method. British Standards Institute, London

Carbon Trust (2011) Closing the gap: lessons learned on realizing the potential of low carbon building design. Report CTG047. Carbon Trust, London

DECC- Department of Energy & Climate Change (2013a) Guidance: Standard Assessment Procedure. https://www.gov.uk/standard-assessment-procedure. Accessed June 2015

DECC- Department of Energy & Climate Change (2013b) Energy performance Certificate (EPC) example. https://www.gov.uk/government/uploads/system/uploads/attachment_data/file/49997/1790388.pdf. Accessed June 2015

EED (2012) Energy efficiency directive, official journal of the European union. Directive 2012/27/EU of 25 October 2012 on energy efficiency, amending Directives 2009/125/EC and 2010/30/EU and repealing Directives 2004/8/EC and 2006/32/EC [OJ L315 p 1]

English Heritage (2011) Energy efficiency and historic buildings—Application of part L of the building regulations to historic and traditionally constructed buildings

EPBD (2010) Energy performance building directive. Official journal of the European union. Directive 2010/31/EU of 19 May 2010 on the energy performance of buildings [OJ L153 p 13]

FLIR (2011) Thermal imaging guidebook for industrial applications. FLIR systems AB

Galvin R (2014) Making the 'rebound effect' more useful for performance evaluation of thermal retrofits of existing homes: defining the 'energy savings deficit' and the 'energy performance gap'. Energy Build 69:515–524

Gorse CA, Johnston D, Miles-Shenton D (2011) Low carbon housing—failings in construction practice. Architect Sci ArS(3):53–67 (Taiwan (ISSN 2219–1577))

Gorse C, Stafford A, Miles-Shenton D, Johnston D, Sutton R, Farmer D (2012a) Thermal performance of buildings and the management process. In: Smith SD (ed) Procs28th annual ARCOM conference, 3–5 Sept 2012, Edinburgh, UK, Association of Researchers in Construction Management, 1413–1422.

Gorse C, Stafford A, Miles-Shenton D, Johnston D, Sutton R, Farmer D (2012b) Do we know they work: shedding light on the actual thermal performance of buildings. Constr Innov Res 3(3):30–33 (CIOB Englemer Ascot, CRI)

Hart JM (1991) A practical guide to infra-red thermography for building surveys. Building Research Establishment. Via the Constriction Information Service

Historic Scotland (2011) Technical paper 10: u-values and traditional buildings -In situ measurements and their comparison to calculated values. Historic Scotland Conservation Group

Johnston D, Farmer D, Brooke-Peat M, Miles-Shenton D (2014) Bridging the domestic building fabric performance gap. Build Res Inform. doi:10.1080/09613218.2014.979093 (Online version). http://eprints.leedsbeckett.ac.uk/802. Accessed June 2015

Künzel H, Kiessl K (1996) Drying of brick walls after impregnation, International Zeitschrift fur Bauinstandsetzen 2, Ed 2. 87–100. Fraunhofer Institute of Building Physics, Fraunhofer

May N (2013) The risks and realities of low carbon building. Good Homes Alliance presentation, London

Melville I, Gordon I (1973) The repair and maintenance of houses. The Estates Gazette Limited, London

Miles-Shenton D, Wingfield J, Sutton R, Bell M (2011) Final report to Joseph Rowntree Housing trust project title: temple avenue project part 2, Energy efficient renovation of an existing dwelling: evaluation of design & construction and measurement of fabric performance. Leeds Metropolitan University, UK http://www.leedsmet.ac.uk/as/cebe/projects/tap/tap_part2.pdf. Accessed June 2015

Oreszczyn T, Mumovic D, Davies M, Ridley I, Bell M, Smith M, Miles-Shenton D (2005) Condensation risk—impact of improvements to Part L and Robust details on Part C. Report number 10—Final Report; A Report to the ODPM building regulations division under the building operational performance framework. Project Reference number CI 71/6/1 (BD2414), University College London and Leeds Metropolitan University

Pearson C (2011) Thermal imaging of building fabric: A BSRIA Guide. BISRIA. www.bsria.co.uk. Accessed June 2015

RICS (2010) Building surveys and technical due diligence of commercial property. Royal Institution of Chartered Surveyors, Coventry (RICS guidance note)

RICS (2013) Surveys of residential property, 3rd edn. Royal Institution of Chartered Surveyors, Coventry (RICS guidance note)

Stafford A, Bell M, Gorse C (2012a) Building confidence: a working paper, report no 8. Energy Saving Trust & Low Carbon Futures. [www.lowcarbonfutures.org] March 2012

Stafford A, Gorse C, Shao L (2012b) The retrofit challenge: delivering low carbon buildings, report no 4. Energy Saving Trust & Low Carbon Futures. [www.lowcarbonfutures.org] March 2012

Stafford A, Johnston D, Miles-Shenton D, Farmer D, Brooke-Peat M, Gorse C (2014) Adding value and meaning to coheating tests. Struct Surv 32(4):ISSN:0263–080X

STBA (2012) Responsible retrofit of traditional buildings. Sustainable Traditional Buildings Alliance, UK

STBA (2014) Moisture risk assessment and guidance: technical appendix. Sustainable Traditional Buildings Alliance, UK (Department of Energy & Climate Change)

Thomas F, Gorse C (2013a) Wall types and changes in construction; Design and good practice. LSI technical paper 2013-TP-0005 Leeds Sustainability Institute

Thomas F, Gorse C (2013b) Wall types and changes in construction; Design and good practice. LSI technical paper 2013-TP-0006 Leeds Sustainability Institute

Usamentiaga R (2014) Infrared thermography for temperature measurement and non-destructive testing. Sensors, 14, pp 12305–12348 www.mdpi.com/journal/sensors. Accessed June 2015

Viitanen H, Vinha J, Salminen K, Ojanen T, Peuhkuri R, Paajanen L, Lähdesmäki K (2010) Moisture and biodeterioration risk of building materials and structures. J Build Phys 33(3):201–224 (Finnish Funding Agency for Technology and Innovation)

Chapter 3
Monitoring and Measuring Building Performance

Christopher Gorse, David Johnston, David Glew, Fiona Fylan, Felix Thomas, Dominic Miles Shenton, Martin Fletcher, Aitor Erkoreka and Anne Stafford

3.1 Integrated Energy Monitoring: Process and Building Performance

The whole-life sustainability of a building should be underpinned with a demonstration of functional value and an awareness of the direct environmental impact. While a great deal of energy and resources are consumed in the construction of buildings, this is marginal when compared to the operation costs and associated energy used during a building's life cycle. Many reports identify the build costs and associated resources to be less than 1% of the whole-life operation costs. The exact energy use of a building can vary widely, depending on the use, energy efficiency of the building and occupant behaviour; thus, a greater deal of attention should be given to understanding the energy used in buildings and how energy efficient operation is achieved.

For some time, the scientific community has been able to measure building performance and provided a perspective on the buildings ability to meet sustainable agenda. Unfortunately, the resource implications associated with measuring energy efficiency in detail required by the scientific community can be prohibitive, preventing mainstream use of energy monitoring. However, through advances in energy monitoring technology and data capture methods, the monitoring of energy use has become more accessible, with some of the cost and time implications reduced or removed. While in-depth scientific monitoring and forensic analysis will always have its place, with relatively basic monitoring equipment it is now possible

C. Gorse (✉) · D. Johnston · D. Glew · F. Fylan · F. Thomas · D. M. Shenton · M. Fletcher · A. Stafford
Leeds Sustainability Institute, Leeds Beckett University, Broadcasting Place, Leeds LS2 9EN, UK
e-mail: c.gorse@leedsbeckett.ac.uk

A. Erkoreka
ENEDI Research Group, University of the Basque Country (UPV/EHU), ETSI Bilbao, Alameda Urquijo s/n, 48013 Bilbao, Spain

© Springer International Publishing Switzerland 2016
M. Dastbaz et al. (eds.), *Building Sustainable Futures*,
DOI 10.1007/978-3-319-19348-9_3

to collect live data on the building fabric, service and occupant energy behaviour; even consumer-focused devices such as "smart meters" are now coming to market.

Scientific investigation will always be required to underpin smart data, but the live smart data that is captured and feed back to the occupant and building manager is becoming more important. If gathered correctly, such data can be used for energy monitoring, financial analysis and forecast, as well as identifying building and service efficiency. Monitoring data can be used to trigger alerts and importantly can provide valuable feedback to the design and construction processes.

Variance between theoretical and actual thermal performance of buildings presents a major barrier to achieving the zero-energy targets and to private investors and the broader financial payback schemes. However, the research groups that have identified the performance gaps have also shown that the aspirational energy performance standards in retrofit and new build projects can be achieved, within acceptable tolerance. Unfortunately, such demonstrations of good practice occur in isolation and are not indicative of the norm; more widespread use of in-use monitoring and building performance evaluation will help to build an understanding of how buildings and their occupants truly behave.

The intelligent use of the existing scientific and basic energy monitoring tools can make a significant step change in understanding building performance. Where valid data exist for different building types, building elements and features, they provide a baseline, defining characteristics and behaviour and improving the ability to predict future performance. Once good data and understanding exists, and prototypes or standards developed, simple checks can be used to align production, enabling the reproduction of the energy standard. Advances in monitoring technology, apps and smart metering provide the transitional step towards checking that buildings are responding as expected, allowing whole building life cycle feedback loops that enable self-commissioning, effective maintenance and performance upgrades.

The future of energy monitoring is relatively clear; through the systematic analysis of energy data, basic energy signatures can be used to inform diagnostic gap analysis and performance. Data can be used to develop and refine predictive tools improving the accuracy of energy performance predictions so that design performance gaps can be closed. Through more autonomous feedback, an insight into variances in observed energy use data can be gained. Such processes will allow for system calibration, continuous commissioning and energy improvement cycles. The data can also be digitally integrated, managed and visualised. However, while many in the industry claim this is already happening, the data being extracted are not always simple and meaningful.

The following chapter outlines some of the basic monitoring and testing tools and techniques that are being used to inform the understanding of building performance.

3.2 Logically Defining the Test and Monitoring Environment

Research has demonstrated that there are gaps in building performance (Gorse et al. 2011), some of which have come about through a failure to understand the design and construction process and other due to a neglect of the usefulness of whole building measurement (Stafford et al. 2014). There are a number of tests that can be applied to building performance; however, knowing their limitations and capabilities, they must be applied systematically and logically.

When setting up a testing or monitoring campaign, the first question that should be asked is "what do you want to characterise?" This question if applied to building fabric determines the context, environment, experimental design and analysis method that should be applied (Fig. 3.1). *http://dynastee.info/ and http://www.leedsbeckett.ac.uk/as/cebe/resources.htm. Free software to develop the logical links can be found on the Xmind site (http://www.xmind.net/).*

Since there are many different objectives when measuring the thermal performance of buildings or building components, a good way to treat this question is to develop a decision tree and logically work to the standards, guides and methods that will help answer the question. By working through a logical structure, it is possible to identify the standard documents that exist, methods that have been developed and requirements of the monitoring to obtain the necessary control of the test or monitoring environment to obtain good quality data and feedback *(a full version of the decision tree referred to in this document can be found on the following sites).*

To obtain reliable results for full-scale, whole building or building element tests, it is essential that the work is conducted within a good test infrastructure, which means having control of the variables that impact on the test and or knowledge of their behaviour.

It is clear that the understanding of designed building components must be deeper than just measuring its heat loss or U-value under steady-state laboratory conditions. The modelling and testing of the dynamic thermal behaviour of the building's components, envelope and whole building must be more representative of real-world environments and accurate. Therefore, to gain a much deeper understanding, it is necessary to conduct tests in the field as well as in the laboratory. Even when conducting experiments in the field, there are the steps that are not always the same. Although many different procedures have been developed to test the dynamic behaviour of building components and buildings *in situ*, few of them have become internationally accepted standards. Indeed, many of these procedures may never result in a standard, since the nature of dynamic testing exposes the same test component to different dynamic conditions. Thus, some procedures and data can vary considerably depending on the conditions under investigation. However, the monitoring and testing procedures are important to develop further knowledge, and understand contributing factors and the relationships between the dependent and independent variables.

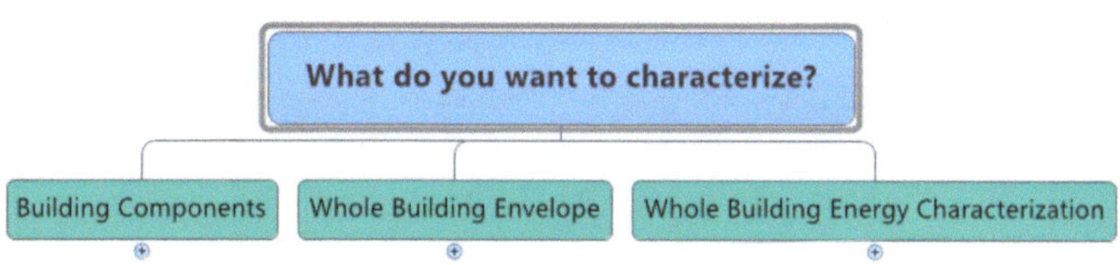

Fig. 3.1 Main branches of a decision tree Gorse et al. (2015) Decision Tree Whole Building Characterisation: IEA Annex 58

3.3 Decision Tree Logic in Building Performance

The logical steps in identifying a good experimental set-up to characterise building and building component performance are shown in Fig. 3.2. From the IEA's Annex 58's perspective (Gorse et al. 2015), once the overarching logic was identified, logical branches were then developed to show the options available. An extract from the branches under building components and whole building element is shown in Fig. 3.3.

As can be seen in Fig. 3.4, once the second level of the logical branch was defined, the question "What do you want to characterise?" is explored again. Following this question, the environment requires further consideration and is explored further by raising the question, is the test environment in the field (in situ) or in a controlled laboratory? What should be remembered is that the different approaches will make different observations and yield different results. Both approaches can be complementary, but resource and test environment constraints may mean that it is not possible to do both laboratory and field tests. What must be remembered is whether the approach will yield suitable results.

Another key area is to explore the type of test conditions. Are the conditions considered to be dynamic, quasi-steady state or steady state? In Fig. 3.4, for example, the decision tree has been designed and exploded to reveal the logic and questions that lead to dynamic testing of building components. Here, some of the authoritative publications and standards are identified.

The first level of the decision tree has three choices: *building components*, *whole building envelope* and *whole building energy characterisation*. These are the main three levels where the different full-scale testing is carried out in the building sector.

Fig. 3.2 Overarching logic of decision tree, an example of logic of the decision tree produced by the International Energy Agency, Annex 58 (Gorse et al. 2015)

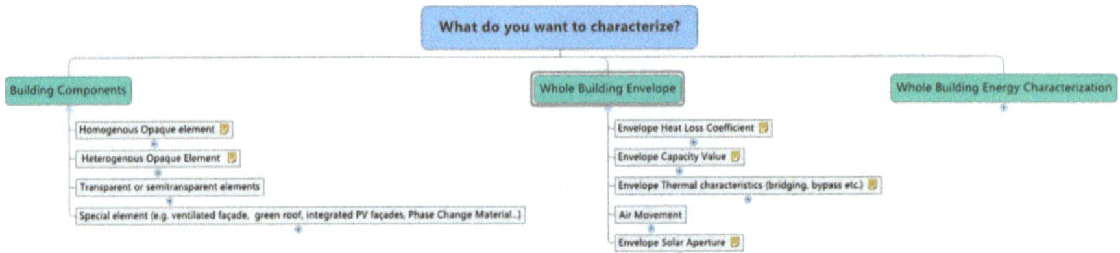

Fig. 3.3 Partial view of building component and whole building branches of the decision tree

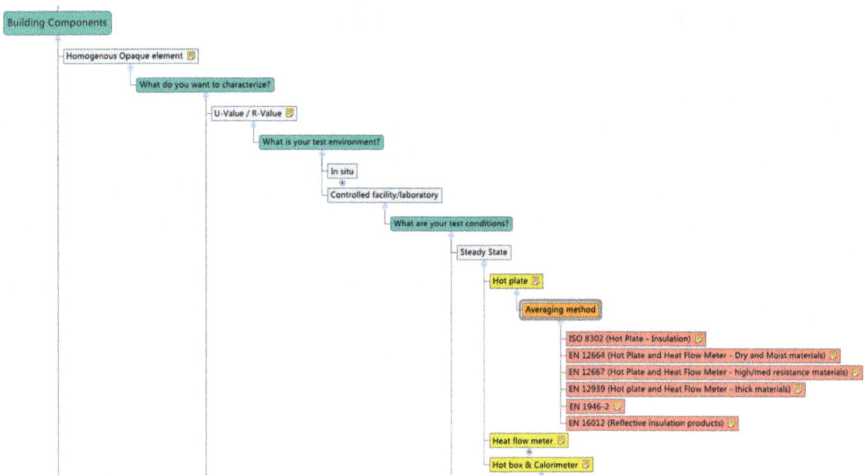

Fig. 3.4 Partial view of decision tree process

The building component branch is focused on how to test a building component in isolation, without considering the effect of the whole building on the building component. The building component branch is focused on the U-value characterisation of walls and windows under well-known standards, but also considers how to test and characterise special building components such as ventilated façades, green roofs, etc. Tests on components are relatively well established. However, the science behind understanding component assemblies as part of the whole building thermal envelope is a developing area.

In the IEA Annex 58 decision tree, the whole building envelope branch is focused on characterising and/or modelling the main energy characteristics of the whole building envelope. The term "characteristic" in this case encompasses the envelope U, C and gA values and also for the buildings envelope special characteristics such as thermal bridging, modelling and characterisation of the air movement through and within the building envelope. These areas are a key concern to thermal transfer within the building envelope.

Finally, the more complicated area of research is where attempts are made to characterise energy used when all factors are considered for the whole building. This general characterisation considers the three main reasons for the energy consumption in buildings: the buildings thermal envelope, the buildings systems and the user behaviour. The understanding of whole building energy use and its characterisation is a new and emerging area of research.

3.4 Building Component: Key Considerations

During the recent decades, much work has been carried out on building component energy characterisation. As can be seen in Fig. 3.3, there are four options inside the main level of the "Building components" branch. The first three options consider the characterisation of "common" building components. They are homogeneous opaque elements, heterogeneous opaque elements and transparent or semi-transparent elements.

The main thermal characteristics tested and modelled include the thermal transmittance value (U-value), the thermal capacity value (C-value) and the solar gain (g-value) or solar heat gain coefficient (SHGC); these are the common building component considerations. Other thermal aspects should also be considered such as the hygrothermal behaviour, thermal bridging, reflection, absorption and transmittance of light and air permeability, all of which impact on performance.

Most building components are already covered by standard test procedures, but many of the newly developed "special" building components cannot be tested within the standards. For example, a ventilated façade or a green roof is not suitable for testing in a guarded hot box since they are passive solar components, and the correct thermal characterisation of these components requires tests to be carried out under real weather conditions or at least with the use of a solar simulator.

3.5 Whole Building Envelope: Key Considerations

In addition to understanding the performance characteristics of individual construction elements and materials in isolation, it is important to appreciate their interaction across the whole building envelope. In order to do this, tests are often undertaken *in situ*. As can be seen in Fig. 3.3, common measures include whole envelope U value, whole envelope C value, whole envelope gA value, envelope special thermal characteristics (thermal bridging) and air movement.

3.6 Whole Building Energy Characterisation

The whole building energy is characterised by methodologies centred on monitoring the main contributors to energy use in whole buildings, namely the building envelope, building systems and users.

When undertaking such studies, the environmental conditions must be understood and assessments need to be made on what tests are and are not applicable (Stafford et al. 2014). For some tests, occupancy is highly significant and may limit the type of tests that are possible or permitted to be undertaken. When undertaking whole building tests or monitoring, it is important to establish whether the properties are occupied, unoccupied or expected occupancy patterns at an early stage.

Following occupancy assessment, further environmental considerations can be explored. In addition to the environment and conditions, the whole building energy characterisation also clarifies the usage of the building, splitting into domestic and commercial properties. It is important to distinguish use, purpose, occupancy patterns, building typologies, system infrastructure as well as defining the rationale for research as all will impact on the findings.

As can be seen, understanding whole building performance can be particularly challenging. The remaining sector of this chapter outlines some of the different approaches, tests and considerations. The explanations are not detailed and should be used as an introduction to each of the topics, rather than a definitive guide.

In-use monitoring of whole building energy and environmental conditions is becoming more common and developing as a science. If done correctly, valuable information can be obtained.

3.7 In-Use Monitoring

In-use monitoring refers to any collection of data from an occupied building, generally including energy and temperature data and often with the purpose of assessing the energy efficiency of buildings. It can also monitor humidity levels to assess for condensation risks, check whether set point temperatures are being achieved to provide occupant comfort or even assess air quality as part of a ventilation strategy. In-use data can be as light touch as taking manual readings of daily kWh usage or as complex as using sensors to log the internal and external environmental conditions and gas and electric use every half hour. The degree of in-use monitoring employed allows for a different type of analysis and information that can be provided about a building. The following section will introduce how relatively intensively collected in-use monitoring can be used to improve the energy efficiency of buildings.

3.8 In-Use Energy Consumption

As discussed, the predicted energy performance of dwellings has until relatively recently been taken for granted to be accurate. However, there is a mounting body of evidence to suggest that buildings are not performing *in situ* as they were originally intended, referred to as the performance gap. This is typically attributed to the underperformance of the building fabric, faults in building services and the influence of occupants.

While the predicted performance of the building fabric and services is commonly derived from laboratory-based testing performed under standardised conditions, energy use predictions attributed to occupancy are often based on assumptions taken from theoretical occupancy patterns and applied to a model (BSI 2008). These occupancy patterns are estimates and so cannot accurately reflect how the building is used in practice; as a result, they are responsible for a degree of the gap between energy use prediction and reality. For example, a home will have differing energy consumption if a family with 4 children and a stay-at-home parent move when compared to a single person who commutes 2 h to work every day. Similarly, a primary school classroom used for occasional lessons will have different energy consumption to a similar classroom that doubles up as the indoor break time recreational space and which also hosts afterschool events every day of the week and yoga classes on the weekend. Thus, the only way to accurately determine the energy consumption of an individual building (and to understand how to make efficiency improvements) is to monitor it in-use.

Accurately identifying energy use in relation to external and internal environmental conditions gives the building manager or occupant the ability to diagnose issues relating to the use of the building and the services. For example, energy use during unoccupied times can be identified, set point temperatures can be refined where over or under heating is occurring, and responsiveness to external temperatures can be assessed. Other useful outputs from in-use monitoring in buildings are to show energy performance deteriorating over time, identify threshold temperatures when specific problems occur, track improvements resulting from interventions and refurbishments and even compare the actual running costs with that predicted by their design stage models. It is also possible to use in-use monitoring to compare buildings to one another, showing how different levels of thermal mass affects heating profiles, how different property types compare in terms of energy efficiency and also ranking and prioritising specific inefficient buildings that might benefit from improvements.

3.9 Monitoring Needs: What to Monitor and Why?

In-use monitoring can be costly and complex; thus, what variables are chosen to be monitored should be a function of what is trying to be deduced. For instance, if space heating demand is being investigated, more is required than simply taking gas meter readings; internal and external temperatures must also be logged in order to get an understanding of heating demand, plus a recording of any secondary heating system energy use is needed. If there are excessive electrical loads that contribute greatly to the amount of incidental gains, such as incandescent lighting, it may also be important to record these too, even the number of occupants in a building at any one time may also be important as could their window opening behaviour. Often, very basic sets of in-use monitoring data are collected, with electricity, gas and household temperature recorded (Fig. 3.5), but more detailed monitoring can also be undertaken (Fig. 3.6) (Aerts et al. 2014; Fabi et al. 2011).

In-use monitoring studies fall broadly into two categories: case studies and large samples. Case studies consider a small number of sites in a detailed way, often using disaggregated data to analyse parameters to a high degree of complexity. Large samples tend to utilise aggregated data such as total electricity use or total gas use for a large number of cases to compare trends and identify outliers. While it is possible to collect large amounts of high-level disaggregated data, these studies

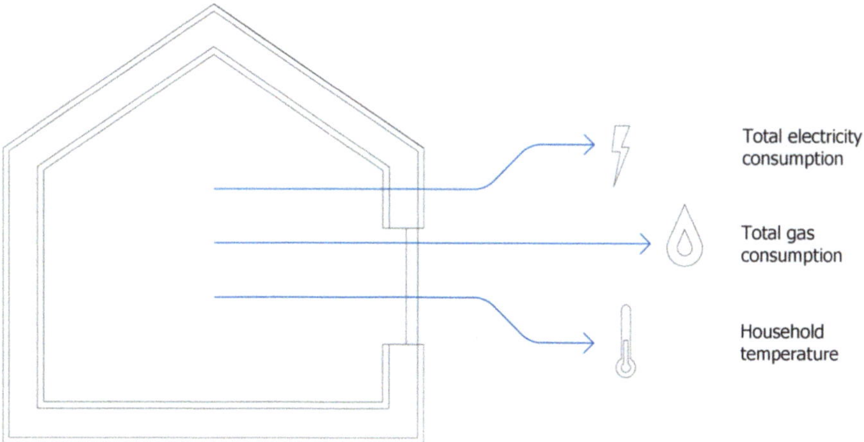

Fig. 3.5 Large aggregated data. Total energy, gas and household temperature monitored (Thomas and Gorse 2015)

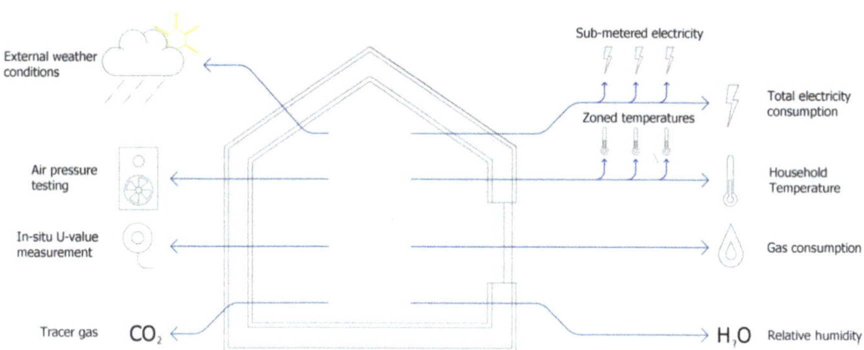

Fig. 3.6 Disaggregated data. Energy and environmental data collected separately and systematically (Thomas and Gorse 2015)

are expensive and require resources that are beyond the capacity of most research projects, and the large amount of individual tests and monitoring equipment can be intrusive for building occupants. An aggregated approach is the cheapest and easiest to implement and can be as simple as collecting meter readings on a monthly basis. The downside to this simplicity is that in isolation, this type of data often lacks explanatory power and requires additional data to add context. A disaggregated approach is far more detailed and robust, with measurements typically taken at shorter intervals and for a wider range of parameters, allowing a greater level of complexity in analysis. The obvious disadvantage is that this extra monitoring often comes with additional costs, as shown in Fig. 3.7.

When considering what to monitor, the analysis techniques that will be employed when handling the data should be considered. If the purpose of the research is to detect and diagnose faults within a case study, the resolution and complexity of

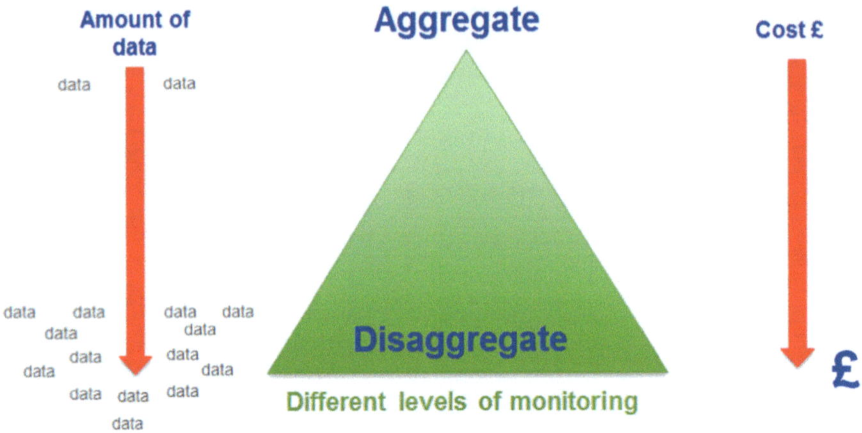

Fig. 3.7 Case study (disaggregated data) vs. large sample (aggregated data) in-use monitoring

the monitoring data are likely to be higher than a study looking for general trends and patterns in domestic water use across regions of the country. By determining the level of complexity required to answer the research question, superfluous data collection can be avoided.

Once the variables and approaches have been considered and a monitoring "wish list" created, practical considerations can be made. The majority of monitoring projects operate under strict financial budgets, limiting the type and amount of equipment that can be used. In many cases, equipment will require professional installation, disruption to residents and ongoing data retrieval and storage charges. These practical constraints may in reality be the limiting factor in determining the type and scale of in-use monitoring projects.

3.10 How to Monitor?

Energy monitoring systems are developing at a rapid pace as demand for energy use information grows from both in the construction industry and the individual homeowner. Currently, these systems can be grouped into two types: standalone and data acquisition. Standalone systems are individual in nature, with internal memory for data storage. The main advantage of standalone monitoring is that they are often a simple, low-cost solution that can be left at a site and collected at a later date. The main disadvantage is that any issues or patterns that arise during data collection will not be discovered until equipment is checked. Data acquisition systems offer an alternative; they are comprised of multiple sensors communicating with a central hub, which may simultaneously store data and transmit to an external storage platform, e.g. via Wi-Fi to the cloud and often with an interface for interrogating the data. This allows regular access to identify issues, though these systems tend to be more expensive to purchase and incur ongoing costs for data retrieval.

The ability to monitor in-use electricity, water and gas is dependent on these utility meters being "pulse-enabled". Commonly, the wide variety of domestic gas meters are not pulse-enabled, even if they have been relatively recently installed. Currently, there is no consensus between meter providers on how they will allow information to be retrieved from their meters or indeed whether this should always be possible and so this should always be checked before purchasing monitoring equipment. If utility meters are deemed not to be suitable for data collection, a separate meter will need to be installed, or where this is not possible in the case of gas, data on heat being used could be collected by installing heat meters on the heating system that may be used as a proxy for the kWh of gas used.

There is heterogeneity between the accuracy and recording range of different items of monitoring equipment. For domestic and commercial building scale energy consumption, resolutions may not need to be too refined and often an accuracy to 0.5°C may be sufficient. Industrial applications may require a greater level of accuracy to identify or predict faults or inefficiency in their processes. Generally speaking, each in-use monitoring installation will be bespoke, which allows individuals to have flexibility; however, it also means that it can be confusing for building owners to know what they need and can also mean it is difficult to compare or validate data derived from different in-use monitoring installations.

3.11 Considerations for In-Use Monitoring

There are a number of considerations that need to be made when undertaking in-use monitoring. These are as follows:

- Financial limitations are a barrier to what can be achieved.
- Conclusions drawn can only be robust if the necessary data has been collected, e.g. space heating energy efficiency cannot be determined from gas consumption alone.
- Building owners should contribute to the aim and design of the monitoring system, grant access to the installation and repair monitoring equipment.
- Ethical issues exist whenever collecting data from individuals; proper procedures in data collection and storage are required.
- Consideration should be given to the frequency of data capture and the capacity to monitor this data; higher frequency will give a finer resolution and strengthen the analysis, but requires more time and resource.

3.12 Understanding Building Energy Use, Occupant Behaviour and Change

A key aim of both new build projects and retrofits is to produce buildings that are highly energy efficient, thereby leading to a reduction in energy use. This contributes to various targets to reducing CO_2 emissions.

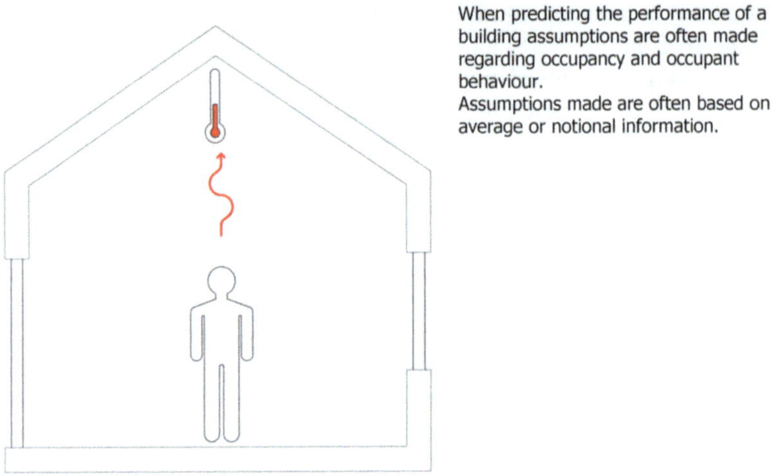

Fig. 3.8 Assumed building performance based on some basic occupancy, environment and energy needs (Thomas and Gorse 2015)

Calculations of energy efficiency are usually required before a project commences. There are often an assumed set of circumstances. Basic occupancy, temperature and energy requirements, such as those illustrated in Fig. 3.8, are often used, but in reality, occupancy, service use and building performance may be much more complicated. For example, Fig. 3.9 illustrates how the heat gains from occupants and animals contribute, as well as how the building is used. Figure 3.10 also shows how the use of appliances contribute and this is without considering the defects in the building (Fig. 3.11) and discrepancies in service performance. Thus, the difference between the predicted energy use and actual demand can differ considerably.

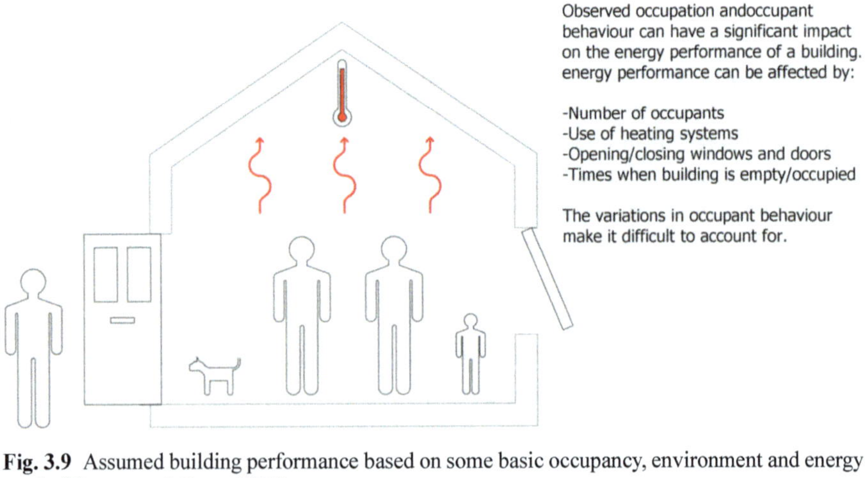

Fig. 3.9 Assumed building performance based on some basic occupancy, environment and energy needs (Thomas and Gorse 2015)

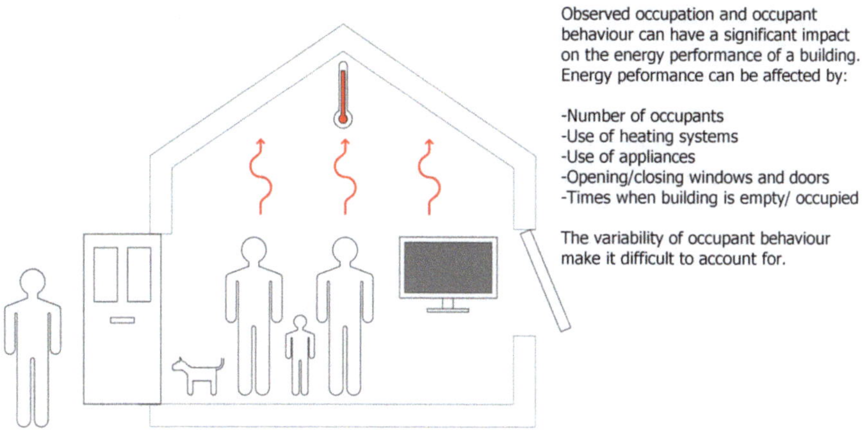

Fig. 3.10 Understanding Occupancy and Use: Consideration for occupancy, behaviour and appliances (Thomas and Gorse 2015)

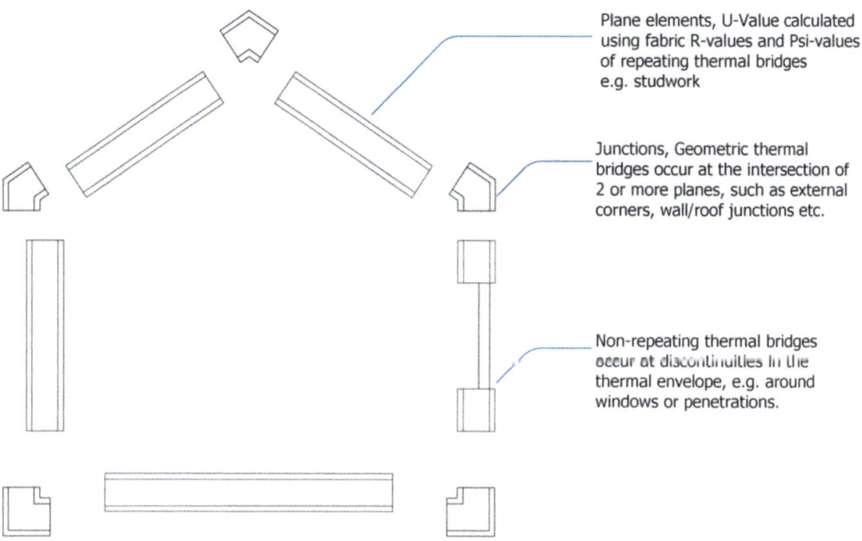

Fig. 3.11 Aspects of the building fabric that can contribute to differences in performance (Thomas and Gorse 2015)

Where, in practice, actual buildings underperform when compared with predicted performance, this had been termed as the performance gap. The performance gap means that the expected savings in both energy use and the cost of heating people's homes are not always achieved. There has been some work to understand why this is the case, although there remain many unanswered questions. We know that living in a better insulated home can increase, rather than decrease, energy consumption (e.g. Hens et al. 2010), and homes with better insulation have higher internal temperatures (Santin 2013). This has been referred to as the "rebound effect" in which people offset greater energy efficiency benefits against greater use of heating system. Sometimes, higher energy use arises from occupants not understanding how to use their heating system effectively (Linden et al. 2006), which can be because they were not instructed in how to use it, or they are reluctant to engage with the new technology. It is clear that a substantial proportion of variation in energy use is due to individual behaviour. Around half of the variation in the amount of energy the people use to heat their homes is due to differences in behaviour rather than the energy efficiency of dwellings themselves (Gill et al. 2010). This can be through occupancy patterns, i.e., the number of hours the house is occupied and the rooms used (De Meester et al. 2013) as well as the temperature the house is heated to.

We therefore need to gain a more in-depth understanding of people's experiences of heating their homes and their decisions around energy saving behaviour. Few qualitative research studies have been undertaken, yet this approach has the potential to provide greater insight into energy use behaviour and the user experience of buildings (e.g. Kim et al. 2013). It is also essential when developing interventions to change energy behaviour. Qualitative research techniques, such as interviews and focus groups, are likely to become increasingly important in informing the design and implementation of future energy efficiency schemes. The semi-structured approach typically adopted allows occupants to discuss areas influencing their energy behaviours not anticipated by the researchers. As such, this approach enables depth of understanding and the ability to develop insight and innovative ways of engaging with occupants and changing their behaviour.

Quantitative research methods can also be used, either instead of or alongside qualitative techniques, and yield data suitable for statistical analysis. Questionnaires are typically used to collect information from occupants on their behaviour. Most of these address patterns of occupancy, and some also include beliefs about energy efficiency and the environmental impact of energy use. It is also useful to include questions derived from psychological models of behaviour. There are several theoretical frameworks used to model behaviour, most commonly social cognitive models such as the theory of planned behaviour (TPB), and these have been applied to energy use behaviour. For example, perceived behavioural control is a strong predictor of the amount of energy savings that are made in domestic households (Abrahamse 2009). These authors found little predictive power in other TPB constructs, however. Therefore, feeling confident in one's ability to be able to reduce the amount of energy used predicts actual energy savings. This is useful, but we need to know more about psychological predictors of energy use and how people respond to interventions to change behaviour. This is an area that deserves substantial future research.

In addition, research with occupants can provide evidence on how housing influences quality of life. Living in a warmer more comfortable home could potentially lead to people feeling happier, and it has the potential to improve any long-term health conditions they have. Our research has also shown that retrofitting an area leads to people believing the area appears more attractive and this increases their sense of community. These aspects are valuable to include when researching the wider impact of increasing energy efficiency.

Our recommendations for what to include when designing research on occupant behaviour, based both on previous research and our experience of researching the impact of occupants on energy use, are shown below.

- Who lives in the home (people and pets) and any regular visitors, such as grandparents providing childcare.
- How the home is used, e.g. when people are in and out of the home, whether rooms are used differently in winter and summer, whether windows are kept open and doors are opened frequently.
- Any health problems that people have that might affect the temperature of the home and the energy they use, together with a measure of their health status.
- Preferred temperature within the home, the reason for this, and reasons why the actual temperature is not the same as the preferred temperature.
- Any life events that mean people might change the energy they use, such as spending more or less time in the home, having a baby, losing their job or retiring.
- Understanding of how to use the heating and ventilation systems.
- Confidence in the ability to use less energy ("perceived behavioural control").
- Beliefs about the advantages and disadvantages of using more or less energy.
- Beliefs about what other people expect in terms of energy use ("social norms").
- How satisfied people are with factors such as how quickly their home heats up, how warm it gets, how drafty it is, how damp it gets, how much it costs to heat, how much noise it lets in.
- How satisfied people are with living in their neighbourhood, e.g. its appearance, how safe it feels and how much they feel they belong.

In addition to people, we also need to consider how the services and renewable energy sources feedback to the building.

3.13 Monitoring of Renewable Technologies

3.13.1 Policy Background

Renewable technologies in dwellings are becoming increasingly common, largely due to government incentives such as Green Deal, Feed-in Tariffs (FiT) and Renewable Heat Incentive (RHI). However, the benefits available under these schemes are based upon dwelling assessment, monitored electricity output and deemed (or in some cases monitored) heat output, respectively, and the incentives offered also change over time as policy is reviewed. This is not sufficient to provide information on whether expected savings or technology performance is being achieved in practice. The Renewable Energy Consumer Code (RECC 2014) sets out clear responsibilities for accredited installers (including general business practices) and requires that reasonable estimates of performance expectations are provided to the customer, but these are generally based upon the customer's historic consumption data (where possible) together with data acquired from laboratory testing of equipment under applicable standards. There is no obligation to monitor the performance of systems in situ.

The most common technologies eligible for the domestic RHI are heat pumps, biomass boilers and solar thermal. Under the RHI tariffs are paid per kWh of renewable heat generated, but this is generally "deemed" (based upon calculated heat output requirements), rather than monitored, in order to avoid the additional cost of metering. Only in special cases (e.g. where there is a secondary heating system, or where the dwelling is not continuously occupied as in the case of a second home) is there a requirement for the heat output to be measured. Any customer may choose to install a meter, and is eligible for additional payments to cover the costs of doing so, but the resulting data acts only as confirmation that the system is working as expected, and the RHI payment is still based upon the deemed output.

In the case of electricity generation (e.g. solar PV and micro-wind turbines), payments are made under the FiTs scheme and are based upon the amount of electricity actually generated. There is also an export tariff that provides additional payment for electricity exported to the grid rather than used on-site. This means that installations must include an import/export electricity meter. Nevertheless, without complex submetering and weather data, this is not sufficient to give more than a very rough indication of technology performance.

3.13.2 Complex Nature of Field Trials: Monitoring Renewable Technology

The above analysis illustrates the basis of the problem with regard to domestic renewable energy technology performance in general. In order to understand the system performance in situ, rather detailed monitoring of a wide range of parameters is required for each individual system. Thus, monitoring becomes expensive and time-consuming if performed with real rigour. DECC has sought to address the issue by commissioning a series of field trials, e.g. the field trials program on domestic heat pumps which was undertaken in two phases by EST on behalf of DECC between 2008 and 2013 (Dunbabin and Wickins 2012; Dunbabin et al. 2013). This was a full-scale research project, with careful attention given to the data produced, but even so many difficulties were experienced, particularly relating to fair comparison of the performances of a wide range of different system types, the analysis of issues relating to domestic hot water (DHW) production and the treatment of system boundaries and heat losses not accounted for under practical metering configurations. Inspection of the two reports produced gives a clear impression of the scale of the difficulties that can experienced in this type of study.

Field trials such as the one referred to in the previous paragraph, and also similar smaller-scale trials undertaken by Leeds Beckett University and other researchers, tend to indicate that while good performance *can* be achieved in practice, the reality falls short of expectations in too many cases. Therefore, it is of vital importance to undertake further detailed studies in the field in order to enhance our understanding of factors affecting real performance in situ.

3.14 Practical Metering Issues: Monitoring Renewable Technology

For renewable heat generation, monitoring involves measuring heat output. In the case of heat pumps and pellet biomass boilers (but not pellet stoves with back boilers, other types of biomass boilers or solar thermal installations), metering guidelines are produced by DECC for compliance with their Metering and Monitoring Service Packages(DECC 2013). Placing of heat metres is vital in order to include all required outputs without unrecorded losses, but in recognition of the practical difficulties in the case of some heat pump configurations, the strategy for heat pumps is to measure heat output from "whatever components are practical to measure", provided that the energy inputs to the same components are also measured. The guidelines do however specify that heat meters should consist of a flow meter together with a matched pair of temperature sensors, and should fall within accuracy Class 3 or better, as defined by Annex MI-004 of the Measuring Instruments Directive (European Commission 2004) Furthermore, the meter resolution must satisfy the relationship:

$$[\text{Resolution of Heat Meter}] \leq 3\% * [\text{Minimum non-zero heat output in two 2minutes}]$$

In practice, this tends to result in a required resolution of 1 pulse per 10Wh, or more frequently, 1 pulse per 1Wh. The latter is almost always required to take account of short DHW draw-offs. At the time of writing, many commercial heat meters are not able to achieve this resolution, and therefore, available compliant solutions are restricted and may be costly. In addition, the time resolution of 2 min for data recording is recommended, resulting in large datasets being generated. At the same time, the specified level of accuracy in measurement is compromised by allowing water-calibrated heat meters in water/antifreeze systems, and not requiring bi-directional heat meters for air-source heat pumps in defrost mode. In these cases, a fixed percentage penalty is applied, or assumptions made regarding losses, when calculating the system efficiencies.

For renewable electricity generation, other practical difficulties apply. The measurement of solar PV system performance requires metering of relatively high-voltage DC power as well as AC output if losses at the inverter are to be characterised. Understanding of system performance also requires measurement of the in-plane solar irradiance, and is enhanced by other measurements such as external air temperature.

3.14.1 *Data Management*

The generation of large datasets, together with the possibility of metering malfunctions, indicates that frequent downloading, inspection and processing of data is good practice. For reasons of convenience and efficiency, and to prevent inconvenience to dwelling occupiers, remote management is usually necessary. Data are therefore logged and stored on-site by some device capable of transmitting data via the mobile phone network (GSM/GPRS modem) or via the Internet. The former can present difficulties relating to poor or variable signal coverage, while the latter presents issues surrounding ownership and the use of equipment and Internet connections. In addition, data losses can occur between sensors/meters and the logging device, in transferring information via wireless radio signal, which may or may not be based on localised networks such as Zigbee (2014). While data downloading and management on a regular, frequent basis may be time-consuming, it does have the advantage of enabling prompt recognition of sensor/meter failures, and can also play a useful part in remote system diagnostics, enabling "action research" style interventions where systems are performing poorly.

3.15 Installation Issues: Monitoring Renewable Technology

The performance of many renewable technologies is sensitive to the details of design and quality of installation. Engagement with monitoring programs can highlight some of these issues and serve as a learning process for installers.

The installation of monitoring equipment can also require careful thought and analysis in order to maximise reliability and minimise disruption or inconvenience to occupiers. At the time of writing, most monitoring programmes are research initiatives, and monitoring systems can therefore be designed and tailored specifically for a given dwelling or group of dwellings. Even so, experience suggests that, too often, not enough attention is paid to the monitoring system installation procedure. Field trials in the past have tended to be most successful when the installation of monitoring equipment on-site has been overseen by a researcher or monitoring specialist. This approach also has the advantage that any individual modifications (such as non-ideal placement of sensors or meters due to practical considerations) are understood and approved from the outset, and their potential effects on the analysis can be estimated.

Nevertheless, as monitoring programmes become larger or more widespread, this approach will rapidly become impracticable. Standards and guidelines for monitoring systems will then become very important, for the purposes of comparability and quality assurance, especially if routine monitoring is undertaken by commercial interests. It has been suggested that in some cases (e.g. for heat meters on heat pumps), the best way of achieving this would be via a requirement for the monitoring equipment to be a standard part of the system as supplied to the customer. This is also likely to be less costly than a retro-fitted monitoring system.

3.16 Social and Behavioural and Renewable Heat Systems: Context

The performance of systems such as micro-wind and solar PV is relatively independent of user behaviour, the main issue being whether the energy produced is used on-site or exported to the grid. However, renewable heat systems can be more sensitive to user behaviour. For example, since the coefficient of performance of heat pumps depends upon the input/output temperature difference, then lower preferred internal dwelling temperatures can lead to improved efficiency. There are also a number of issues surrounding DHW production temperature and storage. Therefore, it can be important to consider monitored systems in the context of dwelling occupancy and usage patterns.

Most users wish to get the best from their systems, but since the principles of operation may be unfamiliar, or even counter-intuitive, they may not have sufficient information or expertise to do so easily. This problem can be exacerbated by poorly designed or complicated control interfaces. An important area of future research will centre on designing controls which are user-friendly and encourage efficient system operation. Intelligent controls, with some element of remote management, may also prove advantageous, especially for the purposes of addressing grid-related issues. These types of innovations are likely to provide further opportunities for routine monitoring of systems.

3.17 Measuring the Performance of the Building Fabric: Aggregated and Disaggregated Performance

Many tools, techniques and methods currently exist that are capable of being deployed in the field to measure a range of parameters relating to the building fabric's *in situ* performance. These tools, techniques and methods can be separated into two very distinct categorises that describe the approach that they adopt to measure building fabric performance, namely:

- An aggregate approach.
- A disaggregate approach.

Those that adopt an aggregate approach are capable of capturing all of the complex inter-related heat transfer mechanisms that occur within a building, such as conductive and convective heat loss, radiation exchanges, bulk air movement and thermal bypassing. However, as they tend to aggregate all of the individual heat loss mechanisms together, the resulting output is a single heat loss figure for the entire building envelope. Consequently, if this type of approach is adopted in isolation, then it is not possible to determine the heat loss attributable to the various aspects of the building envelope, such as background ventilation heat loss or the heat loss attributable to different elements of the buildings fabric (walls, floors, roofs, etc.). Only a very limited number of aggregate (whole building) approaches are currently available. Examples include the coheating test method (Wingfield et al. 2010; Johnston et al. 2013),

ISABELE (Bouchié et al. 2014), the Primary and Secondary Terms Analysis and Renormalisation (PSTAR) method (Subbaro 1988; Subbaro et al. 1988) and the Quick U-value of Buildings (QUB) method (Mangematin et al. 2012). Of the four aggregate methods available, two of the methods (the QUB and the ISABELE method) are currently under development and their use has primarily been limited to research and development purposes. The QUB test has been successfully used and the results compared, within an environmental chamber against the coheating method.

Quick U-Value of Buildings (QUB) is a very simple diagnosis method enabling the measurement of the heat loss coefficient in one to two nights. It is done by measuring the building's temperature response to two consecutive thermal loads (Mangematin et al. 2012). Constant heat is driven into the building followed by a free cooling period, when the supply of heat is terminated. The QUB test has been performed with a range of heat loads and time periods; following a number of short studies, the procedure was found to give some surprisingly reliable results in less than 2 h, and more generally, uncertainties estimated to be around $\pm 15\%$ when the tests are done over a single night. Such dynamic tests and in-use tests, using similar energy profiles are likely to become more common in the future and could be integrated into smart monitoring technologies. Blind tests were conducted by Saint Gobain Recherche, Leeds Beckett University and Salford University as part of the Saint Gobain Energy House project, the results of both the performance of the retrofitted project and the cross-comparison of coheating and QUB tests were encouraging (Saint Gobain 2014; British Gypsum 2014; Farmer et al. 2015).

With the two short dynamic methods, the application of the PSTAR method in the field has been limited, particularly in the UK, with the QUB test showing some encouraging results. However, due to the reliability of the full coheating test, the aggregate method that has seen most significant application in the field, both in the UK and abroad, has been the coheating test method. The only aggregate method that has seen significant application in the field and development, both in the UK and abroad, has been the coheating test method. In the UK, the coheating test method has become established as a recognised *in situ* test method to obtain an estimate of the overall heat loss coefficient (HLC) of an unoccupied dwelling and formed a requirement of the recent Technology Strategy Board's Building Performance Evaluation Programme, which aimed to understand the key factors that influence the in-use performance of buildings (Technology Strategy Board 2010).

The vast majority of the tools, methods and techniques that are currently available adopt a disaggregate approach to measuring building fabric performance. This approach tends to involve measuring the heat loss attributable to a particular aspect or element of the building fabric in isolation, such as the heat loss attributable to a ground floor. Although it is possible to take the results that have been obtained from a wide range of disaggregate methods and aggregate them together, it is not practically possible to measure all of the individual heat loss mechanisms that occur within a building. In addition, it is also only practically possible to take *in situ* measurements from a very small proportion of the total surface area of the building fabric. Consequently, it is very difficult to try and replicate the results that have been obtained using an aggregate approach using disaggregate approaches alone.

Examples of disaggregate approaches include air flow measurement, cavity temperature measurement, heat flux measurement, leakage detection, pressurisation testing, thermal imaging, tracer gas measurement and partial deconstruction of the building envelope.

3.18 Coheating Test Method

The electric coheating test method was first proposed by Socolow (1977) in North America in the late 1970s, with the first documented description of the test method being published by Sonderegger et al. in 1979. The test method was originally devised to determine the net efficiency of domestic heating and cooling systems (Sonderegger et al. 1979). It involved heating the dwelling electrically, using thermostatically controlled portable heaters, to a constant internal temperature and then measuring the reduction in the electric heating required when the dwellings own heating system was operated. The term "coheating" was aptly used to describe the method, as two heating systems were required to be operated during the test. It was also found that by applying this test method, the overall aggregate heat loss coefficient (in W/K) of the dwelling (both fabric and ventilation) could be determined.

The earliest documented use of the test method in the UK was in the mid-1980s (see Siviour 1985; Everett 1985), where it was developed to empirically determine the overall heal loss coefficient (HLC) and response to solar gains of unoccupied dwellings using portable electric resistance heaters only. Since then, the test method has seen very limited application in the field (for instance, see Bell and Lowe 1997) up until the last decade or so, most notably following development and application of the test method by Leeds Metropolitan University (now Leeds Beckett University) as part of the Stamford Brook Project. During this project, a modified version of the electric coheating test method developed by Siviour (1985) and Everett (1985) not only identified a significant performance gap in new housing, but was also used to identify and quantify the party wall bypass heat loss mechanism (see Lowe et al. 2007). Following the Stamford Brook Project, the electric coheating test method was further refined and developed by Leeds Beckett University, resulting in the 2010 version of the Whole House Heat Loss Test Method (see Wingfield et al. 2010). This version became recognised as an established test method in the UK when it was incorporated within the Post Construction and Initial Occupation studies undertaken under the Technology Strategy Boards (now Innovate UK's) Building Performance Evaluation Programme (Technology Strategy Board 2010).

The latest version of the electric coheating test method described by Johnston et al. (2013) is a quasi-static state test method that has been developed to be applied to an unoccupied dwelling in the field. It is considered to be a quasi-static state test method, as the internal temperature conditions within the dwelling are held static, while the external conditions are allowed to vary dynamically in response to the external climatic conditions. The test method involves heating the internal volume of an unoccupied dwelling to a mean elevated homogeneous internal temperature using thermostatically controlled electric resistance point heaters, and then

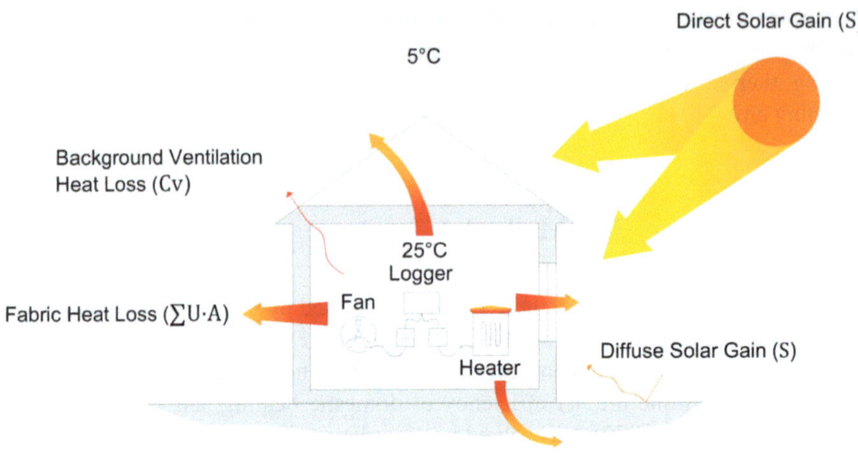

Fig. 3.12 Diagrammatic representation of an electric coheating test [Source: Brooke-Peat 2015]

maintaining this temperature constant for a specified number of days, typically between 7 to 21 days. In addition to the electric resistance point heaters, electrically driven air circulation fans are also installed and strategically positioned within the dwelling to ensure that the internal air is adequately mixed and that there is minimal temperature stratification (see Fig. 3.12). In the UK, the mean-elevated internal temperature is commonly set at 25 °C to ensure that there is a sufficient temperature difference between the inside and the outside of the dwelling (ΔT) such that the primary direction of the heat flow is from the inside to the outside of the dwelling. The direction of the heat flow is also ensured by undertaking the tests during the heating season (October/November to March/April). This normally ensures that at least a 10 K ΔT is maintained throughout the test period. A mean internal setpoint temperature of 25 °C also ensures that the internal temperature lies within the range of temperatures that would normally be expected to occur within the dwelling during occupation. During the electric coheating test, various internal and external parameters are measured, such as internal and external temperatures, solar radiation and the total electrical power input to the dwelling. By measuring these various parameters, the total daily heat input to the dwelling in Watts that is required to obtain a particular ΔT in K can be established.

Central to the analysis of the coheating test data is the assumption that the following energy balance holds true:

$$Q + R.S = \left(\Sigma U.A + C_V\right).\Delta T \tag{1}$$

Where:

Q = Total measured power input into the dwelling (W)

R = The solar aperture of the house (m^2)

S = The total amount of South facing solar radiation (W/m^2)
ΣU.A = Total fabric heat loss (W/K)
C_v = Background ventilation heat loss (W/K)
ΔT = Temperature difference between the inside and the outside of the dwelling (K).

An estimate of the raw uncorrected heat loss coefficient (HLC) in W/K for the dwelling can then be determined by rearranging the energy balance such that:

$$(\Sigma U.A + Cv) = (Q + R.S) / \Delta T \qquad (2)$$

By plotting the daily heat input in Watts (Q) against the daily internal and external temperature difference in Kelvin (ΔT), the resulting gradient of the plot gives the raw uncorrected heat loss coefficient. An example of such a plot is shown in Fig. 3.13. The raw uncorrected heat loss coefficient can then be corrected to take into account external environmental effects, such as solar radiation, by applying multiple linear regression analysis techniques. Further details regarding the techniques that can be used can be found within Johnston et al. (2013). An example of a plot using the multiple linear regression analysis method to account for solar radiation is illustrated in Fig. 3.14.

Although the electric coheating test has been around for some considerable time, widespread application of the test method in the field in both the UK and abroad has been limited. The reasons for test methods limited application in the field include:

- The dwelling needs to be unoccupied during the entire test duration.
- The test duration is relatively long, typically between 1 and 3 weeks.

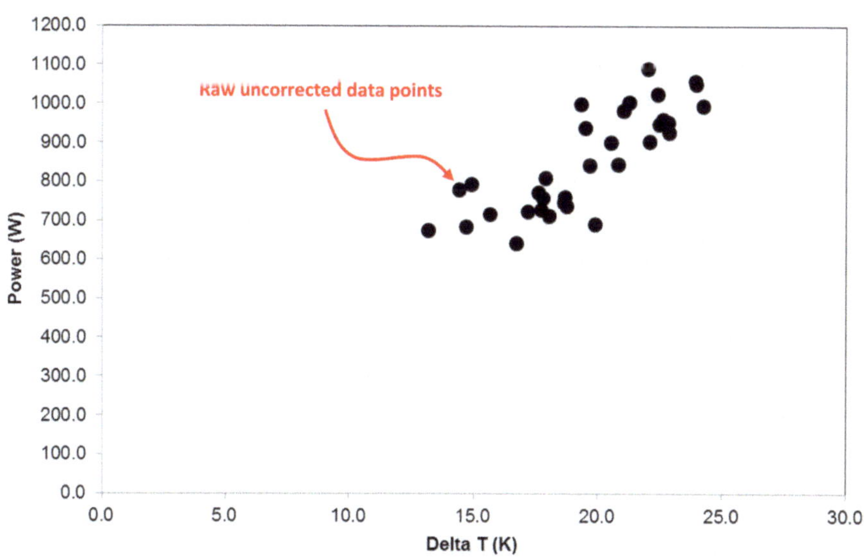

Fig. 3.13 Example plot of the raw data obtained from a coheating test

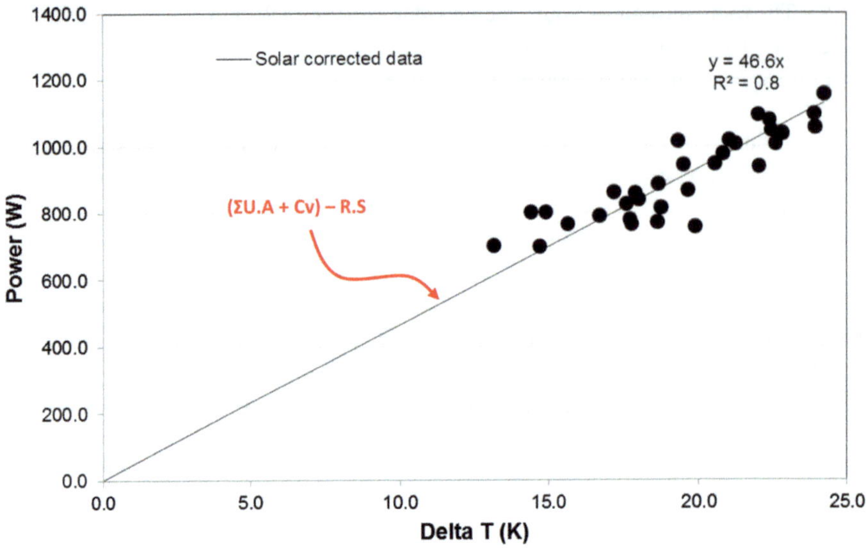

Fig. 3.14 Example plot of solar corrected data obtained from a coheating test

- Access to the dwelling needs to be restricted throughout the test duration to the minimum required to ensure that the test is proceeding satisfactorily.
- All electrical items of equipment need to be isolated during the test, such as fridges, freezers, MVHR units. This may be problematic in the existing dwellings.
- The test can only be undertaken within a restricted time period which may not coincide with dwelling completion. In the UK, this time period is restricted to the heating season, typically between October and March to minimise the impact of solar heat gains on the building fabric and to ensure that a sufficient daily average ΔT is maintained.
- When undertaking the test, there is no guarantee that a confident estimate of the HLC will be obtained, as it is highly dependent upon the environmental conditions experienced during the test.
- The dwelling needs to have sufficiently dried out prior to commencing the test. Otherwise, the rate of heat loss measured could be exaggerated if there is any residual construction moisture in new builds and renovations, or moisture build up on older buildings. The additional advantage of ensure that the dwelling has dried out, before heating the building under test conditions, is that the potential for surface condensation and mould growth is considerably reduced.
- There is currently no recognised international test standard on electric coheating, although a draft CEN test standard is currently under development.
- There is a lack of testers with sufficient experience of undertaking the test method.
- The financial costs, amount of equipment and time required to undertake the test can be prohibitive.

In recent years, some concern has also been raised regarding the reliability and practicality of the electric coheating test method and the level of uncertainty associated with the estimated HLC (see Butler and Dengle 2013). There are various uncertainties associated with the test method that may have an impact on the reliability and repeatability of the estimated HLC. The main areas of uncertainty relate to errors associated with the data analysis, errors associated with the measurement of various parameters, procedural errors and the individual circumstances surrounding the particular test (e.g. the external environmental conditions, the thermal mass and capacity of the building envelope and the moisture content of the individual building materials). Data analysis errors can be minimised by ensuring that appropriate analysis techniques are used and care is taken when analysing the test data. Measurement errors can be minimised by ensuring that the test equipment is suitably calibrated. Procedural errors can be minimised by ensuring that the test method protocol is carefully followed and that experienced testers undertake the test. However, the uncertainties associated with the circumstances surrounding the test are much more difficult to minimise, as they are not only highly context dependent, but more often than not are inter-related. One of the few authors to have explored the uncertainty associated with some of the environmental conditions is Stamp et al. (2013). They concluded that under cold low solar conditions, unbiased estimates of the HLC could be obtained in highly glazed heavyweight dwellings. However, in the same highly glazed and heavyweight dwellings, under high solar conditions, the HLC may be underestimated.

Leeds Beckett University has also undertaken work on the repeatability and reliability of the electric coheating test. The results obtained from a repeat test on a dwelling located in the field, as well as a series of tests at different ΔT's on a dwelling located within a test chamber, revealed no significant difference in the measured HLC. Although the results obtained from these different tests do give a degree of reassurance in the repeatability and reliability of the electric coheating test methodology, it is recognised that further research is required to gain a more thorough understanding of the uncertainties involved in the test method.

3.19 Summary

This chapter has introduced some of the issues that need to be considered when attempting to understand and measure the thermal performance, energy behaviour or occupants and energy demand of building in use. As understanding is developed into the behaviour of buildings, occupants and services, we need to feed this information into models, simulations and designs, so that we can improve predicted performance. This chapter should also be read with the supporting chapters on building performance.

References

Abrahamse W, Steg L (2009) How do socio-demographic and psychological factors relate to households' direct and indirect energy use and savings? J Econ Psychol 30:711–720

Aerts D, Minnen J, Glorieux I, Wouters I, Descamps F (2014) A method for the identification and modelling of realistic domestic occupancy sequences for building energy demand simulations and peer comparison. Build Environ 75:67–78

Bell M, Lowe RJ (1997) The York energy demonstration project. Final Report on a DOE Greenhouse Project, CEBE Research Report No 3, Leeds Metropolitan University

Bouchié R, Alzetto R, Brun F, Boisson A, Thebault PS. (2014) Short Methodologies for In-situ Assessment of the Intrinsic Thermal Performance of the Building Envelope, In Sustainable Places 2014, Nice

British G (2014) Saint-Gobain energy house brochure, energy house salford, British gypsum.com online http://www.british-gypsum.com/case-studies/energy-house-salford. Accessed Feb 2015

Brooke-Peat M (2015) Personal communication. Leeds Beckett University, Leeds

Butler D, Dengle A (2013) Review of Co-heating Test Methodologies. Report F54. Milton Keynes, NHBC Foundation

BSI (2008) BS EN ISO 13790:2008 Energy Performance of Buildings. Calculation of energy use for space heating and cooling. British Standards Institution, Londo

DECC (2013) Metering and monitoring service packages technical supplement: outline of metering and monitoring service package requirements for the domestic renewable heat incentive. Department for Energy and Climate Change, Crown Copyright, London

De Meester T, Marique A-F, De Herde A, Reiter S (2013) Impacts of occupant behaviours on residential heating consumption for detached houses in a temperate climate in the northern part of Europe. Energy Build 57:313–323

Dunbabin P, Wickins C (2012) Detailed Analysis from the first phase of the Energy Saving Trust's heat pump field trial, Department for Energy and Climate Change, London, March 2012. www.gov.uk/government/publications/analysis-from-the-first-phase-of-the-energy-saving-trust-s-heat-pump-field-trial. Accessed Nov 2014

Dunbabin P, Charlick H, Green R (2013) Detailed Analysis from the second phase of the Energy Saving Trust's heat pump field trial, Department for Energy and Climate Change, London, May 2013. www.gov.uk/government/publications/analysis-from-the-first-phase-of-the-energy-saving-trust-s-heat-pump-field-trial. Accessed Nov 2014

European Commission (2004) Official Journal of the European Union. Directive 2004/22/EC of the European Parliament and of the Council of 31 March 2004 on measuring instruments. Available from http://eur-lex.europa.eu/LexUriServ/LexUriServ.do?uri = OJ:L:2004:135:0001:0080:EN:PDF. Accessed Nov 2014

Everett R (1985) Rapid Thermal Calibration of Houses. Technical Report ERG 055 for the Science and Engineering Research Council, Milton Keynes, UK

Fabi V, Andersen RV, Corgnati SP, Olesen BW, Filippi M (2011) Description of occupant behaviour in building energy simulation: State-of-art and concepts for improvements. Proceedings of Building Simulation 2011: 12th Conference of International Building Performance Simulation Association, 2882–2889

Farmer D, Brooke-Peate M, Miles-Shenton D, Cuttle C, Gorse CA (2015) Saint gobain energy house thermal upgrading. Leeds Beckett University, Leeds (unpublished)

Gill ZM, Tierney MJ, Pegg IM, Neil A (2010) Low-energy dwellings: the contribution of behaviours to actual performance. Build Res Inf 38(5):491–508

Gorse CA, Johnston D, Miles Shenton D (2011) Low Carbon Housing—Failings in Construction Practice, Architecture Science, ArS, No.3, June 2011. Taiwan (ISSN 2219–1577) 53–67

Gorse C, Erkoreka A, Fletcher M, Martin K, Sala JM (2015) Optimising test environment and test set up for characterizing actual thermal performance of building components and whole buildings, International Energy Agency: Annex 58. http://www.leedsbeckett.ac.uk/as/cebe/resources.htm. Accessed June 2015

Hens H, Parijs W, Deurinck M (2010) Energy consumption for heating and rebound effects. Energy Build 42:105–110

Johnston D. Miles-Shenton D. Farmer D, Wingfield J (2013) Whole House Heat Loss Test Method (Coheating). June 2013 [Internet] Leeds, Leeds Metropolitan University. Available from: www.leedsbeckett.ac.uk/as/cebe/projects/cebe_coheating_test_method_june2013.pdf. Accessed 14 Jan 2015

Kim MJ, Oh MW, Kim JT (2013) A method for evaluating the performance of green buildings with a focus on user experience. Energy Build 66:203–210

Linden A-L, Carlsson-Kanyama A, Eriksson B (2006) Efficient and inefficient aspects of residential energy behaviour: What are the policy instruments for change? Energy Policy 34:1918–1927

Lowe R, Wingfield J, Bell M, Bell J (2007) Evidence for heat losses via party wall cavities in masonry construction. Build Serv Eng Res Technol 28(2):161–181

Mangematin E, Pandraud G, Roux D (2012) Quick measurements of energy efficiency of buildings. C. R. Physique 13:383–390

Renewable Energy Consumer Code (2014) Renewable energy and consumer code: protecting consumers and promoting renewable, October 2014. www.recc.org.uk. Accessed Nov 2014

Saint G (2014) Saint Gobain Whole House Retrofit http://www.saint-gobain.co.uk/latest-news/2014/february/saint-gobain-whole-house-retrofit-project-achieves-63-energy-saving-at-salford-energy-house/. Accessed Feb 2015

Santin OG (2013) Occupant behaviour in energy efficient dwellings: evidence of a rebound effect. J Housing Built Environ 28:311–327

Siviour J (1985) Experimental Thermal Calibration of Houses. In: Everett R Rapid Thermal Calibration of Houses. Technical Report ERG 055 for the Science and Engineering Research Council, Milton Keynes, UK

Socolow RH (1977) The twin rivers program on energy conservation in housing: highlights and conclusions. Energy Build 1:207–242

Sonderegger RC, Condon PE, Modera MP (1979) In-situ Measurements of Residential Energy Performance Using Electric Co-heating. ASHRAE Transactions, 1979; Vol. 86 (I).

Stafford A, Johnston D, Miles-Shenton D, Farmer D, Brooke-Peat M, Gorse. C (2014) Adding value and meaning to coheating tests. Structural Survey 32(4):ISSN:0263–0080X

Stamp S, Lowe R, Altamirano-Medina H (2013) An Investigation into the Role of Thermal Mass on the Accuracy of Co-heating Tests Through Simulations & Field Results. In Proceedings of BS2013: 13th Conference of international Building Performance Simulation Association (IPBSA), Building Simulation 2013, Chambery, France, August 25–28, pp. 39–46.

Subbarao K (1988) PSTAR—Primary and secondary terms analysis and renormalization: a unified approach to building and energy simulations and short-term testing—A summary. September 1988. SERI/TR-254–3347. Solar Energy Research Institute, Colorado

Subbarao K, Burch JD, Hancock CE, Lekov A, Balcomb JD (1988) Short-term energy monitoring (STEM): application of the PSTAR method to a residence in Fredericksburg, Virginia. September 1988. SERI/TR-254–3356. Solar Energy Research Institute, Colorado

The Technology Strategy Board (2010) Building performance evaluation, domestic buildings—Guidance for Project Execution. The Technology Strategy Board, Swindon

The ZigBee Alliance (2014) What is Zigbee. http://zigbee.org. Accessed Nov 2014

Thomas F, Gorse C (2015) An illustrated review of the factors contributing to differences between designed performance and in-use energy demand. LSi Technical Paper 2015-TP-0001. Leeds Sustainability Institute. http://www.leedsbeckett.ac.uk/as/cebe/resources.htm. Accessed 3 March 2015

Van der Geer J, Hanraads JAJ, Lupton RA (2000) The art of writing a scientific article. J Sci Commun 163:51–9.

Wingfield J, Johnston D, Miles-Shenton D, Bell M (2010) Whole House Heat Loss Test Method (Coheating). [Internet] Leeds, Leeds Metropolitan University. http://www.leedsbeckett.ac.uk/as/cebe/projects/coheating_test_protocol.pdf. Accessed 24 April 2014

Chapter 4
Construction Sustainability Through Visualisation of Building Operation

Farzad Khosrowshahi and Parisa Ghodous

4.1 Introduction

In its broadest definition, sustainability entwines environmental, economic and social concepts. Due to its socioeconomic and environmental significance, construction industry has its fair share of contribution to sustainability agenda (Pitts 2008). In most industrialised countries, buildings account for about 40% of total energy use and around 30% of greenhouse gas emissions (*Autodesk 2008 for US*). The scale of the impact may vary at different phases of project life cycle but all phases play a contributory role. While the operation phase consumes, by far, the biggest share of the life cycle cost and time, the decisions at the design phase are likely to exert widespread implications, affecting all other phases including maintenance and demolition.

Typically, performance evaluation and analysis are made after the design is developed rather than during. Further, in most cases such decisions are made to examine project cost rather than considering the lifecycle impact of the decisions. On occasions where evaluation tools are used, they are applied in isolation rather than examining their combined effect. Apart from adverse impact on project performance, the lack of holistic and integrated analysis of life cycle performance will undermine sustainability priorities (Schueter and Thessling 2008).

A collaborative integrated environment underpinned by visualisation and analytical tools, offers intelligent sharing of coordinated multidisciplinary information throughout the whole project life cycle, and the decisions at the design phase in particular. The consideration of sustainability at the design phase can manifest itself in a variety of ways (Azhar et al. 2009); examples include material selection, site

F. Khosrowshahi (✉)
School of Built Environment and Engineering, Leeds Beckett University,
Northern Terrace, Leeds LS2 8AG, UK
e-mail: f.khosrowshahi@leedsbeckett.ac.uk

P. Ghodous
Computer Science Department (Départment Informatique),
University Claude Bernard Lyon 1, Lyon, France

© Springer International Publishing Switzerland 2016

M. Dastbaz et al. (eds.), *Building Sustainable Futures*,
DOI 10.1007/978-3-319-19348-9_4

selection, management (Hardin 2009), energy demand (Autodesk, Inc 2008), thermal analysis, ventilation and air flow, solar analysis, resource management, shading design, carbon emission, building orientation, daylight analysis and water reaping (Krygiel and Nies 2008).

Through defacto industry standards language such as gbXM, major providers such as Graphisoft and Autodesk offer packages to transform building information to a diversity of sustainable building performance analysis solutions. Particularly, with the development of building information modelling, BIM), the range of technologies that support building sustainability evaluations has been on the increase, particularly during the past decade (Azhar et al. 2008; Kriegel and Nies 2008; Gleeson 2008). Examples include Autodesk EcotectTM, Autodesk Green Building Studio (GBS); Autodesk1 Virtual Environment; Hevacomp, Energy Plus, Delight; Radiance; HEED; Homer; Virtual DOE; Bentley HEVACOM and Bentley TAS; Climate Consultant. EcotectTM, by Autodesk, for instance, offers energy analysis, thermal analysis and lighting/shading analysis. It is 'a complete building design and environmental analyses tool that covers the full range of simulation and analyses functions required to truly understand how a building design will operate and perform' (Autodesk, Inc 2008). Another Autodesk product is Green Building StudioTM (GBS) which is web-based service for energy and thermal analysis, lighting and shading analysis, and value/cost analysis. It helps to evaluate the environmental impact of individual building components early in the design process. Alternative solutions include Integrated Environmental Solutions (IES)1 and Virtual Environment (VE)TM software. It provides a suite of integrated building performance analyses tools for solar, lighting, energy, costs and egress, as well as other variables.

The technology to view an object like a building (or any of its components) 'in time' has long been developed and exploited by practitioners and researchers alike. By using this technology we are able to view a building from different viewpoints. At the turn of the twentieth century this debate inspired artists such as Picasso to depict a multi-perspective view of a face in one image, as if time is frozen at a moment (Fig. 4.1). However, there has been very little work facilitating visualisation

Fig. 4.1 Painting by Picasso (artnet.com)

of a building 'though time'—viewing it as it degrades over time. This objective will require understanding of the time-related behaviour of building materials and components and its representation in visual form.

In this chapter, the overall model of degradation visualisation is presented and its application is shown by focusing on two example areas: lighting and flooring systems. For the lighting system, the visual impact of intrinsic and environmental factors is discussed. For the flooring system, the main challenge proved to be associated with the identification and visual presentation of peoples' patter of movement. This process is mathematically modelled and visually presented via virtual reality modeling language (VRML).

4.2 Building Maintenance

Building maintenance is a complex and costly activity that spans over a large part of building life cycle. Generations of building maintenance approach have consisted of both planned and unplanned maintenance (BS 3811 1984). The planned maintenance is intended as a corrective measure so as to avoid problems prior to their occurrences, although, this is not always the case. The unplanned maintenance can be grouped into corrective and preventative maintenance which itself is further subdivided into scheduled and condition-based maintenance. The predictive maintenance approaches aim at user comfort, though at a cost. The preventative maintenance attempts to reduce the cost of unnecessary recovery through regular inspections. The corrective approach removes the cost of inspection through timely provision of maintenance. On the other extreme is the maintenance on demand, where the focus is on cost saving rather than user comfort. The choice of these approaches is not only governed by cost, on the one hand, but also factors like building classification, visibility and affordability (Akizuki 2000). Other choices are based on decision criteria such as the knowledge of local expertise, regulations and consequence (Bertelsen and Nesje 2000).

Despite their differences, these categories of building maintenance tend to either compromise cost or user comfort. In other words, the corrective measure is taken earlier or later than the expected time. The balance between the two options has highlighted the benefits of the third alternative to building maintenance, namely, the just-in-time approach. However, the successful implementation of this approach is contingent upon the evaluation of the behaviour of building elements over time. Previous attempts to examine this behaviour date back to late 1970s and early 1980s (Arendarski 1978; Thierry and Zalewski 1982). The more advanced approaches include the use of multi-criteria technique and using artificial neural networks and fuzzy set theory (Urbanski and Waszczyszyn 2002).

4.3 Event and Effect Model

A building consists of a definite number of elements. They are typically categorised in a number of hierarchical layers ranging from substructure, internal finish and fitting to wall finish, flooring, ceiling and then walls, doors and then door handles, etc. Building elements degrade over time thus necessitating corrective measures, such as repair or replacement. The deteriorations are primarily due to buildings' interaction with its environment through events acting upon them. In other words, an effect is produced as a result of an event acting upon building elements. These effects are diverse in nature and include colour rendition, surface texture, shape, mortality, erosion, corrosion, loosing reflective, absorption and transmission properties. The events are also diverse including rain, wear, temperature, people, air movement, long-wave radiation, chemical attack, biological attack, fire, man-made disaster and natural disaster.

The effect of events on building elements is manifested through the interaction of the composite parts of the events, namely agents, (e.g. water: rain) on the constituent component that make up an element—the Parts. In other words, it is the water base of the rain that creates corrosion by coming into contact with the metal part of the element. Naturally, if the metal is protected by plastic covering, there will be no corrosion as the water will not come into contact with the metal. This is shown in Fig. 4.2.

There are events that can be categorised as auxiliary events: For instance rain is an event which manifests its effect through water. Also, pollution is an event which acts upon a building through its chemical compounds. However, when the

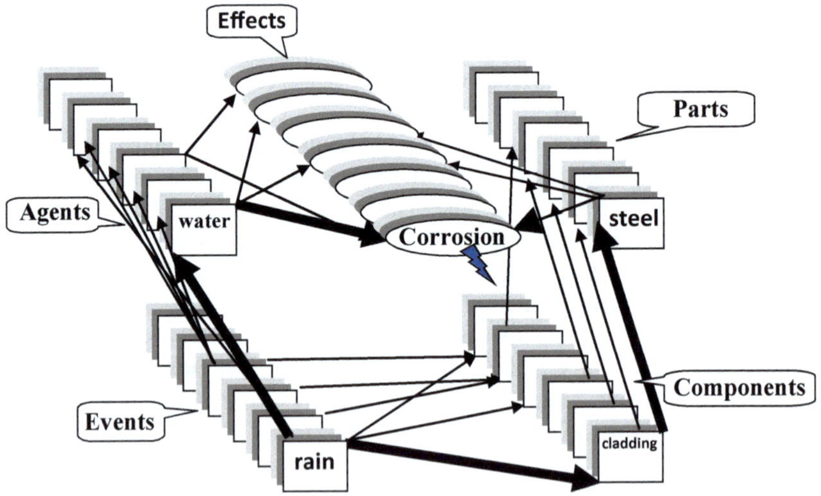

Fig. 4.2 Basic event–effect model

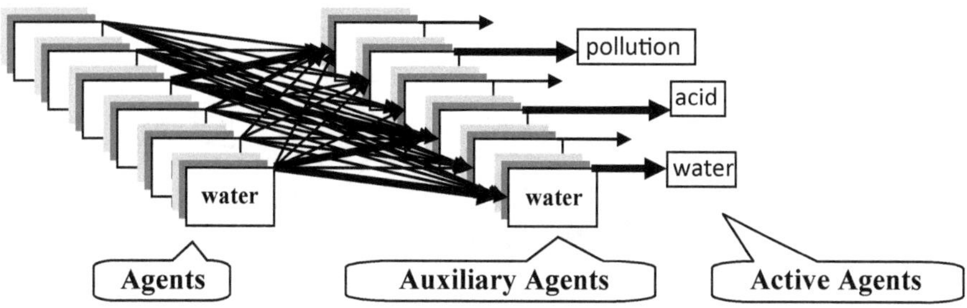

Fig. 4.3 Combined effect of two or more agents. Example: agent-water plus agent-pollution = active agent-acid

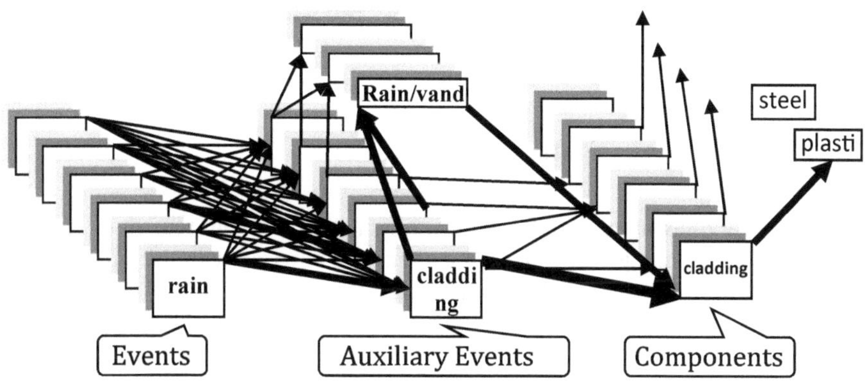

Fig. 4.4 Combined effect of two or more events. Example: event-rain plus event-vandalism-x = active link to component-cladding-element-steel

two events rain and pollution are combined, an auxiliary event—acid rain—is produced which will have a different effect on the building. This is shown in Fig. 4.3.

Similarly, the combination of two or more events can result in the activation of auxiliary elements. For instance, if the steel part of the cladding is covered by plastic coating, no effect will be fired and no corrosion will be expected to occur. But, a type of event, 'vandalism' could cause damage to building cladding and expose the steel part, thus, making it vulnerable to corrosion-effect. Figure 4.4, shows that the introduction of an auxiliary event layer facilitates the incorporation of the impact of combined events. Unlike the previous case where both layers contained the same agents, here, a new Auxiliary event is created.

Figure 4.5, shows the final event–effect model that takes auxiliary events into account.

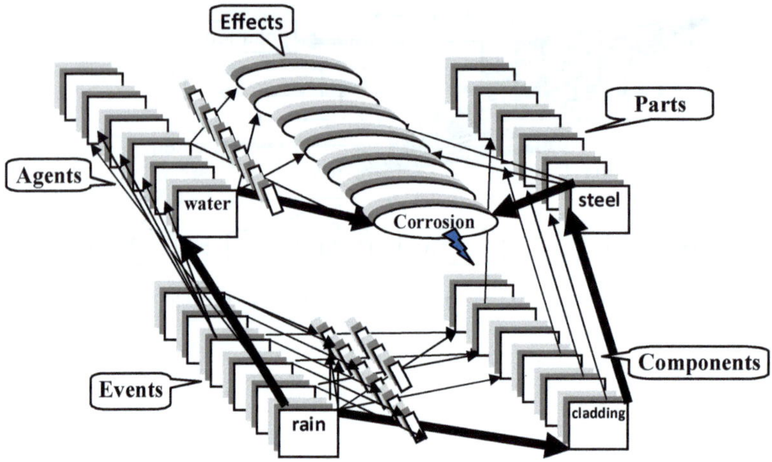

Fig. 4.5 Extended model of events–effects

4.4 Maintenance System Tools

The practical transformation of scientific approach to the management of information, in aid of building maintenance planning, commenced with use of computers during the 1970s. This paved the way for systematic collection of maintenance-related data. By the early 1980s, the use of centralised databases and computer programmes for building maintenance had grown amongst companies which either had dedicated departments or used external maintenance consultants (Pettit 1983; Davis 1990). With the advents of computing and emergence of computer-aided design (CAD), a number of specialised systems were introduced by industry and research community. During this period, the concept of whole-life analysis and life cycle costing formed part of the agenda. Examples of the new generation of facilities management system include FrontLine™, *WLCAT* by BRE (Edwards et al. 2000). *Whole life cost (WLC) Comparator* (Edwards et al. 2000).

Developments in this area resulted in the ability of actors to examine the trade-off between different elements of sustainability, as well as best-value analysis. This was further assisted by rise of World Wide Web environment. *ENVEST2* (http://envestv2.bre.co.uk/account.jsp) which has a strong sustainability angle by examining the environmental influence of building elements. *Tool for Office Building Upgrading Solutions (TOBUS)* (Flourentzou et al. 2002) is applicable to office refurbishment and aims to align the energy and indoor environmental performance with the needs and expectations of office users. With the increasing interest in life cycle considerations, the attention was directed at the post-construction phase. *SUREURO* (Bueren et al. 2002), for instance, offered insight into sustainable refurbishment. *PLACE3S* (U.S. Department of Energy 1996) *is a* system that contains information about the behaviour of element. This is designed for communities to develop a more sustainable urban planning. Another *timely US product is CITYgreen* (http://www.americanforests.org/) *that focuses on* forest preservation, which uses scenario-based approach to modelling. A project with a more focused view of financial strategies, is *INVESTIMMO* (FRANCO 2002), which is a decision aid tool that takes building degradation into account, albeit, in a limited way. Further, its visual examination of time-based behaviour of building elements is simply video clips of their state over time. Another research-based source on environmental assessment is the *Green Building Challenge (GBC)* (http://greenbuilding.ca/). In the USA, much of the practical approach to sustainability is driven by the California Energy Commission *in Sacramento. Their original Guideline Sustainable Building* (California Energy Commission 2000), leading to 'building passport', laid the foundation for future developments.

4.5 Visualisation of Building Operation

Building maintenance is a data-intensive function. While the service life and life cycle data form the basis of decision-making, these data are heterogeneous in nature: they are multi-scale; qualitative versus quantitative; and multi-disciplinary. Their accumulation, structuring and abstraction offer significant challenges. The complexity is exacerbated by the high level of uncertainties surrounding the construction data. These complexities have generated a degree of pragmatism towards providing practical solutions. The development of maintenance solutions commenced with increasing attention to maintenance data. Indeed, the overwhelming volume of building maintenance data has hindered their practical use (Al-Hajj and Horner 1998). Some of the maintenance tools have addressed the need for a more qualitative approach to maintenance data. *EuroLifeForm* (Kirkham 2004) for instance, facilitates application of statistical analyses of data relating to time-related behaviour of building element. *LIFECON1&2* (Saria 2004) did the same but placed the focus on the deterioration of concrete infrastructure behaviour under different climatic environments.

The application of the just-in-time approach to maintenance programming too requires extensive data, as it relies on the identification of the correct time of intervention for each element. This research aims to address this issue by using a visual tool and approach. The visualisation techniques can be effectively used in aid of life cycle decision-making in various ways. In particular, visual-enabled decision-making for early design choices offers the potential to combine astatic desirability with social, economic and environmental objectives. The developments in visualisation technologies have paved the way for the achievement of this objective (Rosenman et al. 2007), however, some challenges still remain (BSI 2012).

The combination of human vision, the expert knowledge and computation tools enables conversion of large volume of diverse date into information and knowledge. Information visualisation is a new but growing area of science. One definition of information visualisation is given by Card et al. (1999) as 'The use of computer-supported, interactive, visual representations of abstract data to amplify cognition'.

Already, there has been a diversity of efforts to address sustainability issues involving a form of visual aid (Kamat and Martinez 2001; Dawood and Mallasi 2006). Other tools have addressed focused areas such as building design and analysis of construction equipment (Lipman and Reed 2000), as decision tools (Flanagan 2002) and for planning and scheduling (Haymaker and Fischer 2001). The use of the visualisation tools for building operation phase has been less frequent. EPIQR (Jaggs and Palmer 2000) use multimedia representation of cases to offer refurbishment decisions for apartment building. EPIOR was then combined with TOBUS to develop XENIO (http://env.meteo.noa.gr/xenios/) which offers a multimedia to achieve cost-efficient renovation; *INTELCITY was a research effort* to develop sustainable cities; and *LIFETIME* (www.rte.vtt.fi/QuickPlace/lifetime_public/Main.nsf) places the emphasis on design management and maintenance planning.

4.6 Time-Based Visualisation Model

The general model of the 4D visualisation of building degradation is shown in Fig. 4.6. The three main parts of the model are the data management, simulation and presentation. The database containing the behaviour of building elements is separate from that containing the state of the current building objects. The simulation part interacts with the aforementioned event–effect model and relies on the time-related data relating to building elements. For the most part these are provided by the manufacturers of the products. Typically, these behaviours are represented by a probability distribution function (PDR), such as that shown in Fig. 4.7.

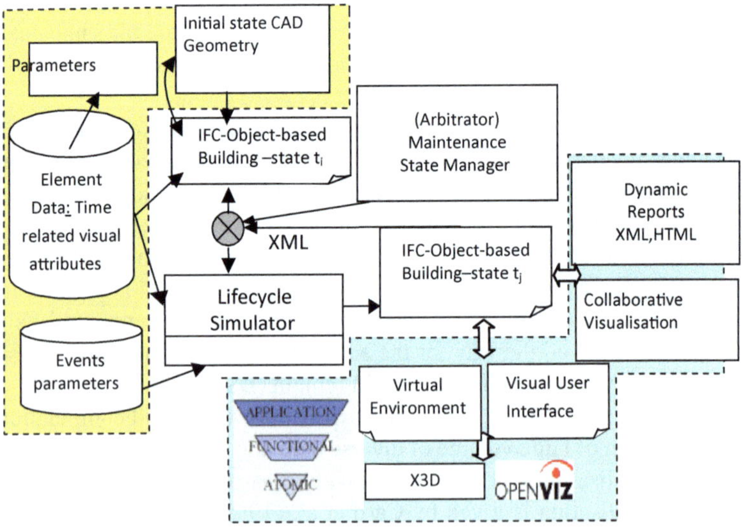

Fig. 4.6 Generalised simulation model

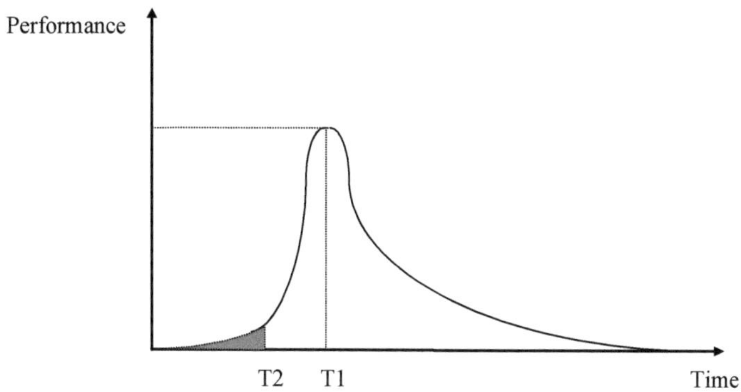

Fig. 4.7 Generalised life span model

The presentation part—visual user interface (VUI) can be facilitated through various means. With the development of sophisticated BIM solutions and their increasing ability to interact with interoperable software, the VUI part needs not be developed independently (Holness 2008; Stephen 2009). Nonetheless, the propose model considers VUI as a separate autonomous part. This independence is essential to ensure scalability, extensibility and reusability.

4.7 Lighting System

Light is a sort of electromagnetic wave which is generated by the elevation of electrons between two different orbits with positive potential energy difference. The main components of a light (luminaire) are the bulb (producing light), reflector (all sources reflecting the light) and shade or filter, (controlling the degree of transparency). Different types of light source are included into atmospheric, spot and normal (Sorcar 1979).

Atmospheric: atmosphere is a source of light (in a dark night, without the moon, still remains some light produced by the atmosphere). This is similar to the inside of a neon lamp.

Spot: a focused source of light that is small compared to the surrounding.

Normal: here the size of the light source is comparable with surrounding objects.

4.7.1 Light Source Behaviour

The performance of a light source is assessed in terms of its lifespan and lifestyle. For the most part, both of these behaviours are explained by their respective probability distribution functions (PDF). The x-random variable in Fig. 4.8 reflects the probability of death (the area of the curve between $X=0$ and $X=\infty$ is 1) (Laplin 1990).

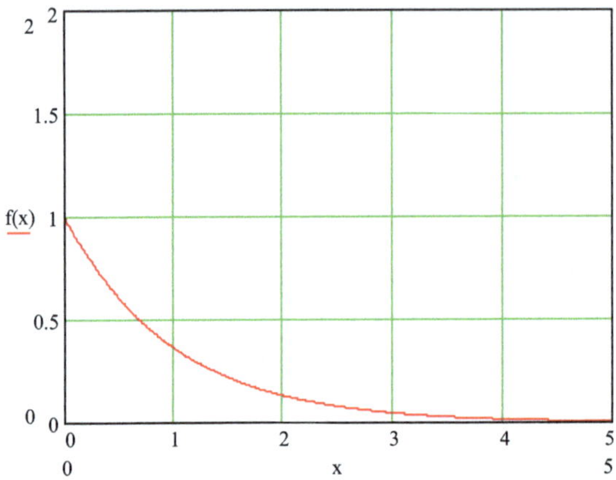

$$k = 1$$

$$F(x) = \begin{cases} ke^{-kx} & if x \geq 0 \\ 0 & if x < 0 \end{cases}$$

$$\int_0^\infty f(x)dx = 1$$

Fig. 4.8 Exponential representation

The illumination performance (lifestyle) of a light source follows an exponential form, though better represented by a normal distribution curve. This is shown in Fig. 4.9.

There are several variables that impact on the performance of a man-made light source. However, there are only a handful of variables with appreciable effect. These variables are as follows:

Life: Each luminaire has a lifespan, which is calculated based on statistical average lifetime of the same type of luminaires.

Age: the age of a light source changes the colour of the illumination, typically shifting towards yellow.

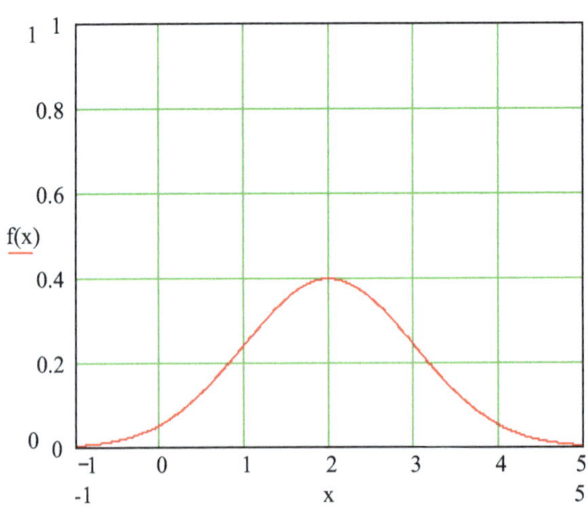

$$f(x) = \frac{1}{\sqrt{2\pi\sigma}} e^{\frac{-(x-\mu)^2}{2\sigma^2}}$$

$$\sigma = Variance$$

$$\mu = Mean$$

$$\int_{-\infty}^{\infty} f(x)dx = 1$$

Fig. 4.9 Normal PDF for illumination reduction

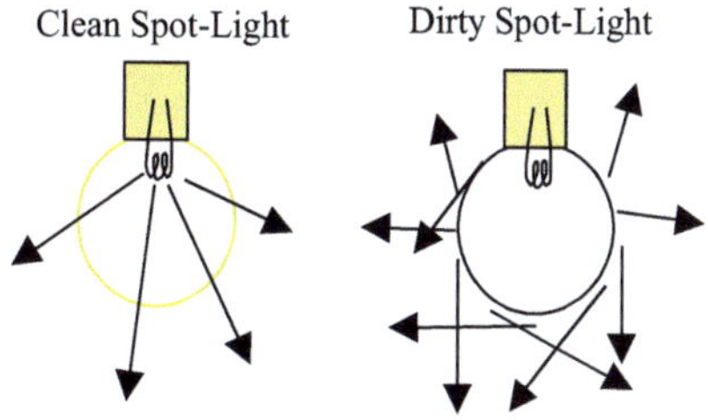

Fig. 4.10 Clean and dirty spotlight

Dirt and Dust: the pollution in the environment can change the reflection factor of the reflector, the shade colour and the transparency factor. As shown in Fig. 4.10, sometimes the pollution converts them into a normal light, as the dusts can reflect or filter the light.

As in the case of atmospheric light, pollution affects the impact of light by increasing the pollution, the atmospheric light will also increase. However, this parameter is not time-dependant but at any given time, different pollution factors can generate different atmospheric effects.

Usage: the pattern of usage (e.g. frequency, length, etc), of a luminaire tend to impact on the life–death status of the bulb.

Room Temperature: the room temperature can increase or decrease the life of a light bulb. In high- and low-temperature places, the lifetime is shorter than normal.

Voltage: a slight increase or decrease in the expected voltage can reduce or increase its life.

The incorporation of the above variables, such as temperature (*T*) and voltage (*V*), requires the identification of the mean value (μ) which basically is the number of life hours (*life*).

$$\mu = life \times \left(1 - C_v \times \frac{v - Stdv}{v}\right) \times \left(1 + C_T \times \frac{StdT - T}{StdT}\right)$$

$$C_v = \text{Constant Stdv} = \text{Standard Voltage (e.g. 220 v)}$$

$$C_T = \text{Constant StdT} = \text{Standard Temperature } (25^\circ \text{ c})$$

The algorithm for the implementation of the above within the overall visual degradation model for the lifespan representation is shown in Fig. 4.11.

The visual representation of the illumination behaviour (lifestyle) requires three separate set of expressions for red, green and blue (RGB) parts. An example of the algorithm for incorporation of the dust factor into the lifestyle is in Fig. 4.12.

The lifetime variance function that includes all adjustment factors (AdjFactor) is therefore, decomposed into constituent RGB models are shown in Fig. 4.13.

The simulation of light-source lifespan requires a different PDF. This is typically based on a Beta-PDF, expressed as follows:

```
Death status of lamp I at the time T
    Probability_of_Death    =    PDF_FNd(I,T)    *
        Environment     function     based     on
        Maintenance factor (I);
    if  Probability_of_Death  >=  RND(I,T)  then
        Bulb is in broken state;
    Else
        Bulb is not in Broken State;
End of lamp death status.
```

Fig. 4.11 Generalised lifespan algorithm

```
Color of Component I at the time T
    Red  =Initial_Red_Colour*PDF_FNr(I,T)
       *DustFactor;
    Green=Initial_Green_Colour*PDF_FNg(I,T
       )
       *Dust_Factor;
    Blue =Initial_Blue_Colour*PDF_FNb(I,T)
       *Dust_Factor;
End of Colour calculation.
```

Fig. 4.12 Visualisation algorithm dust factor

$$f(x)=\frac{\Gamma(\alpha+\beta)}{\Gamma(\alpha)\Gamma(\beta)}\times\left(\frac{x}{c}\right)^{\alpha-1}\times\left(1-\frac{x}{c}\right)^{\beta-1}\times\left(\frac{1}{c}\right)$$

$$Mean=\frac{\alpha}{\alpha+\beta}$$

$$Variance=\frac{\alpha\beta}{(\alpha+\beta)^2}$$

$$\int_{-\infty}^{\infty} f(x)dx=1\Rightarrow c=1$$

The incorporation of external variables such as temperature and voltage will require the following adjustments:

$$Mean = Life\times(1-Useage)$$

$$Variance=\left(\frac{Life}{2}\right)\times\left(1-C'_v\frac{v-Stdv}{v}\right)\times\left(1+C'_T\frac{StdT-T}{StdT}\right)$$

$$C'_v=\text{Constant}$$

$$C'_T=\text{Constant}$$

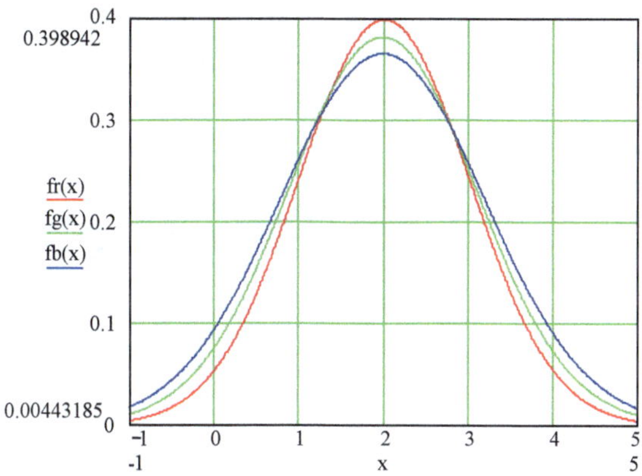

$$\sigma_R = \left(\frac{life}{2}\right) \times AdjFactor_R$$

$$\sigma_G = \left(\frac{life}{2}\right) \times AdjFactor_G$$

$$\sigma_B = \left(\frac{life}{2}\right) \times AdjFactor_B$$

Fig. 4.13 RGB channels normal PDFs

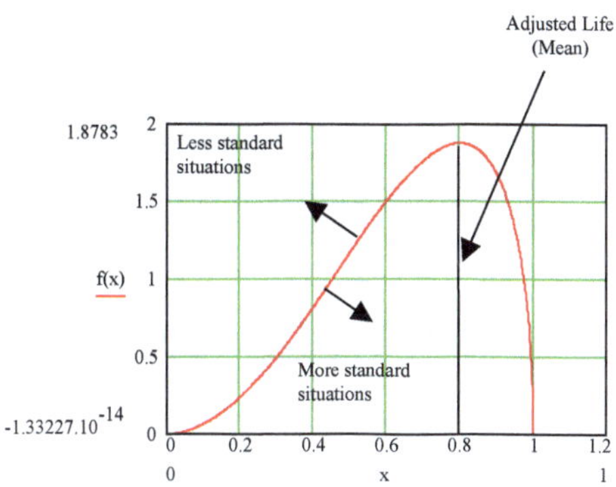

$$f(x) = Life \times \frac{\Gamma(\alpha+\beta)}{\Gamma(\alpha)\Gamma(\beta)} \times x^{\alpha-1} \times (1-x)^{\beta-1}$$

$$\begin{cases} \dfrac{\alpha}{\alpha+\beta} = Life \times (1 - Use) \\ \dfrac{\alpha\beta}{(\alpha+\beta)^2} = \left(\dfrac{Life}{2}\right) \times \left(1 - C'_v \dfrac{v - Stdv}{v}\right) \times \left(1 - C'_T \dfrac{StdT - T}{StdT}\right) \end{cases}$$

Fig. 4.14 Lifespan model

Extending the above to include all variables, the lifespan representation is given in Fig. 4.14.

4.7.2 *Visual Representation of Lighting Systems*

Whatever the choice of VUI, it needs to be able to interact with the simulated data. Here, for ease of use and for demonstration purpose, VRML is used, as it provides easy transformation of statistically generated data into visual form. VRML has clear definition for spot, point and directional lights. It also provides parametric control of positions, state and performance (ISO/IEC 14772-1:1997 1997). The basic model starts with Fig. 4.15, showing two light sources (one creating a shadow) and a table in a room.

Fig. 4.15 Base model using VRML

Fig. 4.16 Visual impact of aging

Fig. 4.17 Visual impact of environmental factors

Figure 4.16, shows how VRML parameters can be changed to reflect the visual impact of aging of the light sources.

Figure 4.17 demonstrates the combined visual impact of aging, dust, grease and dirt, all implemented through the interaction between the statistical data and VRML.

4.8 Flooring Systems

There has been very few works on life cycle performance of flooring systems, particularly with visual representation. But examples do exist (Capper et al. 2012). The main challenge associated with realistic visual representation of the degradation of building flooring system is to determine the pattern of flooring usage, which can be used to map the level of usage.

4.8.1 Behavioural Pattern

The realistic depiction of the degradation of building flooring systems requires understanding of peoples pattern of behaviour within buildings. This is because not all parts of the flooring systems degrade equally. Ground breaking works in this filed include those by Sommer (1969) on personal space and Hall (1966) on proxemics. Spatial behaviour is influenced by the physiology, personality, social group membership and culture of an individual and the environment in which s/he operates. While ergonomics, looks at human physiological process to the performance requirement of work tasks, anthropometrics, involves the detailed anatomical measurements of the human being and behavioural capabilities. (Morgan et al. 1963; Dreyfuss 1967).

Human factor analysis suggests that the human movement patterns are the product of anthroposophy. From anthropological perspective, two approaches have been identified for analysing human spatial behaviour:

- Activity system (Chapin 1968); addressing the organisation of the sequence of activities in a setting.
- Behaviour Settings (Barker 1968); examining the relationship between the setting and a recurring pattern of behaviour.

For the purpose of this research, it is imperative that human pattern of movement is presented in a formulated manner, preferably with mathematical representation. To this end, the behavioural setting has been used. Anthropology, explains human behaviour on two basis:

- Anthroposophy; dealing with human wisdom in his environment.
- Anthropocentric; dealing with human centric tendencies.

The two principles combined suggest that humans tend to seek wisdom and place themselves at the centre of their world. Subsequently, while there is a centric tendency to form colonies, humans desire to distance themselves from each other. The application of the above description on the behavioural pattern of an individual can be mathematically represented as follows:

$$f(x,y)=\frac{1}{\sqrt{2\pi}}e^{-\frac{(y-x)^2}{2}}$$

x is the random variable and y determines the centrality of the behaviour.

Figure 4.18, shows the distribution patterns for 1, 2 and 3 species. (centrality selected at *2* which is almost in the middle of the arbitrary scale from 0 to 5 m where:

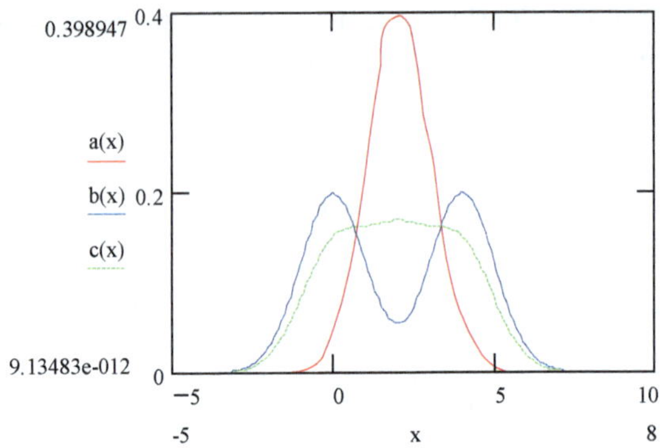

Fig. 4.18 Probability of 1, 2 and 3 creatures

$$a(x) = f(x,2)$$

The distribution pattern is affected by the number of subjects. The following expressions represent graphs *b* and *c* for two and three subjects respectively.

$$b(x) = \frac{f(x, 0) + f(x, 4)}{2}$$

$$c(x) = \frac{f(x, 0) + f(x, 2) + f(x,4)}{2}$$

The model is then expanded to represent the population or colony of subjects, by integrating form *n* number of subjects, when *n* approaches infinity. The expression representing the colony is as follows:

$$g(x) = \int_{y=i}^{j} \frac{f(x,y)}{y} dy$$

The expansion of the above over *n* subjects as *n* approached ∞ (within the arbitrary span of $i=1$ to $j=5$) is shown by the curve in Fig. 4.19, which shows properties similar to normal distribution.

The term 'anthropy' is coined to refer to this concept. The plan view and the three dimensional (3D) representation of an anthropy is shown in Fig. 4.20.

The application of the above principal to single-entry and open-ended corridors are shown in Figs. 4.21 and 4.22 respectively.

An example of the plane view of a combination of an anthropy and two open-ended corridors is shown in Fig. 4.23.

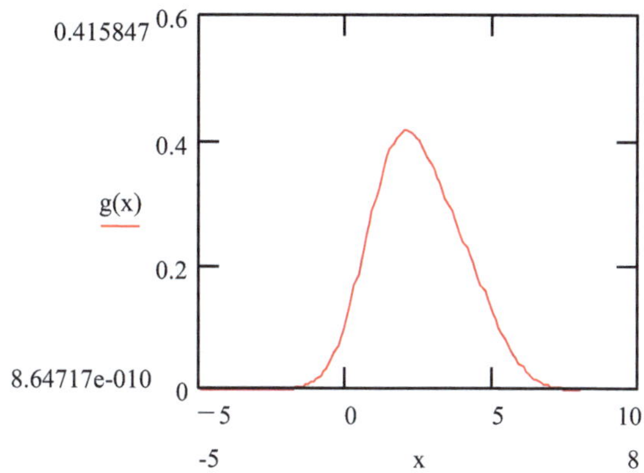

Fig. 4.19 Shows g(x) for $i=1$ and $j=5$

4.8.2 *Visual Representation of Flooring Degradation*

Having identified the density pattern of the use of a flooring system, the next challenge is to create a visual representation of the degradation of the flooring system. Figure 4.24, shows a sample of a brand new carpet and the same in a heavily used state.

The realistic representation of the degradation involves weighted texture mapping of the patter of usage from the new to old carpets. This is shown in Fig. 4.25.

4.8.3 *Anti-Anthropy*

In real life situations, flooring systems are filled with various objects which tend to influence the way the carpet is degraded. Tables, for instance, tend to protect the surrounding carpet. Therefore, the real life scenarios are made up of a combination of anthopies and anti-anthropies. Figure 4.26, is a plan and virtual views of superimposition of an anti-anthropy, created by a round table, on an anthropy.

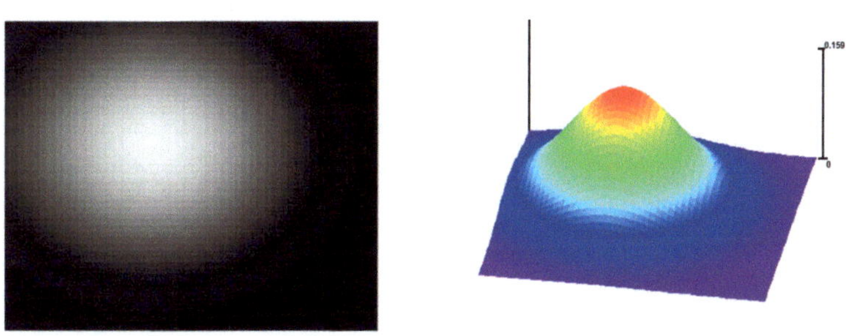

Fig. 4.20 Plan view and 3D: probability of n subjects

Fig. 4.21 Probability of appearance in a single-entry corridor

Fig. 4.22 Probability of appearance of *n* subjects open-ended corridor

Fig. 4.23 Movement density pattern room with two doors

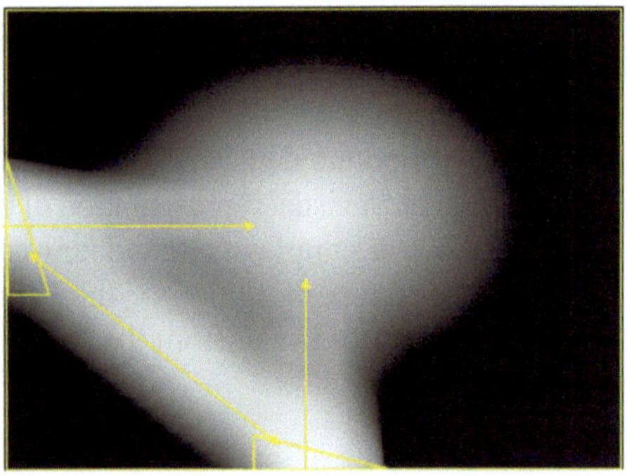

Fig. 4.24 Used and unused carpet texture

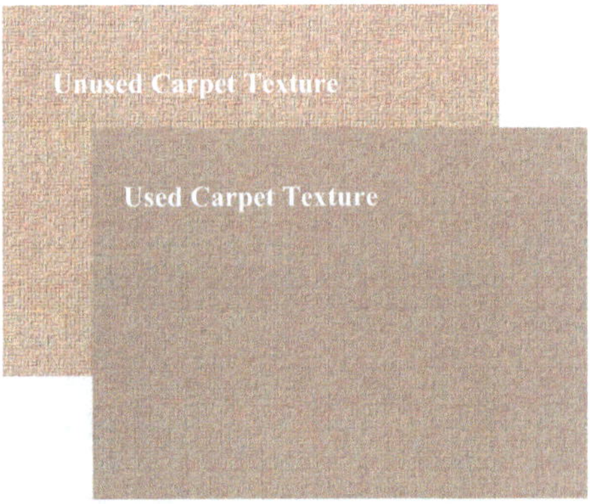

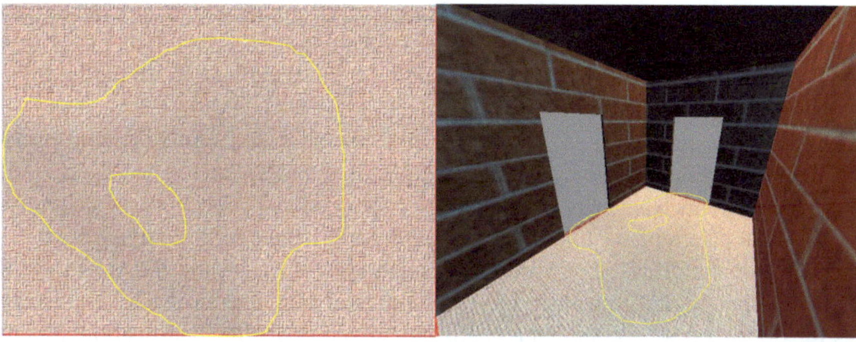

Fig. 4.25 Mapped used carpet texture

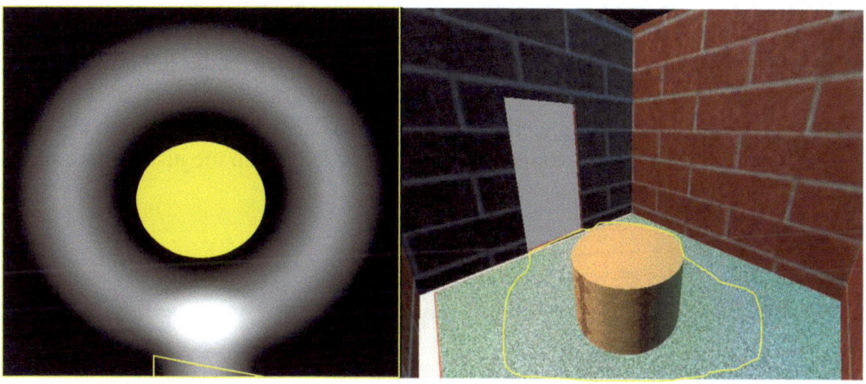

Fig. 4.26 Movement density pattern and actual room with a table

4.9 Conclusions

The chapter reiterated the importance of an effective building maintenance programme in life cycle costing, life cycle services and facilities management provisions. The significance of building maintenance and its impact on sustainability has been highlighted and argued that the current maintenance solutions tend to either compromise the interest of end users of the building or they are uneconomical, as they generate waste. Therefore, just-in-time solution has been offered as a method for overcoming these shortcomings. However, the achievement of just-in-time solution will require experiencing the building twice, which is deemed to be impractical. Subsequently, it is proposed that the first experience of living in the building can be based on a virtual environment. The achievement of this task required the development of the visual simulation of building degradation over time.

This process entailed the development of the overall model which consists of data management, simulation and visualisation. It also required event–effect model that would simplify the complex rules governing the impact of events on the building via its components.

The chapter demonstrates the practicality of the proposed model by focusing on the lighting and flooring components of the building maintenance. The behaviour of these components have been examined and expressed mathematically, which interact with the data representing the time-based performance of building elements as they degrade. For the lighting system, different types of light source have been identified and the impact of intrinsic and environmental parameters has been evaluated. As for flooring systems, the main challenge proved to be the identification of density movement of people defining the pattern of usage and determined the way the flooring system is degraded. To this end, the behaviour setting view of anthropology was exploited. A mathematical model was generated to represent this behaviour. Further, PDFs were used to generate the mathematical equivalence of the lifetime function of the items. They describe the behaviour of each item through time and enable the computation of the percentage damage or change in the status of the item.

The proposed visual approach will provide an integrated approach to the evaluation of the status of building elements. This approach has the potential to facilitate development of an optimum maintenance program, based on the just-in-time methodology.

References

Akizuki S (2000) Maintenance classification for strategic maintenance of government facilities in Japan, in the Facilities Management and Maintenance. Providing facilities solutions to business challenges—moving towards integrated resources management. Proceedings of the Brisbane 2000 CIBW70 International Symposium on Facilities Management and Maintenance, ed. Then, Danny Shiem-Shin, ISBN; 1 864354976, 577–580

Al-Hajj A, Horner MW (1998) Modelling the running costs of buildings. Constr Manage Econ 16:459–470

Arendarski J (1978) Trwalosc I niezawodnosc budynkow mieszkalnch. Arkady, Warszawa

Autodesk, Inc (2008) Improving building industry results through integrated project delivery and building information modeling [WWW Document]. http://images.autodesk.com/adsk/files/bim_and_ipd_whitepaper.pdf

Azhar S, Hein M, Sketo B (2008) Building information modeling (BIM): benefits, risks and challenges. Proceedings of the 44th ASC Annual Conference, Auburn, Alabama, April 2–5, 2008

Azhar S, Brown J, Farooqui R (2009) BIM-based sustainability analysis: an evaluation of building performance analysis software. Proceedings of the 45th ASC Annual Conference, Gainesville, Florida, April 1–4, 2009

Barker RG (1968) Ecological pyschology: concepts and methods for studying the environment of human behaviour. Stanford University Press, Stanford

Bertelsen D, Nesje A (2000) Service life management of building and infrastructure—methodology. SINTEF Civil and Environmental Engineering, Trondheim, SINTEF publication, Norway. www.sintef.no/

Bueren EM van, Bougrain F, Knorr-Siedow T (2002) Sustainable neighbourhood rehabilitation in Europe: from simple toolbox to multilateral learning. In: Martin De Jong W, Lalenis K, Mamadouh VD (eds) The theory and practice of institutional transplantation. experiences with the transfer of policy institutions, GeoJournal series. Kluwer, Alphen aan den Rijn, pp 263–279, ISBN 1-4020-1049-4/

California Energy Commission (2000) A new energy standard for state buildings, 142 Minna Street, San Francisco CA 94105, March 2000

Capper G, Matthews J, Lockley S (2012) Incorporating embodied energy in the BIM process. In CIBSE ASHRAE Technical Symposium, Imperial College, April 18–19, 2012, London, UK

Card SK, Mackinlay JD, Shneiderman B (1999) Readings in Information visualization: using vision to think. Morgan Kaufmann, Academic Press, Harcourt Science and Technology Company, San Diego

Chapin FS Jr (1968) Activity systems and urban structure: a working schema. J Am Inst Plan 34(34):11–18

Dawood N, Mallasi Z (2006) Construction workspace planning: assignment and analysis utilizing 4D visualization technologies. Comput-Aided Civ Infrastruct Eng 21:498–513

Dreyfuss H (1967) The measure of man: human factors in design, 2nd edn. Whitney Library of Design, New York

Edwards S, E Bartlett E, Dickie I (2000) Whole life costing and life-cycle assessment for sustainable building design, Nov 11, (2000), BRE Digest 452, Stock Code 32991, ISBN 1860814417

Environmental management. Life cycle assessment. Principles and frameworks, BSI, 2006. http://shop.bsigroup.com. Accessed 25 June 2012

Flanagan L (2002) Lifecycle assessment as a decision making tool. In: Khosrowshahi F (ed) Decision making in urban and civil engineering. Faculty of the Built Environment, University of Central England, Perry Barr, Birmingham, pp 211–218

Flourentzou F, Genre JL, Roulet CA (2002) TOBUS software—an interactive decision ai tool for building retrofit studies. Energy Build 34:193–202, ISSN 0378–7788

Franco G (2002) INVESTIMMO: a decision-making tool for long-term investment strategies in housing maintenance and refurbishment The national conference on "Urban Maintenance as Strategy for Sustainable Development" held in Naples on 29th November 2002

Gleeson J (2008) Computer-aided green design [WWW Document]. http://www.architectureweek.com/0330/tools_1–2.html

Hardin B (2009) BIM and construction management. Wiley, Indianapolis

Haymaker J, Fischer M (2001) Challenges and benefits of 4D modelling on the Walt Disney concert hall project, Working Paper #064, CIFE, Stanford University, Stanford, CA

Holness G (2008) Building information modeling: gaining momentum. ASHRAE J 50(6):28–37

Jaggs M, Palmer J (2000) Energy performance indoor environmental quality retrofit—a European diagnosis and decision making method for building refurbishment. Energy Build 31:97–101, ISSN 0378–7788

Kamat VR, Martinez JC (2001) Visualizing simulated construction operations in 3D. J Comput Civil Eng 15(4):329–337

Kirkham RJ, Alisa M, Pimenta da Silva A, Grindley T, Brøndsted J (2004) Rethinking whole life cycle cost based design decision-making. Proceedings of the ARCOM annual conference, vol 1, September 2004, pp 91–104, ISBN 0953416194 (2 vols)

Kriegel E, Nies B (2008) Green BIM. Wiley, Indianapolis

Krygiel E, Nies B (2008) Green BIM: successful sustainable design with building information modeling. Wiley, Indianapolis

Lipman R, Reed K (2000) Using VRML, in Construction Industry Applications, Web3D—VRML 2000 Symposium, Monterey, CA, February 21–24, 2000

Pettit R (1983) Computer aids in the management of maintenance, building maintenance information report, RICS BMI, London

Pitts A (2008) Future proof construction—future building and systems design for energy and fuel flexibility. Energy Policy 36(12):4539–4543

Rosenman MA, Smith G, Maher ML, Ding L, Marchant D (2007) Multidisciplinary collaborative design in virtual environments. Autom Constr 16(1):37–44

Sarja A (2004) LIFECON project technical progress report, VTT Technical Research Centre of Finland, 7th December 2004. http://www.ril.fi/dman/Document.phx/~public/LifeTime/259,1,LIFECON project technical progress report

Schueter A, Thessling F (2008) Building information model based energy = exergy performance assessment in early design stages. Autom Constr 18(2):153–163

Sommer R (1969) Personal space: the behavioural basis of design. Prentice-Hall, Englewood Cliffs

Sorcar PC (1979) Rapid lighting design and cost estimating. MacGraw-Hill, New York

Stephen R (2009) Green building XML schema: a building information modeling solution for our green world [WWW Document]. http://www.gbxml.org/about.htm

Urbanski P, Waszczyszyn Z (2002) Neuronowa predycja stopnia zuzycia technicznego budynkow mieszkalnch, XLVIII Konferencja//naukowa Komitetu Inzynierii Wondnej PAN I Komitetu Nauki PZiTB, September 2002, Krynica, 365–372

International Standard ISO/IEC 14772-1:1997 (1997) The VRML consortium incorporation

Chapter 5
Sustainable Facilities Management (SFM) Delivering the Optimum of Cost and Value

Jim W. Ure and Azad Camyab

What is SFM?
The concept of SFM is simple: it is facilities management (FM) that makes a primary commitment and contribution to sustainable development, as defined by Brundtland in 1987: *'meeting the needs of present generations without compromising the ability of future generations to meet their own'* (Brundtland 1987).

Clearly, SFM must be compatible with, and support progress to, sustainable development. To that end, SFM is defined as follows:

It is a facilities management process that optimises financial, environmental and social factors in support of the primary purpose of the organisation.

This chapter addresses SFM under the three main themes of financial, environmental and social.

5.1 Introduction

While organisations continue to step up their corporate response to be sustainable, facilities managers will be increasingly expected to deliver a support service that provides the optimum of cost and value. The FM sector is now expected to deliver a sustainable service that is an integral part of FM rather than a 'bolt-on' enhancement to traditional FM.

With the application of SFM, they can deliver both simultaneously, more from less: for example using fewer resources, better supporting staff needs, and reducing all forms of waste in pursuit of more sustainable operation. SFM is not a cost burden but a cost benefit. If it costs more than the value it produces, it is not sustainable.

J. W. Ure (✉)
ABS consulting (UK) Limited, London, UK
e-mail: jwure@absconsulting.uk.com

A. Camyab
Pearlstone Energy Limited, London, UK

© Springer International Publishing Switzerland 2016

M. Dastbaz et al. (eds.), *Building Sustainable Futures*,
DOI 10.1007/978-3-319-19348-9_5

Simply, it is an optimisation exercise. However, the optimisation of the many competing factors is not always as easy as it sounds. The starting point is to identify the relevant factors and then rank their importance to the organisation's business and the wellbeing of the people who serve it.

While the realisation of SFM is not as easy to achieve as the concept may suggest, those who rise to the challenge will reap substantial reward and deliver benefits to all stakeholders. The opportunity for facilities managers to lead the way in the delivery of sustainable design, construction and operation of their own buildings and infrastructure is enormous. They also have a unique and real opportunity to influence progress to sustainable operation in their organisation's core business.

With this in mind, it is clear that the FM sector has a substantial responsibility in the delivery of sustainability for all stakeholders. FM training and career development should therefore provide a thorough understanding of the implications of sustainable solutions and their impact on the supply and demand chains.

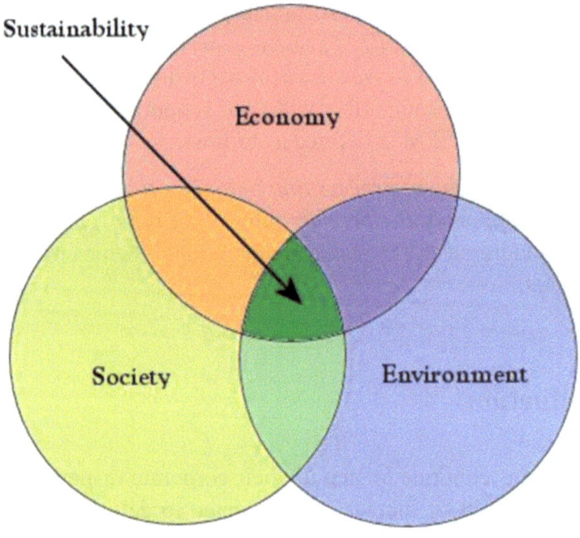

SFM must:

- Deliver an environment that provides a positive feeling of wellbeing for the workforce, enabling them to maximise their contribution to the primary purpose of the organisation.
- Encourage and support boardroom aspirations for improved corporate social responsibility (CSR) and corporate responsibility (CR) while delivering increasing value and wealth for the shareholders or, in the public sector, ever-improving service to the community.
- Demonstrate that the environmental impact of operating the facilities is the minimum needed to deliver the business goals.
- Optimise operating cost and value delivered.
- Create and maintain positive impacts for all other stakeholders.

The Vision can therefore be formulated as:

Sustainable facilities for sustainable organisations resulting from facilities management that embraces design, construction, operation and reuse as a seamless, holistic and continuous process. Organisations and all their stakeholders will be committed to providing better working conditions that encourage social progress while reducing environmental impact, at appropriate cost.

5.2 The Opportunity

There are few professions and market sectors associated with property and construction that have experienced such growth in recent years as has FM. Coupled with that growth, FM has enjoyed increased influence on building design, building operation and other non-core business requirements. The background and experience of facilities managers also provides them with a broad understanding of the issues that influence the success of the core business. The services and activities for which they are responsible have substantial impact on the environment and on the wellbeing of the building occupants.

FMs are therefore uniquely placed to bring the concept of sustainable development to their organisations through SFM. It is by our actions that we show our commitment. If this seems far-fetched, consider for a moment the influence a good caretaker can have in a good school.

Put simply, most of the opportunities for an organisation to improve its own operations—social, ecological or economic—are commonly presented during the operation phase of any facility—*the all-important building life cycle stage following the frantic and relatively short-lived 'design and build' stages.*

Opportunities for change are not confined to the decisions taken by owners and designers during the design phase. These 'procurement' decisions are clearly vital to successful operation of the facility, and FM influence should be involved from the outset of a project. Many, if not most, opportunities to become more sustainable arise in the operation stage of the building life cycle rather than at the procurement stage.

Many organisations will already be developing or operating some form of environmental management strategy. This is the first step to sustainable business conducted from sustainable facilities. But the step from sensible environmental management to sustainable operation is enormous. It requires a long-term commitment to improve environmental, social and economic impacts, without giving undue priority to anyone.

Most organisations that are perceived as wholesome and caring have achieved this important perception by addressing the environmental, ethical, economic and social influences of their business. SFM provides a holistic structured approach to develop and implement what enlightened organisations have realised intuitively as sound business sense for some time.

As with environmental management, a strategy to achieve sustainability must be dynamic. It must be capable of embracing new ideas and technologies as these develop. Facilities managers who understand this, and who can implement a progressive migration strategy, should enjoy a prosperous and satisfying future.

The key message is therefore we require buildings that are *fit for purpose, fit for people and fit for the planet.* This vision will be delivered by SFM.

The place to start in making a contribution to sustainable buildings is to apply what we already know. It is not necessary to wait until we have all the answers before we start. Improving *energy efficiency* will, for the foreseeable future, play a major part in reducing environmental impact, and the opportunities here are enormous.

5.3 Serving the Business and Society

Buildings and their operation often form a substantial part of the facilities manager's workload, but only in so much as they are a necessary to serve the business. A successful SFM strategy limits the environmental and social impacts to enable an organisation to meet or exceed its business plan. This should include occupying the minimum space necessary to conduct its business.

As organisations migrate from environmental management to sustainable development they will take greater account of the social impacts of their activities. The facilities management profession has therefore an unparalleled opportunity to become more involved with the effect that buildings will have on their neighbours. This will be more than simply complying with noise and pollutant regulations. It will mean finding ways to integrate with the community and to seek common goals.

Essential components of a robust SFM strategy include the following:

- **Carbon and energy management (CEM)** policy and strategy that is endorsed at the organisation's executive level, agreed by senior management and shared with all stakeholders.
- **Optimal operating strategies (OOS)** using continuous commissioning **(ConCom)** to ensure sustainable operation of facilities. This requires a thorough understanding of the building life cycle and having a strategy that will prepare the organisation to reduce environmental impact as opportunities arise.
- **Post-occupancy evaluation (POE)** using overall liking score **(OLS)** provides structured occupant feedback which will form an essential part of the SFM. It enables facilities managers to take informed decisions about the working environment created by the facilities they operate.
- **Performance-based service provision (PBSP)** results in the optimum of cost and value for the occupier by providing incentives for the occupier and its service partners to identify, implement and sustain savings and performance improvement in the operation of buildings.
- **Energy monitoring, optimising, targeting and reporting (EMOTR)** identifies savings opportunities and measure the savings in real time.

Implementation of the above techniques and components requires a thorough understanding of building life cycle which is outlined in Fig. 5.1.

Let us consider each of the above components in more detail.

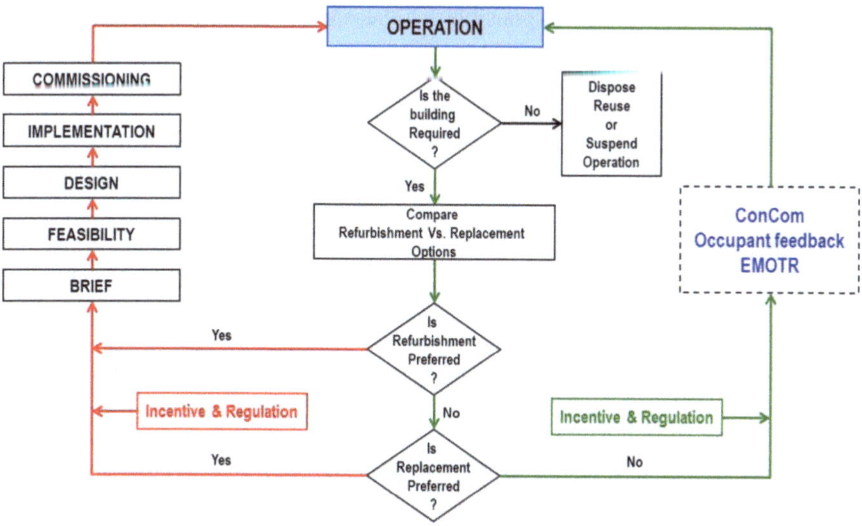

Fig. 5.1 The building life cycle

5.3.1 *Carbon and Energy Management (CEM)*

Successful carbon and energy management (CEM) requires a clear policy statement supported by an implementation strategy that includes challenging but achievable targets.

The primary requirement of a CEM strategy is to ensure that the organisation can realise its business objectives at the optimum of cost, environmental impact and value while meeting and where relevant, going beyond, statutory requirements. It also takes account of the changing needs of the business over the longer term and the impact that such changes will have on energy cost, energy consumption and carbon emissions. There are two main components to consider as outlined below:

Minimising Demand

The simplest way to minimise demand is to ensure that energy-efficient plant and equipment of appropriate capacity with effective controls is specified when refurbishing existing, or fitting out new, premises. This will be achieved by the following:

- The preparation and regular review of specifications for systems, components and equipment that are energy efficient in operation and in their embodied energy.
- Carbon efficient procurement, design, installation, maintenance and operation of energy consuming services and equipment.
- The introduction of commissioning management as part of the design and construction process of all new and refurbishment projects from feasibility stage.
- Rigorous physical commissioning prior to handover of projects with significant involvement from those responsible for the operation and maintenance of the existing estate.
- The preparation of building log books in accordance with the requirements of the building regulations prior to the handover of projects.
- The selection of service partners and contractors that have demonstrated a thorough understanding of the organisation's requirements for energy and carbon efficiency.
- The implementation of a process such as continuous commissioning to provide feedback from operation to design and to ensure that the energy consumption and carbon emissions of facilities are limited to those necessary to operate the facility effectively.
- The preparation and implementation of an energy and carbon monitoring and targeting strategy plus refining the process over time to provide meaningful sustainability benchmarks.
- Introducing, and keeping under review, procedures for identifying and eliminating avoidable energy consumption and carbon emissions.

Controlling Demand

Designing for minimum energy demand will not deliver energy and carbon efficiency unless the strategy includes a rigorous process for controlling demand. Therefore, fundamental requirements for controlling demand that form an integral part of the strategy are as follows:

- Determine and publish standards for operating all energy consuming services and equipment.
- Use regular formal occupant feedback to operators and designers on the performance of systems and the working environment.
- Apply continuous commissioning as the process for delivering the optimum of cost and value during operation with regular feedback to design.
- Conduct an ongoing energy and carbon awareness campaign and promote good housekeeping to enable all employees, service partners and contractors to make a contribution to energy efficiency.
- Provide awareness workshops for employees, service partners and contractors with repeats as necessary.
- Include energy and carbon efficiency awareness in employee induction training.
- Implement energy and carbon monitoring from which challenging but realistic reduction targets can be set and monitored.
- Introduce energy and carbon key performance indicators and review as necessary.

Energy Procurement

It is important to ensure that energy procurement forms an integral part of the strategy and is not carried out as an independent activity. Predicting unit consumption and demand profiles help to ensure that savings can be made by selecting tariffs that match consumption profiles.

Financing Energy Efficiency

The options for financing energy efficiency and supply can be summarised as follows:

> **Option 1**
> Capital expenditure—The organisation owns, operates and maintains building services plant and equipment.
> **Option 2**
> Contract energy management—The organisation enters into a contract with a supplier to finance, design, operate and maintain building services plant and equipment charging the occupier on an ongoing basis for the service provided.
> **Option 3**
> Operating leases—The organisation purchases the benefits from a product rather than the product itself through an operating lease thereby using revenue rather than capital to fund projects. Operation and maintenance of the leased product could be provided by the occupier or the leaser.

Option 2 transfers the capital and operational risk to the supplier, and the supplier will take account of carrying this risk in the contract price. This option will only be viable if the overall operating cost to organisation is less than it would be with

Option 1. Option 2 is particularly relevant for complex specialist plant, such as CHP, enabling the organisation to transfer risk to a supplier that has greater experience in managing the risk.

Option 3 enables organisation to reduce its revenue spent while benefiting from the savings provided by an energy-efficient system or product without capital expenditure.

With all options, it is important to ensure that the organisation takes full advantage of any tax benefits, government funding or other incentives for CEM.

Regulations and Incentives
An essential part of the strategy includes understanding and being prepared for current and future regulations and incentives. Increasing regulation in recent years has made keeping up-to-date an onerous task. Also, the incentives provided by government to encourage organisations to reduce their carbon emissions are substantial, and it is essential that the organisation makes full use of these.

5.3.2 Optimum Operation Strategy (OOS) by Continuous Commissioning (ConCom)

Let us now consider why buildings should consider commissioning in the first place.

There are two primary reasons for commissioning buildings and their services. These are:

- To meet legal obligations
- To meet the business need

The cost of commissioning should be no more than is necessary to satisfy these two reasons, and to achieve this, a process that optimises cost and value is required. This means that the commissioning process must be designed to meet the needs of the business that operates from the building and not relate to the building as an end in itself. The deliverable should therefore focus on meeting the needs of the building occupants and the business rather than the assets.

The author's experience is that the commissioning of buildings is rarely carried out properly in the UK. Poor commissioning is often the result of inadequate planning and a lack of time. Typically, this results in increased energy consumption, occupant discomfort and shortened plant life.

Recent research concludes that statutory requirements for commissioning buildings in the England and Wales are inadequate and a reliance upon regulation will result in a lost opportunity to avoid substantial energy consumption and therefore carbon emissions. Including the requirement for rigorous commissioning in the building regulations will improve matters, but once a building is in operation, it will largely remain the responsibility of FM to deliver OOS. (Lau 2014)

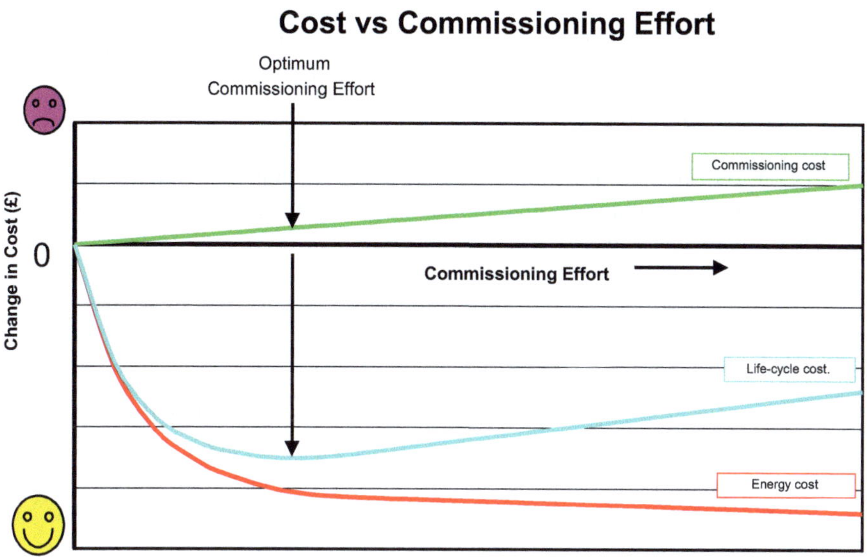

Fig. 5.2 Cost vs. commissioning effort

An increase in commissioning effort will reduce energy consumption and its associated cost as shown by Fig. 5.2. To realise the opportunity for energy efficiency, 'optimum commissioning effort' will be achieved when the increase in commissioning effort costs more than the resulting energy cost savings. Figure 5.2 illustrates this point of 'diminishing returns' for the building life cycle cost (cumulative 'present value' costs of the building over its life).

Over the building life, increased commissioning effort can optimise occupant satisfaction, plant control and maintenance costs as well as energy costs, in accordance with their priority in support of the main purpose of the organisation. Consequently, it is often more appropriate to commission beyond the point of optimum effort for energy efficiency and thereby add value from these other benefits.

What is ConCom?

- Continuous commissioning (ConCom) is the process developed and used by ABS, an energy efficiency company in the UK, for optimising energy efficiency, operating cost and occupant comfort.
- It provides a strategic approach to the operation of buildings that delivers the optimum of cost and value to the occupier. It enables the energy consumption of building services to be limited to that required to meet the occupier's business needs and to provide a safe and comfortable environment for the occupants.
- ConCom's key characteristic is its holistic and ongoing nature.

Through ConCom's holistic and ongoing approach, buildings are able to deliver what is required of them in the most cost effective and sustainable way without compromising quality.

Holistic:

ConCom involves the whole organisation from top to bottom, and ConCom addresses all factors affecting the project's success: technical, motivational and managerial.

Ongoing:

ABS embeds the ConCom principles in the client's overall strategy and day-to-day operations.

How does ConCom work?

- Through a series of activities, ConCom determines a building's ideal and actual provisions and delivers solutions that eliminate avoidable waste and optimise the performance of all soft (i.e. people, management, etc.) and hard (i.e. technical equipment, building management system (BMS), etc.) assets.
- ConCom provides two levels of recommendations:
  - Level 1: Recommendations that get the estate operating at its most efficient using the existing systems and technology with little or no capital expenditure.
  - Level 2: Projects that require capital expenditure to increase energy savings or to ensure that the systems can operate in accordance with the design intent and the client's requirements.
- ConCom acts as the all-inclusive energy efficiency and carbon management programme and includes the following:
  - Technical surveys
  - Post-occupancy evaluation—overall liking score (OLS)
  - Energy monitoring and targeting (M&T) spreadsheet
  - Carbon and energy management (CEM) policy design
  - Motivational workshops
  - Training workshops
  - Carbon footprint (CF) spreadsheet

ConCom provides the fundamental delivery process for SFM as it incorporates a range of services including energy and carbon management, facilities management, feedback, regulation compliance, building services and training.

Goals it Delivers

Typical SFM Goal List	Delivered
Cut energy costs (in a sustainable/long term way)	✓
Optimise current energy usage/ existing systems	✓
Ensure (measure)return on capital investments	✓
Acquire data for measuring, monitoring and reporting	✓
Acquire quantitative support for decision making	✓
Reduce CO2 emissions and environmental impact	✓
Comply with environmental and energy related legislation	✓
Mitigate risks associated with increasing energy prices	✓
Increase asset value	✓
Demonstrate Corporate Social Responsibility (CSR)	✓
Gain positive publicity and PR	✓
Gain staff support on energy and carbon reduction efforts	✓
Improve working environment (Greater occupant safety, comfort, satisfaction, = retention & productivity)	✓

The ConCom approach:

Delivers these goals by helping the organisation adopt a **holistic** approach in tackling energy and carbon reduction.

Is designed to strategically bring together, coordinate and optimize all soft and hard assets of an organization.

Doesn't only focus on the technical factors of the project but also on the motivational and managerial ones:

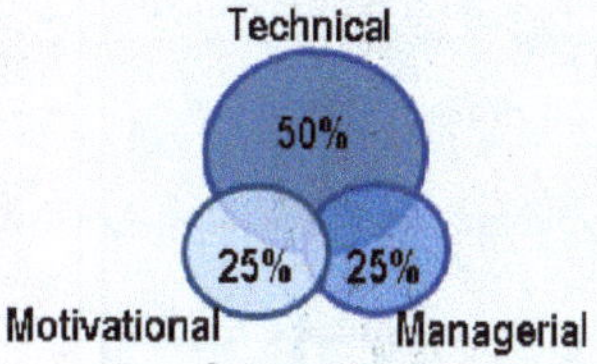

More specifically the ConCom process as shown in figure 3:

- Identifies main areas of inefficiency or ineffectiveness
- Identifies priorities for improvement
- Ensures that the existing systems are delivering the best service they can at little or no capital spend
- Identifies capital projects that meet an organisation's financial criteria and are compatible with the estate's strategy

✓Establishes and delivers an energy and carbon management strategy

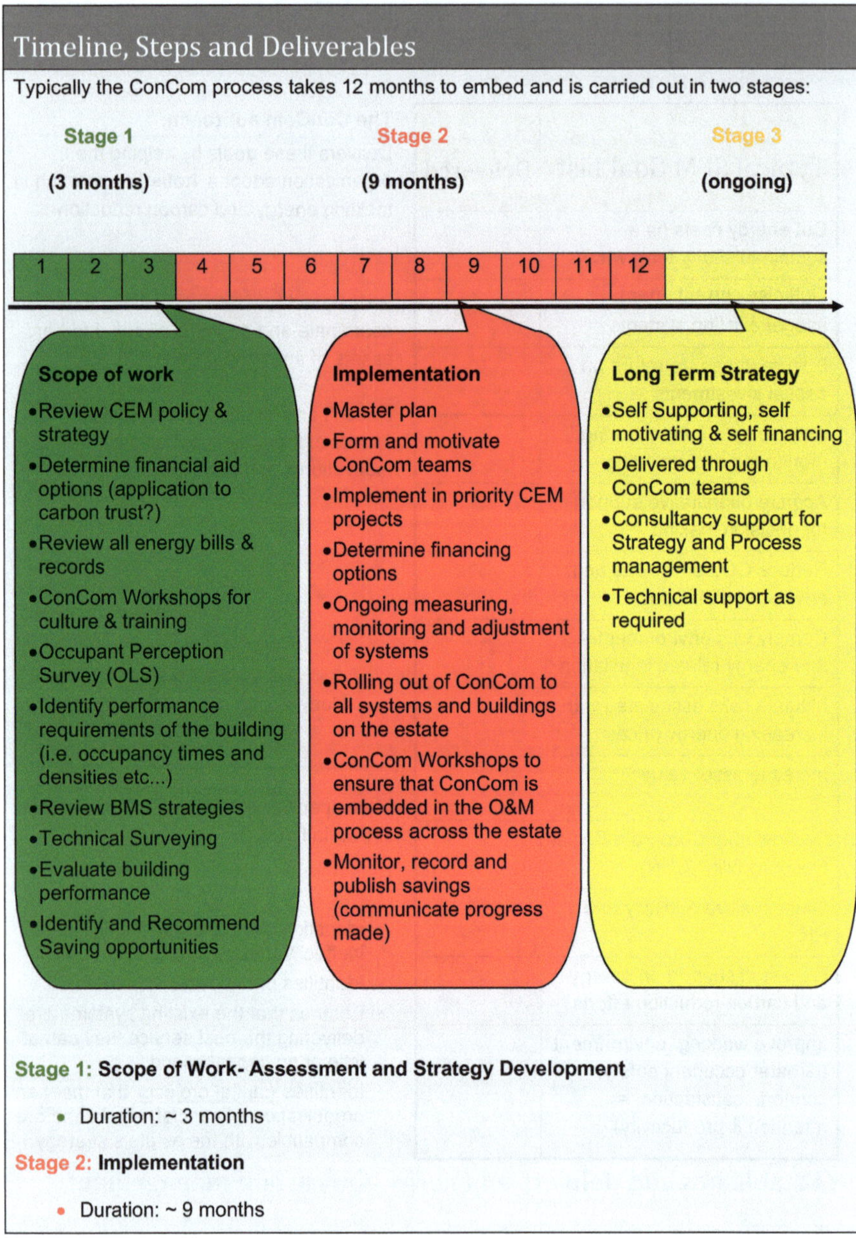

ConCom—Some Examples

In this section, examples of some buildings that have been studied by the author are discussed. Figure 5.4 shows three buildings where the actual consumption is significantly greater than it needs to be. In all cases, it was possible to reduce this consumption to at or near to the good practice figure by commissioning the building services systems and tuning them to meet the occupier's requirements. The financial and environmental savings are considerable. The expenditure and environmental impact will be contained within that needed to meet the occupier's needs by regularly checking the building by the process of ConCom.

Figures 5.5 shows the results of an analysis of three buildings in which some form of ConCom is included in the operating strategy. This provides powerful evidence that ConCom can deliver substantial cost savings. In addition, there are other worthwhile benefits of greater occupant satisfaction, reduced maintenance costs and reduced environmental impact. In all cases, these buildings are consuming less than typical benchmarks, and with more rigorous ConCom, their consumption would be less than good practice benchmarks.

None of the six buildings in Figs 5.4 and 5.5 was designed with energy efficiency as a priority. They are all conventional air-conditioned office buildings. The difference in energy performance between the buildings in Figs. 5.3 and 5.4 is the way in which they are managed. The management strategy for the buildings featured in Fig. 5.4 is managed with energy efficiency in mind and benefit from the application of elements which are part of the ConCom process. These findings support previous research that it is not the application of technology that results in energy-efficient buildings but the ability of those that manage them.

Figure 5.6 shows the result of the application of ConCom to a building that was designed for low energy use. Unfortunately, commissioning was not carried out properly prior to occupation. This resulted in significant occupant complaints

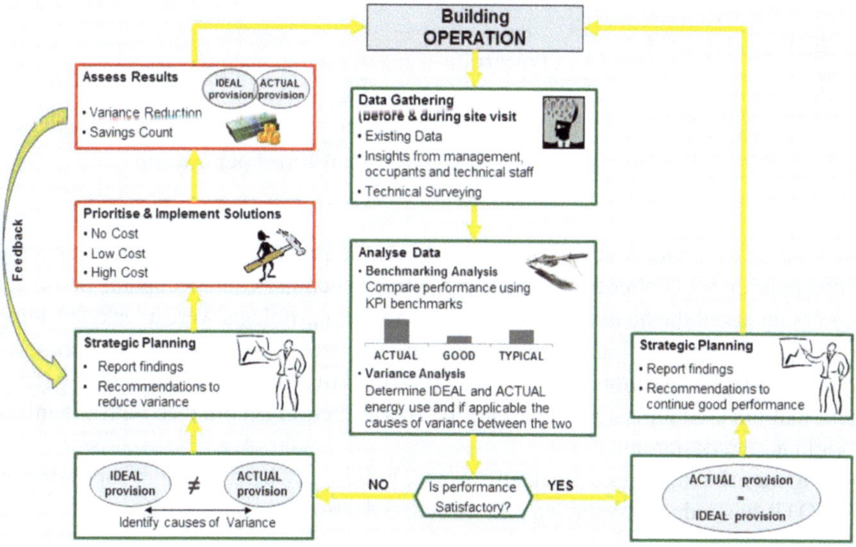

Fig. 5.3 The ConCom Process

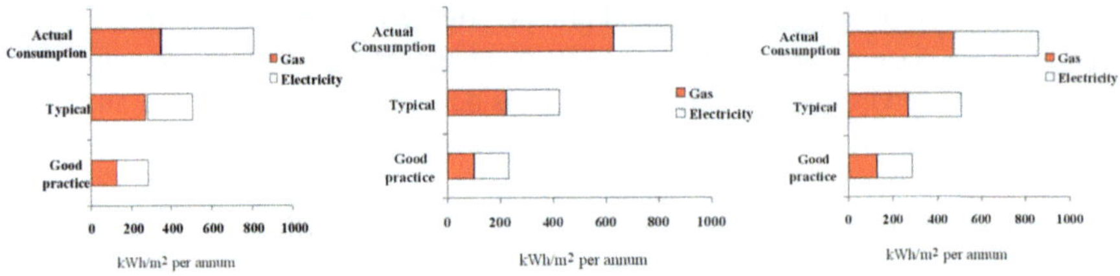

Fig. 5.4 Energy use in three buildings which were poorly commissioned

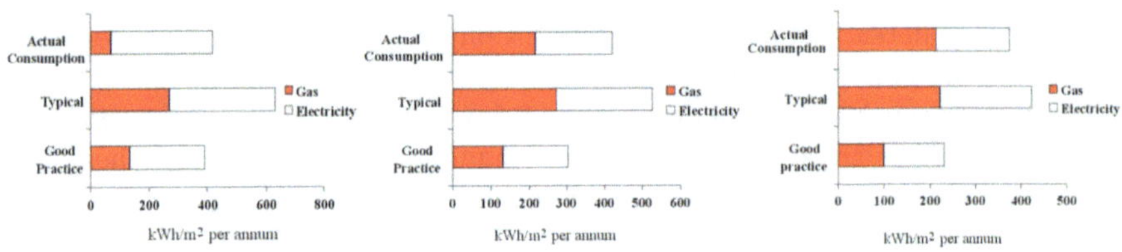

Fig. 5.5 Energy use in three buildings in which ConCom was applied

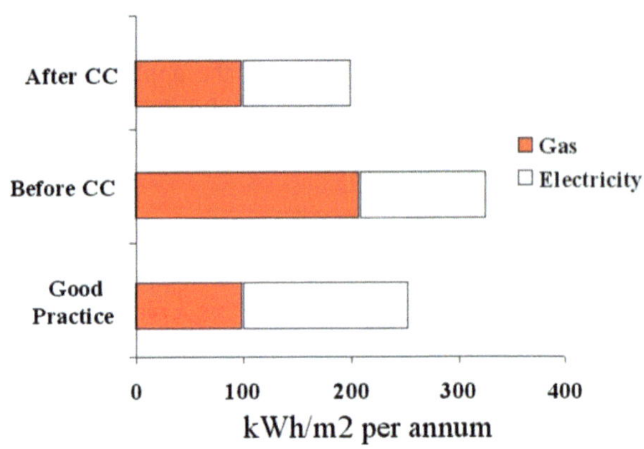

Fig. 5.6 Energy savings with the application of ConCom

of poor heating and draughts. After applying the process of ConCom, the energy consumption was better than the good practice benchmark and occupants no longer complain about the heating and draughts. The main problems were a leaky building envelope, inadequate balancing of the heating system and poor commissioning of the controls. Furthermore, the process confirmed that there were no design defects and that once commissioned the building performed in accordance with the initial brief and the design intent.

Figure 5.7 summarises the results from 2 buildings where ConCom supported by EMOTR has reduced sustained energy cost reduction by up to 50%.

Energy Monitoring, Optimising, Targeting and Reporting (EMOTR)
EMOTR is an important part of ConCom as it provides real-time information to gain a clear understanding of current performance and desired next steps.

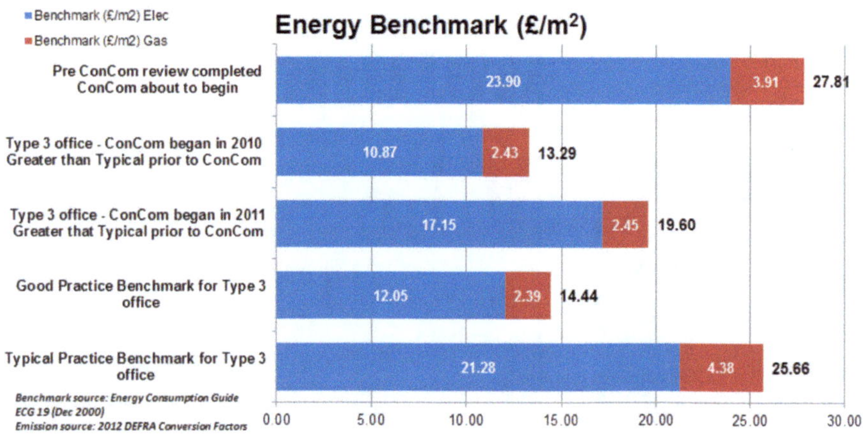

Fig. 5.7 Energy cost benchmark

The purpose of EMOTR is to enable those responsible for energy efficiency to have real-time information that provides a detailed understanding of consumption trends, identification of waste and measurement of savings as a result of intervention. It includes the following:

> Energy: Gas, electricity, oil, solar, wind and water with further analysis to identify dynamic and controllable loads. Other emission sources such as transportation, business travel and waste can be included.
> Monitoring: Electronic display of real-time consumption, carbon and cost (kWh, tCO2, £) via the Internet.
> Optimising: Limiting consumption to that needed for the occupier to deliver its business plan.
> Targeting: Benchmark performance and set challenging but achievable targets.
> Reporting: Regular reports on consumption, carbon and cost trends with exception reports of deviations from targets plus live energy data on display dashboards in reception and other areas to increase awareness.

Examples of the benefits of EMOTR are shown in Figs. 5.8, 5.9, 5.10, 5.11.

Figure 5.9 shows a significant increase in water use at large office building during a 7-day period. This was found to be caused by a passing ball valve. It was investigated and rectified within 8 hours of being identified on the EMOTR platform. This is just one simple example of the benefits of real-time monitoring.

Figure 5.10 shows the air handling unit (AHU) supply and extract fans operating incorrectly as identified on the EMOTR platform and the results following rectification.

Issue: Extract fan consuming more than twice that of the supply fan.

Correction (1): Control fault corrected on May 7 reducing consumption by some 60 % while improving air quality in the office.

Issue: System operating at weekends.

Correction (2): Control repaired and reset avoiding weekend operation by 16 May.

Figure 5.11 shows typical daily electricity use characteristics for the same day of the year before ConCom (red) and 1 year after ConCom (green).

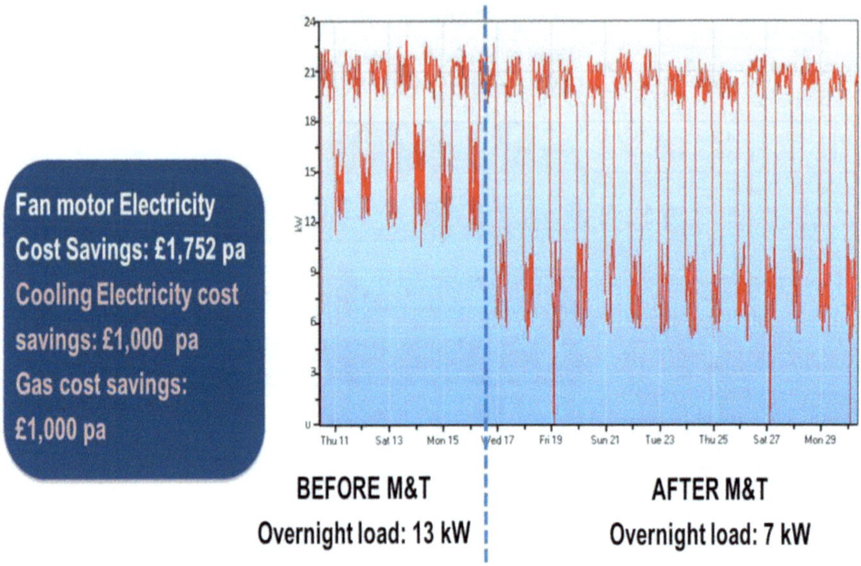

Fig. 5.8 Savings Measured with EMOTR

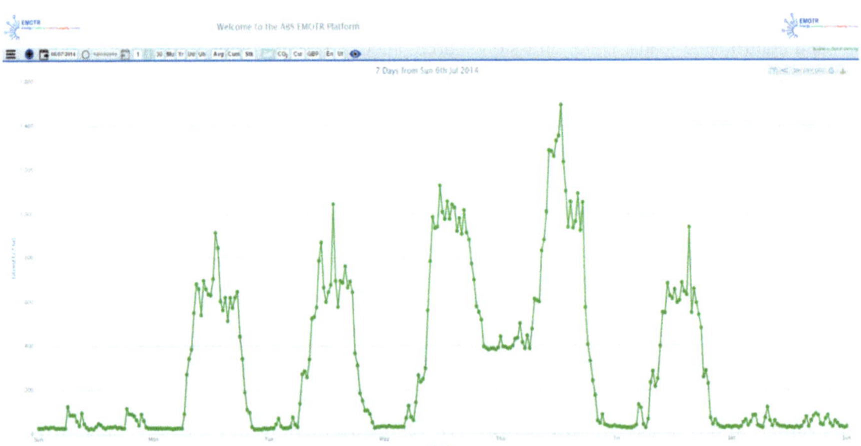

Fig. 5.9 Waste Identification with EMOTR

5.3.3 Post-Occupancy Evaluation (POE)

Structured occupant feedback is an essential part of SFM. It enables facilities managers to take informed decisions about the working environment created by the facilities they operate. The author uses a process known as overall liking score (OLS) for POE.

Fig. 5.10 EMOTR identified faulty operation of the AHU

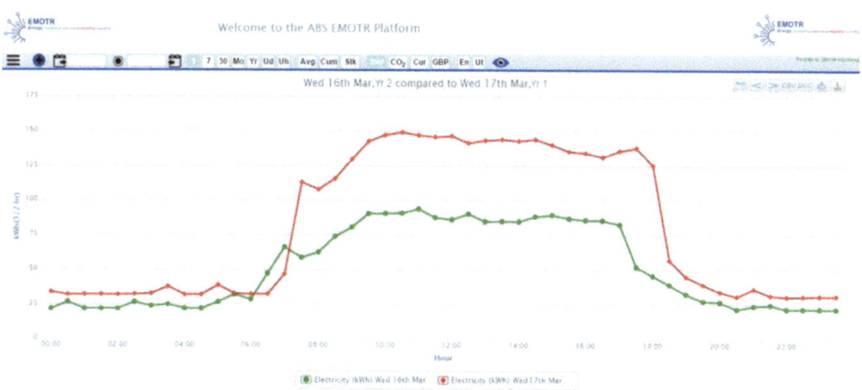

Fig. 5.11 EMOTR energy saved following 1 year of ConCom

POE has many definitions. Some definitions include an evaluation of the procurement process as well as the performance of the building in operation. Others focus on the condition and performance of the assets. This guidance on POE in this chapter has concentrated on the performance of the working environment based on occupant feedback as defined below:

POE is the systematic evaluation of the working environment and building performance based on the perception of the occupants.

This section describes the occupant feedback process developed by the author and used to inform building performance and occupant satisfaction through the application of continuous commissioning.

Defining the Brief and Scope

The key objective is to match the occupier's expectation, and thus, the best starting point is to define the brief and the scope of the evaluation. The evaluation should include an assessment of the building's technical performance (objective viewpoint) as well as its performance as perceived by the occupants (subjective viewpoint). Therefore, a post-occupancy evaluation should include the following:

(a) An occupant perception survey (e.g. overall liking score approach[1] or similar) and
(b) A technical review of the issues identified by the occupants in the survey.

It is appropriate to re-emphasise that although all aspects of facilities management are not covered in this chapter, they have a significant relevance for the users of a building and can influence their overall perception. For example, if users occupy a building that provides a high-quality physical environment, but the support services (soft services) and the management of the organisation are poor, it is likely that occupants will be dissatisfied. In order to validate an occupant's review of the building services, it is suggested that the occupant perception surveys include questions about general management and work satisfaction (e.g. 'Do you like your colleagues?' and 'Are you satisfied with the company's approach to staff development?').

A suggested brief/scope could cover the following:

- Provide a detailed measure of how people feel about their working environment.
- Understand occupant needs and preferences in order to fine-tune the performance of a building following occupation.
- Assess the success or otherwise of building procurement and construction performance, measured against the set objectives (time, cost and quality) and functionality.
- Evaluate financial performance of occupancy (planned against actual).
- Determine a clear idea of the areas that need improvement (i.e. thermal comfort, lighting, management policies, etc.).
- Identify and prioritise any improvements that are relevant and will inform the maintenance, operation and carbon management strategies (e.g. identify opportunities for cost and energy reduction and prioritise tasks for building recommissioning).
- Ensure that energy-saving initiatives increase or sustain the quality of services provided.
- Prepare benchmarking and comparison with best practice and market norms as well as benchmarking with performance overtime if possible.
- Provide the base benchmark on which to measure the success of the working environment going forward.
- Engage the occupants and demonstrate that their views are important, establishing organisation wide support towards energy cost and CO2 reduction and other FM initiatives.
- Provide simple low-cost solutions for immediate adoption where possible.
- Recommend a plan of action in order to address the survey findings.

[1] The overall liking score (OLS) approach in an analytical approach that measures how people feel about their work environment. Through a questionnaire, it assesses all the major areas of importance in a work environment and evaluates how much occupants like, and how important they find, each particular aspect.

Fig. 5.12 Delivering post-occupancy review

To deliver this brief, it will be necessary to carry out company-wide surveys and structured interviews. In addition, there will be an element of desktop technical assessment on the maintenance regime. All this should be evaluated in a report with recommendations for improvements (see Fig. 5.12 for a diagram of the process).

The purpose of defining the brief and scope of the review is to provide clarity, leaving little room for misplaced expectations. It is also to explain the limitations of the review (Fig. 5.13).

Planning and Delivering the POE

A typical programme for the POE using OLS is shown below:

As the above illustrates, the suggested POE process takes 6 weeks and is carried out in two stages:

Stage A: POE Strategy Development Duration: 2 weeks

Tasks:

A.1 Obtain an understanding of the working environment and the client's specific needs and goals

- Visit the premises to meet client representatives who will be involved with the POE programme.
- Agree the tasks of the individuals involved.
- Determine the sample to be surveyed.
- Obtain general information on the building such as occupant density, net floor area, gross floor area, energy cost and consumption, age of the building, specific problems and previous reports.

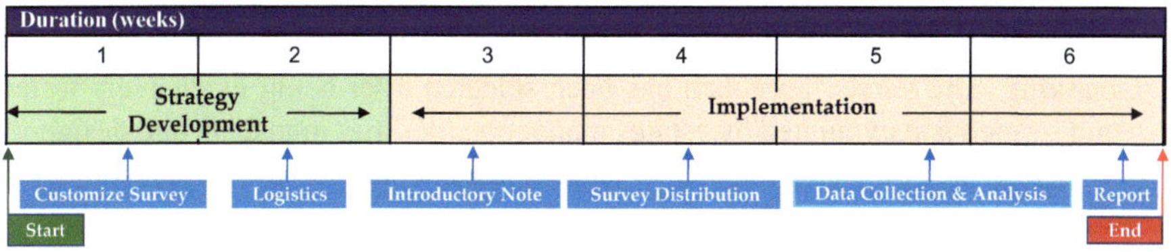

Fig. 5.13 Timeline of the post-occupancy review

A.2 Customise standard questionnaire to the particular needs of the client

- Brief client on survey format—The structure of the survey should be simple, and it should not take more than 10–15 min to complete. There are many survey formats, but the one we have developed and used is the OLS approach. It provides a multi-choice format that rates both liking and importance of each aspect examined and has space for the respondent to include comments (Fig. 5.14):

The derived individual overall liking score for a building or a building's aspect reflects how much respondents like and how important they find each aspect of their working environment.

- The questionnaire should cover all aspects of the working environment including:
  - Thermal comfort
  - Acoustic comfort
  - Lighting (daylight and electric)
  - Air quality
  - Amenities—quality of the office (attractiveness, office services, work space)
  - Level of personal control
  - Quality of general and energy management.

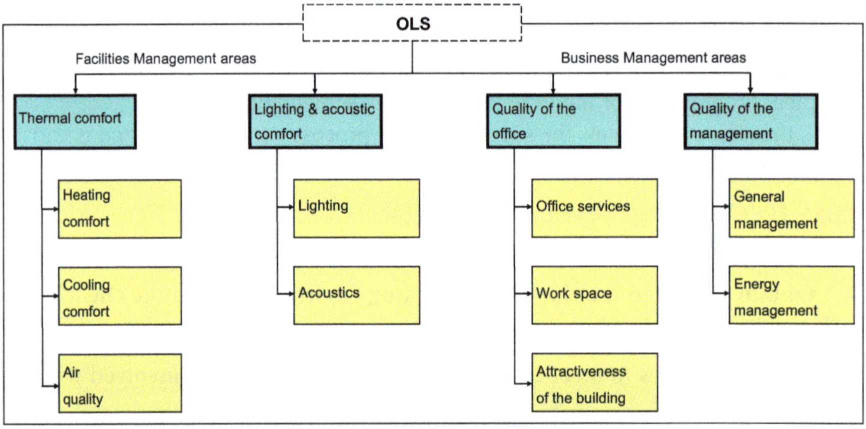

- Include questions that reflect the client's specific needs and goals. The OLS survey consists of the 45 core questions which should be customised in order to meet the client's specific characteristics and requirements. However, caution should be used when choosing to change and omit questions as the core of the questionnaire should remain relatively unaltered to provide consistency from survey to survey, to maintain data integrity and to provide consistent benchmarking. The survey approach has been selected after being extensively tested and refined and evaluated by estates managers, facilities managers and designers to ensure accuracy and relevancy. Clients are encouraged to add questions but advised not to take questions out.

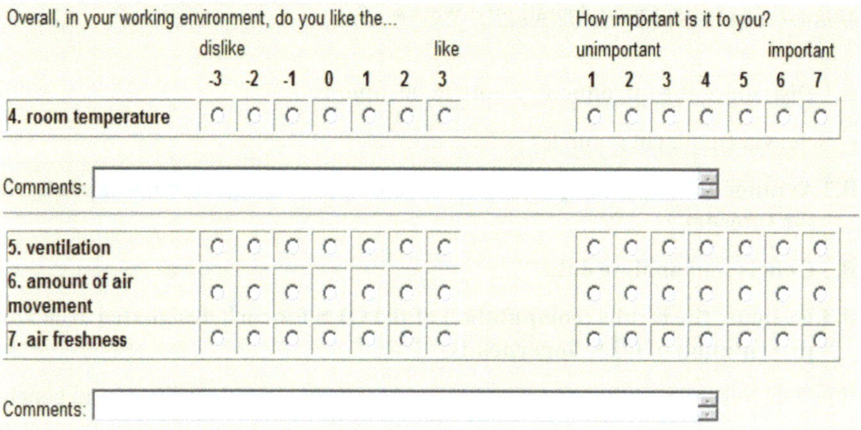

Fig. 5.14 Sample format of an occupant perception survey based on the overall liking score (OLS) approach

- Select the stakeholders that will be surveyed. The extent of the survey will be dependent on the management policies of the organisation and the role of the facilities, estates or project team. The extent can range from totally inclusive, where every user in an organisation is invited to respond to the survey, to a randomly selected sample of respondents. It is important to note that the wider the sample, the more representative the respondents' perception of satisfaction levels and the issues affecting them will be. An indirect impact of these satisfaction levels is the perceived productivity of the respondent.

A.3 Determine the method of distribution and collection

- The questionnaire can be distributed in paper format or electronically via the Internet.

A.4 Prepare an overall methodology and schedule for the project assigning tasks as necessary to all parties involved

- Figure out logistics (i.e. who will be the client communication representative? who will help with distribution of the surveys, the introductory note and the report? etc.).
- Appoint an appropriate member of client's staff as the liaison officer.

A.5 Prepare an introductory note for distribution to occupants

- Key points:
  - Explain how it serves the company's strategy.
  - Identify to which company initiative this survey belongs.
  - Give details on the company's initiative, its goals and the steps it involves.
  - Provide information on the survey.
  - Encourage contribution.

Stage B: Implementation Duration: 4 weeks
Tasks:

B.1 Distribute the introductory note to occupants

- Send via email, put reminder posters in break rooms, etc.

B.2 Arrange for the OLS questionnaire to be made available to occupants via the Internet

B.3 Collect and analyse data

B.4 Evaluate the results (computation of the OLS for each designated area and presentation of OLS fingerprints)

When the results have been received, they must be evaluated, analysed and benchmarked against internal criteria (and, if appropriate, external criteria). If there is information on the satisfaction levels from the previous environment, this should be compared with the results of the current post-occupancy survey.

Examples of OLS Surveys The most important input to the brief for a new project comes from formal and anecdotal feedback from occupants of facilities that are to be replaced by the new project. Such information can also be used for new facilities where the specific occupier is unknown, but the type of occupier is known. For example, occupant feedback from an office can provide valuable input to the brief for an office development that will not have a tenant at the briefing stage. The feedback does not have to be specific to the occupier, but it does need to be specific to the type of occupancy.

Formal occupant feedback is particularly helpful to facilities managers as it identifies areas of concern that are often caused by poor performance of building services engineering (BSE) systems and their associated controls. The results of occupant feedback using overall liking score (OLS) for 11 UK offices are shown in Fig. 5.11. These show that respondents perceive BSE and controls to be major sources of dissatisfaction. None of these offices would have been designed or operated with the intention of achieving poor scores for BSE. Therefore, if the cause of the poor scores can be identified, strategies for improvement can be developed and implemented (Fig. 5.15).

Even in the best office, the scores for BSE are negative and perceived by occupants as sources of dissatisfaction. In both best and worst cases, the response to the business culture questions is positive, showing respondent support for the business and its management. This gives some confidence that the negative responses are genuine: Why would respondents who support the business and like the management exaggerate negative aspects?

The fingerprints for best and worst offices are shown in Figs. 5.16 and 5.17. They have the following similarities:

- Full air conditioning that was nearing the end of its economic service life
- Town centre locations in SE England surrounded by busy roads
- Shortage of funding for investment in improvements to the BAT

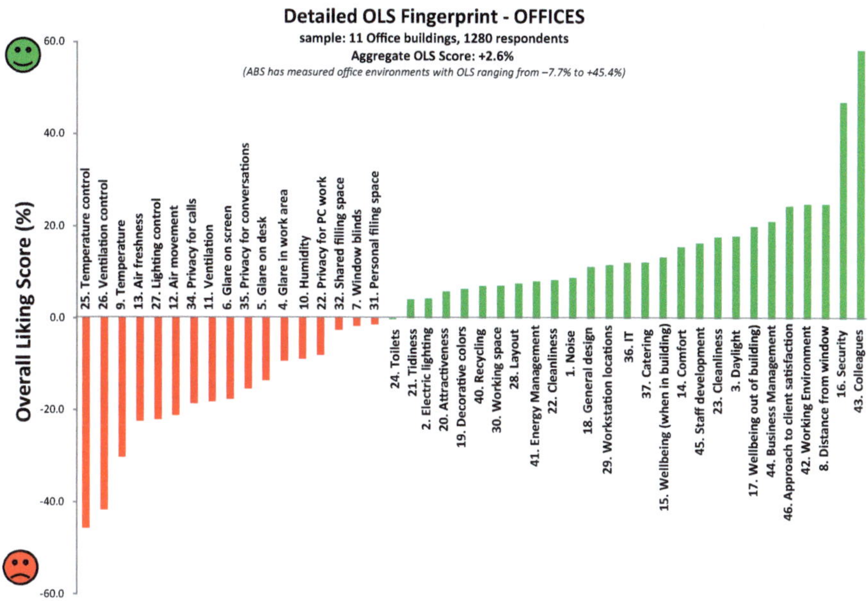

Fig. 5.15 Occupant feedback from a sample of 11 office facilities in the UK

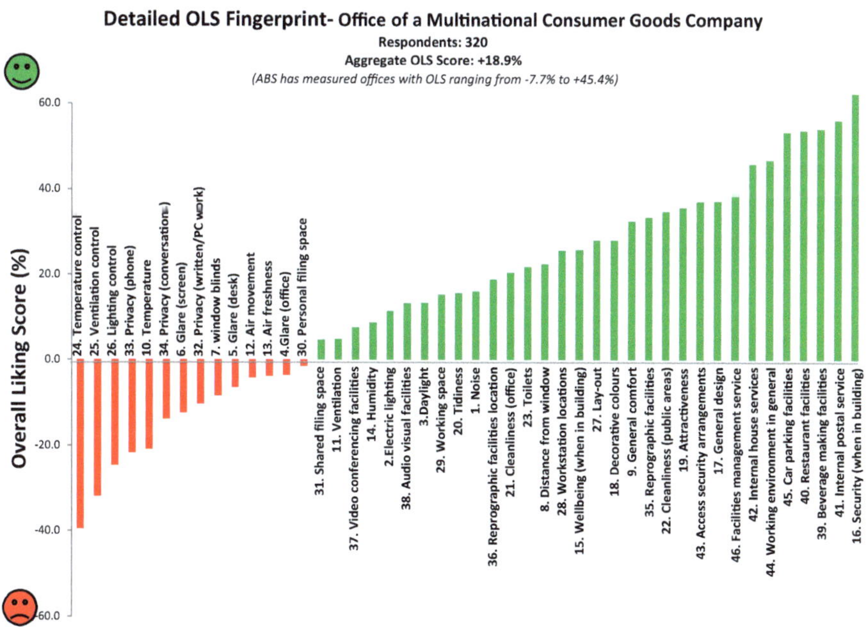

Fig. 5.16 Occupant feedback: highest score in the sample of 11 office facilities in the UK

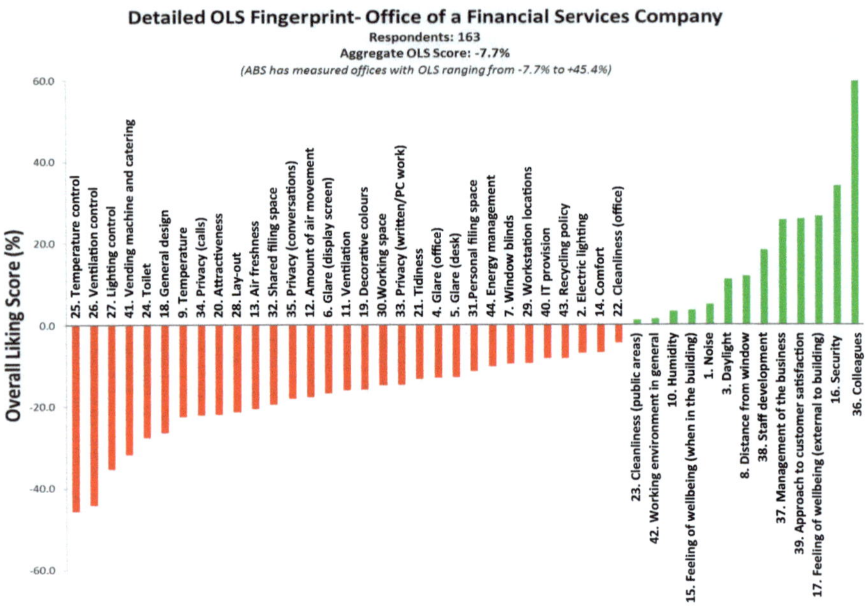

Fig. 5.17 Occupant feedback: lowest score from the sample of 11 office facilities in the UK

- Old BSE that was obsolescent and poor records
- Prescriptive input maintenance contracts with no specific requirement for operation or energy efficiency
- Inadequate technical expertise of the FM and maintenance team
- No stated carbon and energy management policy or strategy

When the survey results were presented to management, a plan for improvement was implemented and funding provided for an improved operating strategy using ConCom.

5.4 Performance-Based Service Provision (PBSP)

In this section, performance-based maintenance is used to outline the advantages in PBSP.

Performance-based maintenance results in the optimum of cost and value for the client by providing incentives for the client and the contractor to deliver cost savings and improved service. The process provides an opportunity to migrate from a prescriptive input specification to one in which performance is measured by what really counts—output. This requires a rigorous approach to subjective as well as objective customer feedback. Key performance indicators (KPIs) form an essential part of the measurement process using a technique known as overall liking score (OLS). The result is a partnership culture, which encourages continuous improvement in service delivery and reduced operating costs.

Experience suggests that performance is best when it is monitored and rewarded. But the big questions are as follows: how to define performance standards; and how to ensure that reward actually does drive value up and cost down.

Traditional contracts define the contractor's input—what he does. Performance contracts define his output—what he achieves. Doing is easier to specify than achieving. The process outlined here is designed to enable a maintenance contract to evolve by mutual consent of employer and contractor, from a doing, or input basis, to one in which a significant element of reward is based on measured achievement.

Maintenance is 'done to' the physical assets or facilities. But those facilities exist only to fulfil the occupier's business plan. In almost all cases, that fulfilment is achieved by delivering services to the people who form the business. Performance measurement must therefore include the satisfaction of the building's occupants. Key performance indicators (KPIs) can be determined for the contract from the results of OLS or other type of occupant survey. The end product is a partnership culture which encourages continuous service improvement and reduced operating costs.

Why Maintain?

There are two primary reasons for maintaining buildings and their services. These are:

- To meet legal obligations
- To meet the business need

In carrying out maintenance, it is essential to spend no more than is necessary to meet these two needs, and to do so, a process that optimises cost and value is required. This means that the maintenance system must be designed to meet the needs of the business that operates from the building and not relate to the building as an end in itself.

The quality of maintenance has an effect on energy efficiency. Of the ten buildings in Fig. 5.18, only building B consumes less energy than the UK government's recommended good practice standard. Building B also had the greatest occupant satisfaction and scored highest on the occupants' liking for temperature, air quality, lighting, noise and overall comfort. The building was maintained to a high standard with a maintenance regime that focussed on the occupants' needs. These results support the argument that performance-based maintenance creates financial, environmental and occupant satisfaction benefits that together are greater than the sum of the individual parts.

Meeting the business objective is the primary purpose of an organisation, and its people, the primary support, provide the means of doing so. Buildings and other secondary support facilities are there to enable the occupants to maximise their contribution to the primary purpose as illustrated in Fig. 5.19. Maintenance is one important component of the secondary support facilities.

The building, its service and the quality of maintenance have a substantial effect on how occupants feel, which in turn affects their productivity. Therefore, an understanding of the building life cycle will help those responsible for maintenance and operation to prepare an operating strategy that will minimise downtime and maximise occupant satisfaction.

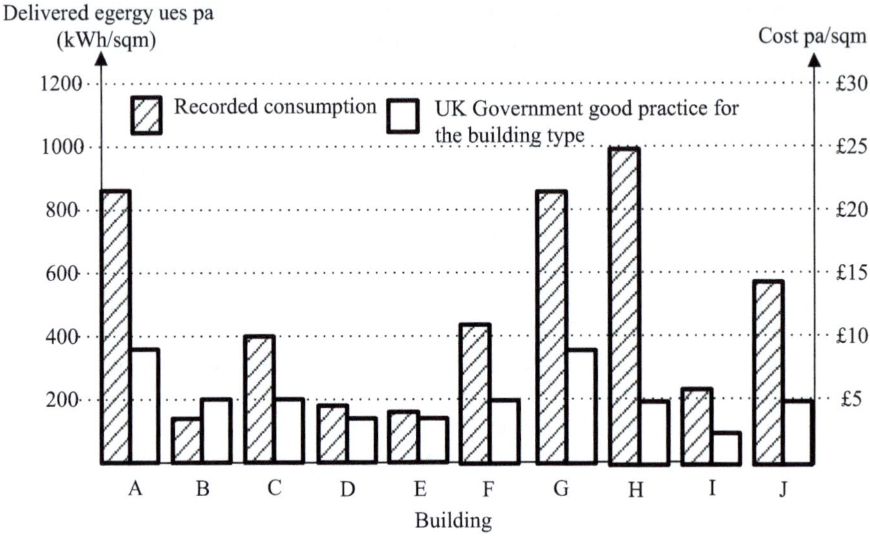

Fig. 5.18 Comparison of ten buildings

Fig. 5.19 Maintenance in support of the core business

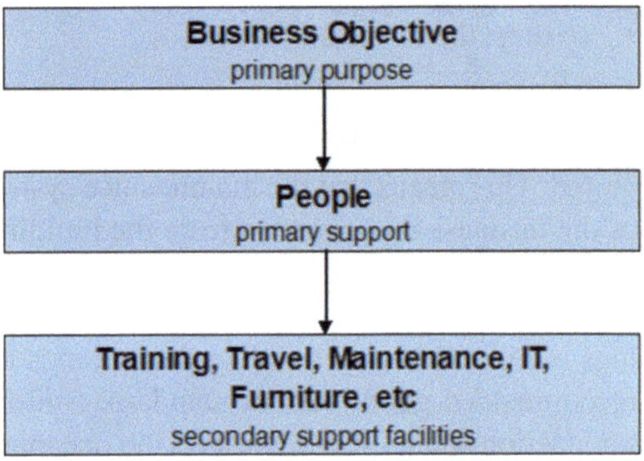

Figure 5.20 shows the main stages of the building life cycle and how they interact. At each decision point, there is an opportunity to reduce environmental impact which usually reduces operating cost and can often enhance value. A rigorous performance-based maintenance strategy will enable a continuous improvement process to be adopted that takes account of minor as well as major refurbishment and replacement. The result will be reduced obsolescence by the replacement of components, as well as complete systems, with the latest technology. To succeed, the process requires innovative maintenance management that can understand and apply technological innovation. Understanding the life cycle and the interaction of the various stages provides the basis for a maintenance and operation strategy that optimises all relevant factors in accordance with their priority.

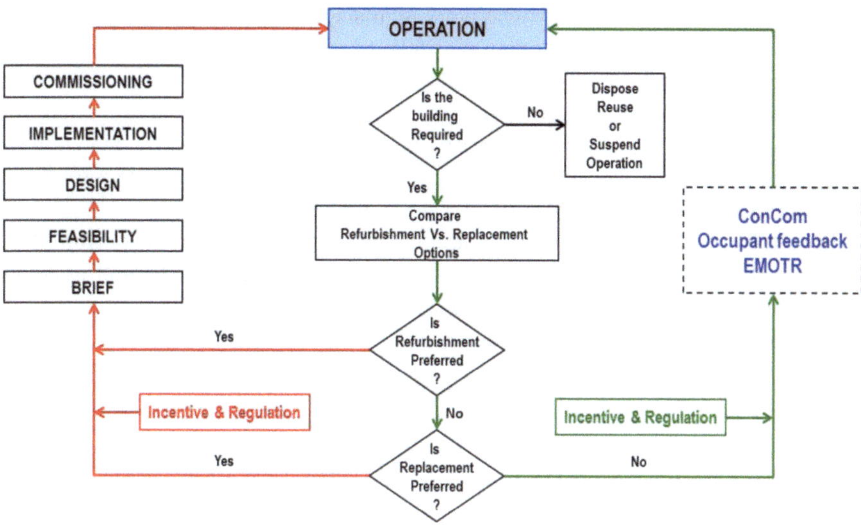

Fig. 5.20 The building life cycle

5.5 Maintenance Contracts

Performance-Based Contracts

Traditional maintenance contracts require the contractor to perform certain planned tasks which are defined by a specification and an asset register. The contractor is paid when the tasks are shown to have been completed. This basic arrangement works but has certain fundamental limitations:

- The need for maintenance of any installation varies according to local conditions including the following:
  - Air quality
  - The quality, age, history and condition of individual systems and items of plant
  - The performance demands on the installation

A standard, planned maintenance specification always contains elements that are either uneconomic (too much service), inadequate (not enough service), or inappropriate (wrong service) for parts of the installation. Furthermore, these factors change with the age and condition of the plant.

- To optimise the overall performance, the cost of maintenance and the management of asset value, there must be a mechanism to adjust the planned maintenance programme and to identify and decide upon special maintenance, repair or replacement issues. Within a traditional contract, the only formal way of doing this is by the costly process of routine inspection and audit.

But the contractors' technicians who work regularly on the plant are the people who know its condition and whether they are under or over-servicing it. They know how many clean filters are routinely replaced (or not replaced despite the specification); they know the sound of a pump, motor, fan or compressor that is in need of attention. They are the ideal routine inspectors and auditors. But they work for the contractor. In the adversarial world of contracting, their job is to perform their employer's contracted tasks and to achieve or better profit targets.

The solution to the problem of optimum operational and economic performance of the building services installation is therefore a simple one—a contract in which the motivation of the contractor and his employees is closely aligned with the client's objectives and operating strategy.

That is a performance contract.

Strategies for Introducing Performance as a Basis of Reward

Contractors are motivated by potential profit. They are not necessarily spurred to greater and better efforts by disproportionate risks and penalties. The essential mechanism of a performance contract is therefore to enable the contractor to improve his profit (not turnover) by sharing the economic advantage he achieves for the client. That advantage may be improved performance and it may be optimisation of asset value, but it is most easily recognised if it is a direct saving in cost.

In a performance contract, the emphasis is on 'output' - what is achieved, rather than 'input' - what is done. 'Output' is less easy to define than 'input' which can be specified as a set of tasks. The starting point for 'output' definition is a set of *key performance indicators*. These are a measure of the elements of performance that are most important to the client. They may relate to the achievement of specified internal conditions, the limiting of downtime, or response times for call-outs, time from reported defect to completed repair, etc. Each client and each building may have different performance needs.

It is possible to create a performance contract from a standing start. But the performance criteria and the rewards or penalties are difficult to define and agree on an equitable (and therefore workable and motivating) basis between the parties. The best way to start is often with a traditional 'input' specification contract which over a period of time migrates to a performance-based output contract. In this case, the contract sum will be based on the input requirements.

The contract includes a statement of the client's real 'output' needs in the form of a set of *key performance indicators*. It also includes a mechanism for the contractor to propose changes that improve performance and/or achieve cost savings by moving away from the specification towards an output definition. Savings are shared. Mutually beneficial changes are adopted; others are not. The result is a virtuous circle of mutual benefit. This whole process is best achieved within what is described as a *partnering* relationship of mutual trust and common interest. Figure 5.4 illustrates the concept of migration (Fig. 5.21).

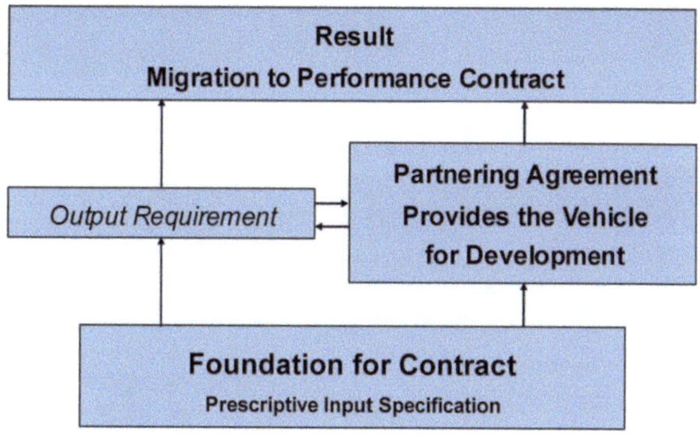

Fig. 5.21 Performance contract strategy

Existing Buildings—Existing or New Contracts
An added advantage of a strategy that starts with a conventional contract is that it can be applied to an existing contract as well as a new one. It is essential, however, that a full audit is made of the documentation, performance and asset condition under the existing contract before any changes are made. It is also a precondition that the incumbent contractor is open to performance-based contracts and to a potential partnering relationship.

Real advantages can be achieved by taking stock of existing maintenance contract arrangements for any reasonably substantial building with a view to obtaining economic and performance benefits by moving towards a performance-based contract.

References

Lau S (2014) Mind the gap: a comparative case study analysis of commissioning in commercial buildings in England and California. MSc. Imperial College London

World Commission on Environment and Development's (the Brundtland Commission) report Our Common Future (1987) (Oxford: Oxford University Press, 1987)

Part II
Exemplars in Sustainable Design and Built Environment

Chapter 6
Delivering Sustainable Exemplars

Guy Nevill and Jo Wright

6.1 Introduction

M&E and environmental engineers, Max Fordham and architects FCB Studios are recognised as pioneers in environmental and sustainable design. The two practices have worked together for over 20 years and, both independently and together, have a long heritage of designing low-energy, passively designed and sustainable buildings.

This chapter describes the particular experiences of the same core team of architects and engineers over the period from 2002 to 2014. They worked closely together delivering a run of buildings which have been recognised as sustainable exemplars, written up by the architectural and engineering press and now often studied by architectural students.

The post-occupancy studies carried out have informed both the approach to and the design of subsequent projects. The contribution of Bill Bordass to this process cannot be underestimated. He has provided the structure for the post-occupancy reviews and we benefitted from his accumulated wisdom, in particular from his work with the Usable Buildings Trust on the PROBE Studies (Online).

The first building covered in this chapter is Heelis—the central office for the National Trust. That will be followed by the Woodland Trust Headquarters and The Hive Library and History Centre in Worcester.

The latest building the team worked on together was an art gallery and residential block in Cyprus for the A.G. Leventis Foundation, which at the time of writing has just begun the post-occupancy review process.

We hope that this chapter will help to highlight some of the key factors in delivering sustainable exemplars and perhaps form the basis for a framework for

G. Nevill (✉)
Max Fordham LLP 42-24 Gloucester Crescent, London NW1 7PE, UK
e-mail: g.nevill@maxfordham.com

J. Wright
FCB Studios, Bath Brewery, Toll Bridge Road, Bath, BA1 7DE, UK

© Springer International Publishing Switzerland 2016

M. Dastbaz et al. (eds.), *Building Sustainable Futures*,
DOI 10.1007/978-3-319-19348-9_6

you to build on for your own projects. At the very least, it enables us to share our experiences!

6.2 Heelis, New Central Office for the National Trust

This headquarters building for the UK's largest charity, completed in 2005, is a highly sustainable workplace. It demonstrates that despite similar budgets, significant improvements can be achieved over the performance of typical commercial buildings. The two-storey, open-plan building provides 76,500 ft^2 of office space, meeting rooms and workshops for up to 500 staff, plus a public shop and café.

The relocation of the central administration functions of the National Trust brought together staff under one roof for the first time, as part of the process of honing the organisation for the new millennium. Previously, operations were dispersed across London and the south west in a variety of adapted accommodation in six separate locations. Sustainability is at the heart of the National Trust's mission and the project brief was to develop the most sustainable building possible within the available budget.

The building is located on the site of Brunel's Great Western Railway works, to the north-west of Swindon town centre where, at the height of activity in the late nineteenth century, 14,000 people were employed. In form and scale, the building makes reference to the tough industrial architecture of the engineering works which once covered much of the site. Its 'deep plan' form is a contemporary interpretation of the historic workshop buildings where natural light and ventilation were the only options in terms of environmental control.

Natural ventilation and use of daylight were two of the key drivers to the design of the building, both to reduce energy use and also to provide an excellent working environment. The form of the building has been strongly influenced by this.

The trapezoidal-plan form synthesises the geometry of the adjacent nineteenth-century buildings with 'solar geometry'. This includes north-south–oriented roof pitches for solar energy collection and controlled natural light, and a 'desire line' (the route the public is drawn to) which traverses the site from the south-east.

The sawtooth roof form, the primary elevation and the 'fifth façade', provides an even distribution of daylight and ventilation across the plan depth, while a series of atria connect the two storeys, creating a real sense of the entire organisation sharing a single volume. The repetitive, ridge-and-furrow profile allows for north light to be optimised and also provides a south-facing photovoltaic collector area.

The resultant form intriguingly has gables on three elevations, with a colonnade to the south addressing the external public space. The south façade, which opens to the new public landscape and gives access to the publicly accessible parts of the building (shop, membership recruitment and café), combines large areas of glazing with cast aluminium grilles, resonant of the foundry which once occupied the site. Perforated aluminium shading provides a sense of enclosure to the colonnade which runs the length of the building.

The Building achieved a BREEAM 'Excellent' rating (Figs. 6.1, 6.2, 6.3).

Fig. 6.1 The colonnade provides shading to the south façade

Fig. 6.2 **a** The building outline is orientated to the site, but the roof form is orientated to the solar geometry. **b** The model shows the array of roof cowls and rooflights. (FCB Studios)

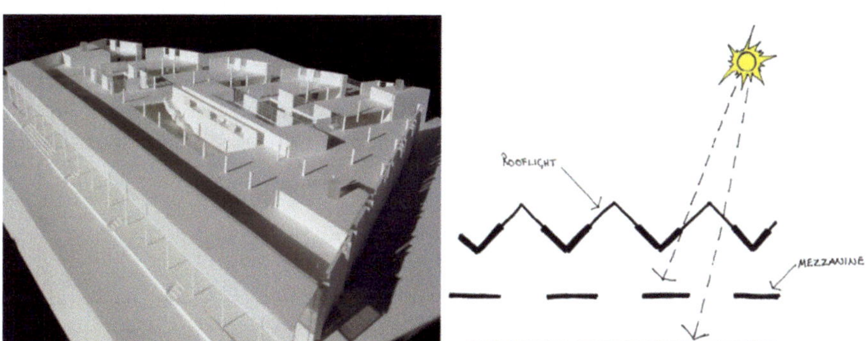

a b

Fig. 6.3 **a** Model showing the first floor with the central atrium, two courtyards and the mezzanines to the north side of the office. **b** The concept sketch from the interview shows daylight to both floors

6.2.1 Landscape

The site encompasses a significant area of additional, publicly accessible landscape in addition to the car parking associated with the building (at a rate of less than one space per three staff). Secure bicycle parking is provided for both staff and the public.

The landscape incorporates simple blocks of birch trees, typical of post-industrial landscapes, which complement the blue engineering bricks of the elevations. Along the south colonnade, a wide planter mediates between floor level and the surrounding landscape, with lavender, echinops and alliums. The planter edge provides a long, south facing seat, with further brick and oak seats inhabiting the new landscape between the central office and the adjacent building.

A hornbeam hedge with a simple galvanised fence encloses the northern boundary. Elsewhere, a variety of planting enhances biodiversity and generates year-round visual interest.

Within the plan-form, two courtyards provide more intimate landscapes and enhance views from both communal and desk areas. With access directly from tea points and meeting rooms, these spaces act as external 'rooms'. Again, emphasis has been placed on year-round interest.

A sustainable urban drainage scheme has been incorporated to reduce the storm-water load on existing infrastructure.

6.3 Procurement

As a charity, the National Trust did not want to spend its funds on an office building. Therefore, the building was institutionally funded and then leased back to the National Trust on a 25-year lease.

The National Trust appointed a contractor (based on RIBA stage C information) to develop the design under a design-and-build contract, with the design team novated to the main contractor.

At times, there were conflicts within the brief, as the building had to meet both the requirements to be a sustainable head office, recognisable as being for the National Trust, while also meeting British Council of Offices standard specification for the funders.

6.4 Briefing

The initial brief to the design team was to deliver a shell-and-core building to accommodate 470 staff at a density of 1/12.5 m^2, with public access to a shop and café for a budget of £ 135/ft^2.

The aspiration was to provide a flexible working environment to suit the Trust or other future occupants while minimising the environmental impact of the Trust's operations.

The National Trust's brief included a guiding principle to deliver a building that 'must achieve high quality benchmarks in all aspects of sustainable design, and should seek to be innovative in at least a few key high profile areas'.

Furthermore, there was a request to consider additional sustainability measures, over and above the requirements of Building Regulations, for which a 'payback' period of less than 25 years could be demonstrated (i.e. within the terms of the lease).

To help demonstrate these sustainability measures, the design team developed a tool which became known as the 'The Sustainability Matrix' (Max Fordham LLP 2010). It sets out possible strategies evaluated as best practice/innovative/pioneering and was also used to highlight the potential areas that could provide suitable payback. On this basis, the project board approved a number of design and specification enhancements including the following:

- Photovoltaics
- Enhanced thermal insulation
- Lighting control system
- Wintertime mechanical ventilation with heat recovery
- Propane chillers
- Lime mortar

6.5 Natural Ventilation and Thermal Mass

The central office was designed as a predominantly naturally ventilated building with mechanical ventilation provided to meetings rooms and throughout the office during the heating season. The building combines BMS-controlled natural ventilation with exposed thermal mass to both the first floor slab and the roof. This enables automated night cooling for the summer months, with additional manually operated openings, to create comfortable conditions for the majority of the spaces for most of the year. Densely occupied or internal spaces (meeting rooms, quiet rooms, etc.) are mechanically ventilated with comfort cooling. Chilled water is supplied from 'Earthcare' chillers which use hydrocarbons with a lower global warming potential than more conventional fluorocarbons.

Air enters the building around the perimeter and the two courtyards via windows and leaves via the roof cowls. The openings are automatically controlled, but all have local over-rides to provide some degree of control to the occupants. All of the automated openings are behind external grilles which provide security. These insulated opening panels comprise a low-level door-sized panel for significant air movement and a much smaller one at high level to admit more controlled quantities of air when the weather is cooler.

The roof vents open into giant, roof-mounted cowls (known as 'snouts') which promote stack-effect and wind-driven ventilation (ensuring a negative pressure at roof level whatever the wind direction) while providing rain protection to the opening vents. In addition, five of the cowls contain mechanical punkah fans which can be used to encourage air movement on still days.

A simple mechanical ventilation system with heat recovery introduces air via the raised floor throughout the building and extracts from central locations. This displacement system maintains air quality and thermal comfort during the coldest winter months when opening windows would risk cold draughts.

The building achieved an air tightness of 5.5 $m^3/h.m^2$ of façade at 50 Pa (Fig. 6.4).

6.6 Daylight and Sunlight

The sawtooth roof forms the most significant facade. It provides an even distribution of daylight and ventilation across the plan depth. As mentioned earlier, a series of voids connect the two storeys creating a sense of the whole organisation sharing a single volume.

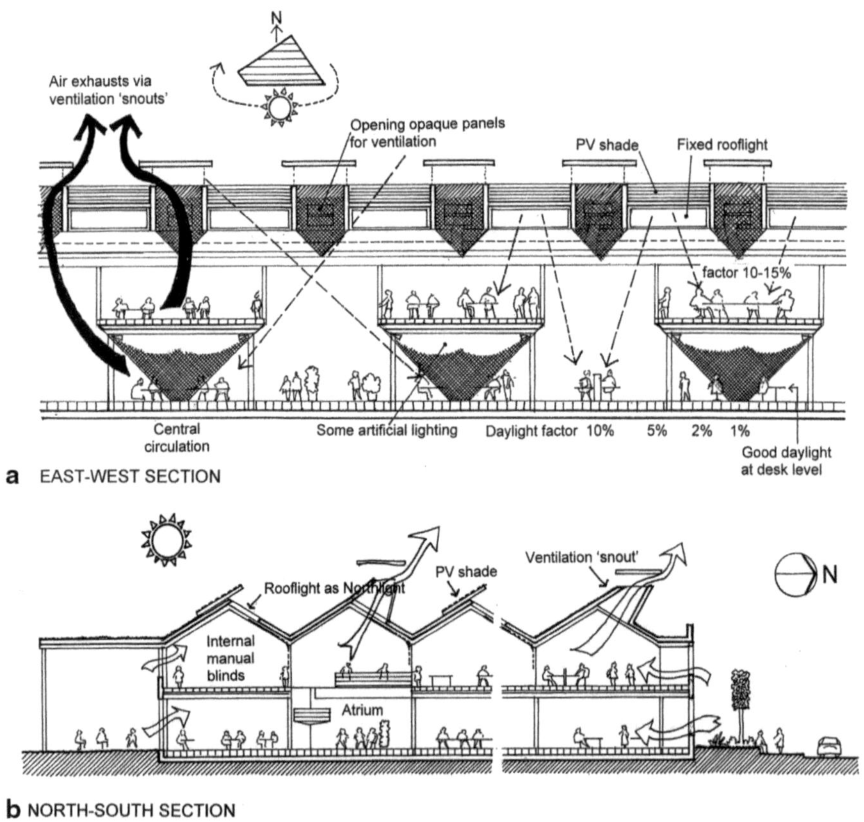

Fig. 6.4 **a** Both floors are well day-lit from the rooflights. PVs and roof cowls prevent direct sunlight from entering. **b** Air enters around the perimeter via automatic openings and leaves via the roof cowl. The PVs shade the rooflights, turning them into northlights

Extensive daylight modelling was undertaken using virtual models together with testing a 1:50 scale model in the UCL Bartlett Artificial Sky. The building has been designed to provide a minimum daylight factor of 5 % to as much of the regularly occupied areas as possible, minimising the use of artificial light by means of a fully dimmable lighting control system which responds to external conditions.

The BCO guide required a minimum light level of 300 lx. This can be met with 5 % (the daylight factor) of an external light level of 6000 lx. Over the year, this level of daylight is available for around 85 % of the working hours 9 am–5 pm, thereby meeting the majority of the lighting requirement using natural daylight only.

The floor-to-ceiling height of 3.7 m at ground floor is unusually high and designed to ensure that both natural light and ventilation work efficiently throughout the plan depth. The mezzanine first floor fingers extend into the main body of the office, allowing daylight to penetrate from the roof to the ground floor. They are orientated north-south to allow the best and most even distribution of daylight.

The external envelope has been carefully considered to avoid potentially harmful solar gain: the brick elevations incorporate cantilevered brick fins, and the roof mounted 'snouts' and photovoltaic panels eliminate high angle south and lower angle east and west sunlight. The combination of a projecting canopy and perforated shading avoid sunlight penetrating the south façade.

An internal environment devoid of sunlight might lack visual warmth and vibrancy—we resolved this through the roof glazing to the central atrium which includes some south facing glazing (with low thermal transmission glass), allowing sunlight to animate this 'breakout' space for much of the day (Figs. 6.5, 6.6).

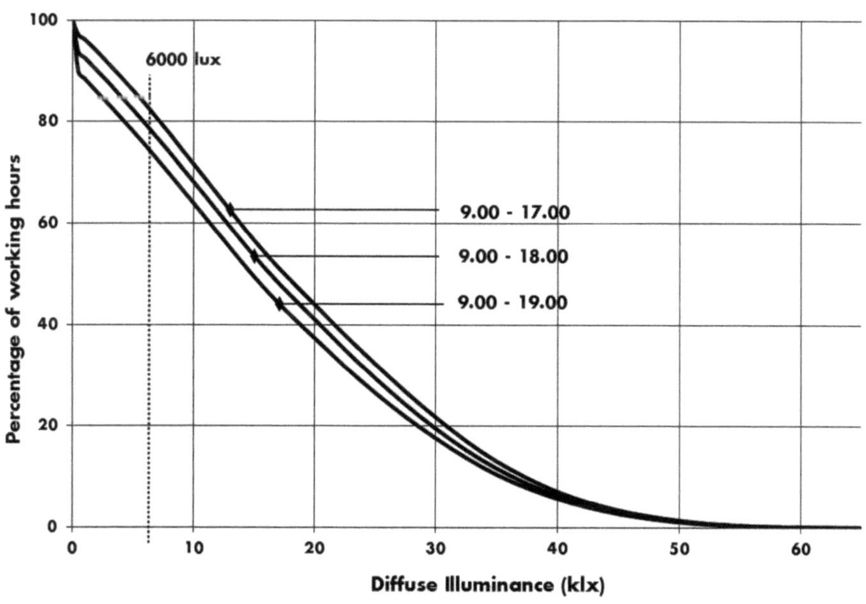

Fig. 6.5 Availability of daylight during working hours

Fig. 6.6 **a** Sunlight is allowed to enter the main atrium to animate the space. **b** The office is well day-lit, but note the dark perimeter wall

6.7 Artificial Lighting

The use of daylight was a key driver to the design so it was important that the artificial lighting scheme could adapt to natural light levels. The artificial lighting scheme was designed to balance fluctuations in external conditions using fully dimmable, ceiling-mounted fluorescent lighting. This provides the minimum light levels of 300 lx to all working surfaces. PIR sensors ensure that only the occupied areas are lit.

It became apparent on completion that due to the triangular form of the building, the rooflights did not extend right to the perimeter. This meant the walls were not well day-lit and were in fact quite gloomy. Despite the desk light level being high enough, wall-washing lights to the perimeter walls would have made a significant difference—potentially allowing the desk light level to be reduced.

The automatic system also did not allow any user interaction or localised control. As a result, some occupants perceived the set light levels as being too high while others found it too low with little means of doing anything about it.

6.8 Materials and Embodied Carbon

The external walls which take their geometry from the adjacent Victorian buildings are clad in a variety of blue bricks, echoing the historic poly-chromatic brickwork. The rippling 'curtain' of brickwork cantilevers to form shading fins that protect north-east and north-west facing glazing.

Heelis was the first commercial building to use batch-mixed lime mortar which reduces cement content (and thus embodied CO2). It also proved to be better suited to the engineering bricks specified than a conventional cement-based mortar, and it will facilitate the future recycling of the brickwork.

The roof covering is a proprietary, standing-seam aluminium system, the majority of which is covered by PVs cantilevered to shade the rooflights. Ventilation cowls are formed in aluminium cassette panels riveted together off-site and craned into position.

The aluminium used in the glazing systems and in the cast grilles to elevations is 92 % recycled and fully recyclable in future, which justified the specification of this energy-intensive material.

The courtyard spaces, which are only glimpsed from outside the building, contrast with the exterior materials, with north and south elevations clad in oak.

The internal environment is reminiscent of the engineering sheds adjacent: a steel frame with precast concrete infill to first floor and roof creates a simple and robust setting. Within the public areas, the blue brick paving used externally is brought inside, while a bespoke carpet made from undyed Herdwick (native British sheep) wool is used throughout desk areas and meeting rooms. The carpet is very durable and long-lasting and this commission helped support the commercial viability of this fleece from a herd saved from the 2001 Foot and Mouth outbreak.

The atrium provides a contrasting environment, with its south-facing wall and stair clad in a variety of timber from National Trust estates, which runs through to the courtyard adjacent. The oak floor continues the theme, creating a warm 'heart' at the centre of the plan, giving access to primary meeting rooms and providing breakout and staff café facilities.

Rob Jarman, Head of Sustainability at the National Trust, said '*I think it's absolutely amazing…the exterior is so appropriate to the rest of the site and the interior is so cool, light, quiet, and completely devoid of smells of new build such as the usual off-gassing from new carpets*'.

6.9 Energy and Thermal Modelling

Thermal comfort targets were the CIBSE standard; dry resultant temperature should not exceed 25 °C for more than 5 % of working hours or 28 °C for more than 1 %; based on weather data from 1997 and CIBSE DSY.

Extensive thermal modelling was carried out using TAS Software to test the amount of ventilation and thermal mass required to meet these conditions.

Following the novation of the design team, the contractor naturally wished to limit construction costs and challenged the inclusion of thermal mass in the roof and the need for the roof cowls. The extent of thermal modelling carried out enabled the design team to robustly defend the design and keep these elements in Fig. 6.7.

Fig. 6.7 The array of roof cowls and PVs provide shading to the rooflights

6.10 Low- and Zero-Carbon Technologies

The National Trust secured £ 300,000 of grant funding from the Department of Trade and Industry's major photovoltaic demonstration programme. Photovoltaic panels are located on the south-facing roof pitches, combining energy collection with shading to the north facing roof lights.

The 83-kWp system consistently generates enough electricity annually to offset approximately 10 % of the electrical load of the building. The efficiency of PVs has greatly increased since they were installed at Heelis, and when the time eventually comes for the existing PVs to be replaced, a comparable PV area will result in at least a doubling of the generation capacity.

6.11 ITC Strategy

The server room contains the central servers for the whole of the National Trust, not just this building, and so was of a significant scale. The room was designed to use free-cooling wherever possible, bringing in fresh air directly from outside whenever the external temperature is below 14 °C.

What became clear in the months after completion was that the electrical load of the IT servers was greater than predicted. Together with the associated cooling requirements, it comprised close to 50 % of the entire building electrical load. The base IT electrical load barely changed day or night as the servers would not respond to actual IT demand—i.e. they stayed 'on' regardless of whether they were being used or not; also demanding constant cooling.

As IT technology has developed, the ability of servers to ramp up and down in response to actual user demand has improved. Due to the relatively high turnover for IT equipment, this scenario may improve over time.

6.12 Commissioning and Handover

There were interface issues between the building management system and the window actuators which meant the controls were not properly commissioned at handover, nor had the graphics pages for the BMS been completed. Fortunately, although the window controls were not operating correctly, there was sufficient thermal mass and ventilation to keep the building comfortable.

Staff were informed of the emerging design for the new building throughout the design and delivery process and the move-in was carefully orchestrated.

A simple occupant guide was written by Max Fordham, including energy do's and don'ts to highlight energy-efficient practices and to maintain the working environment at optimum conditions. This was explained to all staff as part of the induction process. To allow sufficient bedding-in time, the facilities manager instituted a '100-day' rule. Complaints and queries were noted by them but not acted on unless they persisted after the first 100 days while staff settled into and became used to the building.

Staff education is maintained via regular newsletters. As the turn of the season approaches, the newsletters remind staff about how the building's automatic operation is likely to change—windows opening automatically, for instance, as warmer days arrive. During summer, staff understand that the building will be slightly cooler in the mornings to allow for comfortable conditions later in the day and so a 'cardigan-culture' has developed which encourages staff to take responsibility for their own comfort as the internal temperatures change.

6.13 Post-Occupancy Evaluations

FCB Studios funded an independent post-occupancy evaluation (POE) using the cash prize awarded with the 2006 RIBA Sustainability Award. Undertaken by Bill Bordass of the Usable Buildings Trust, this identified issues including energy use at variance with design targets and allowed the team to 'fine-tune' performance. The information from the POE was fed back to the industry over a number of presentations and articles (Nevill 2007; Bunn 2007).

6.13.1 Energy Consumption

The major unexpected energy consumption was that the combination of the café, server room and IT equipment accounted for at least half of the overall energy use of the building and was outside of the control of the design team. These included the IT servers being on 24 h a day and the dishwashers in the kitchen running all day using a steady stream of hot water. This highlighted that the design focus on a delivering a highly sustainable and low-energy office building had unknowingly only considered half the picture.

Despite these issues, a study of the energy use showed that the building performed favourably in comparison with other recently completed office buildings. The study was also valuable in that it reinforced the idea that continual monitoring and adjustment would be required by the building users in order to maintain and improve on energy use.

6.13.2 Comfort and the User Experience

Temperature was monitored during 2006 (an exceptionally hot year) and found to meet the thermal comfort targets, despite the controls only being partially commissioned at that stage. This demonstrated that the base design was robust.

As part of the POE, a building user survey captured the response of staff to the building. The building scored very highly on lighting and perceived effect on health, while the scores for heating, cooling and acoustics were poorer than the overall data set—partly due to the norm being narrow plan buildings. However, when compared with large floor-plate deep plan buildings of which Heelis was one, it came out best and also achieved the highest 'forgiveness factor' (where people so like a building they forgive some degree of discomfort) of any building surveyed by the Usable Buildings Trust at that time.

6.13.3 Key Awards

RIBA Award—2006

British Council for Offices Innovation Award—2006

Architects' Journal 100 Sustainability Award—2006

Building of the Year and the Sustainability Award at the British Institute of Facilities Management Awards—2006

British Construction Industry Awards—Building Award—2006

6.14 Woodland Trust HQ, Grantham

Design of the new headquarters for the Woodland Trust commenced in February 2008. This followed a competitive interview process that selected Feilden Clegg Bradley Studios, Max Fordham, Atelier One and Grant Associates as the design team. The building occupies a green-field site in Grantham, Lincolnshire, just a few hundred metres from the Trust's previous rented workspace. The Trust required a flexible, new office designed to accommodate 200 staff in open-plan accommodation built to suit rental values in Grantham which at that point stood at £ 12.50/ft^2.

The building form, which emerged from early design discussions, comprises a spiral that ascends seamlessly from landscape to three-storey apex. The higher portion of the building, which accommodates the open-plan work space, is orientated north-south and comprises a 15-m-wide block situated towards the middle of the site. A wing of service and cellular space descends from the three-storey volume along the western boundary, culminating in a single-storey bike store. This encloses the central woodland glade which is both entrance court and breakout space. The building communicates the mission and values of the Woodland Trust, while providing an economical, exemplar workplace that meets BCO and institutional requirements.

The building is designed to be naturally ventilated, and to provide excellent levels of daylight throughout, minimising running costs. Whilst the 15 m-wide floor plate is open-plan, most of the cellular support spaces are located along the western side of the building, avoiding the need to externally shade the more challenging west elevation.

The elevations are inspired by woodlands. Windows to the north and south elevations are conceived as three storey, vertical strips, maximising the opportunity for daylight and high-level vents and interspersed with vertical strips of timber cladding. The windows are aluminium clad timber glazed with clear solar-controlled glass. They incorporate insulated spandrel panels coloured in the same dark charcoal grey used for external frames and flashings, with some spandrels are coloured silver grey, echoing the bark of silver birches.

The roof is a single 11.5° pitch, clad in profiled fibre cement sheet, descending from the north elevation to the southernmost point and designed to allow future addition of photovoltaic panels. A single ridge is created over the meeting room wing where the roof 'folds' on a shallow pitch to form the bicycle store.

The building achieved BREEAM 'Excellent' (Figs. 6.8, 6.9).

Fig. 6.8 **a** Large vertical panels of glazing to the north elevation with the external timber baffle to the window openings at roof level. **b** South elevation courtyard with external shading to the windows. (Peter Cook?)

Fig. 6.9 Sketch showing like-building form and the south-facing sloping roof which could take future PVs. (FCB Studios)

6.15 Brief

As a charity, the Woodland Trust required a construction solution driven by cost and building functionality—providing an environment which enabled the staff to work as effectively as possible. The brief demanded a highly innovative and sustainable building built to a budget that reflected the rental values in Grantham. Design options were evaluated in terms of impact on cost and efficiency. Both the structural solution and the research undertaken on the 'Concrete Radiators' (see later) was required to demonstrate that the Woodland Trust would not be paying a premium for the building compared to a more conventional steel/concrete solution.

The brief was also informed by the lessons learnt from Heelis; keeping it as simple as possible; natural ventilation and thermal mass; use of natural and artificial lighting and lighting controls; paying close attention to catering and IT. The Sustainability Matrix was also used again but not followed as rigorously.

6.16 Daylight and Sunlight

The relatively shallow 15 m plan depth of the main office floor plate, and 4 m floor-to-floor heights allows an exemplary level of daylighting, maintaining an average daylight factor of approximately 3 % throughout. This is further improved by the verticality and even distribution of the glazing and the addition of light shelves to the south façade which reflect light onto the soffits and thus further into the plan depth. The design allows some controlled sunlight into the building, particularly through the roof lights over the main stair. This is intended to add an extra layer of 'delight' or 'sparkle' to the internal environment without providing unacceptable glare or potentially harmful solar gain. Elsewhere in the building, glare blinds allow manual user control of sunlight (Figs. 6.10, 6.11, 6.12).

Fig. 6.10 Sunlight is allowed into circulation areas to provide delight without causing problematic glare or solar gain. (Peter Cook)

Fig. 6.11 External louvres and internal lightshelves to the south elevation help to control and improve the quality of daylight. (Peter Cook)

Fig. 6.12 The office floorplate has very good daylighting levels and artificial lighting to the circulation routes. The concrete radiators are painted white to help with daylighting. (Peter Cook)

6.17 Artificial Lighting

Taking the lessons learnt from Heelis, we decided to move from a centrally controlled ceiling mounted system to one with more occupant interaction. Floor-mounted up/downlights are located adjacent to the desks and are entirely user controlled. A low background level of lighting is provided to corridors and circulation routes. Wall-washing lights pick out the internal elevations. These fixed lights are timeclock and central switch controlled.

The result of this is well-day-lit spaces with a background level of lighting and local task lighting that the occupants are very good at limiting the use of. Only about 5 % of task lights are put on regularly and only 30–35 % on dark winter mornings. Most staff work happily at desktop light levels well below the published standards, sometimes dropping as low as 50–75 lx without complaint (Bordass and Burgon 2014a).

6.18 Natural Ventilation and Thermal Mass

The ventilation strategy for the building relies on opening windows at each level and using both cross-ventilation and stack effect ventilation by exhausting at high level on the third floor. Voids in the floorplate allow the air to rise up the building. The openings on the top floor at the building apex are screened by a timber baffle so that a negative pressure is always maintained regardless of wind direction (Figs. 6.13, 6.14).

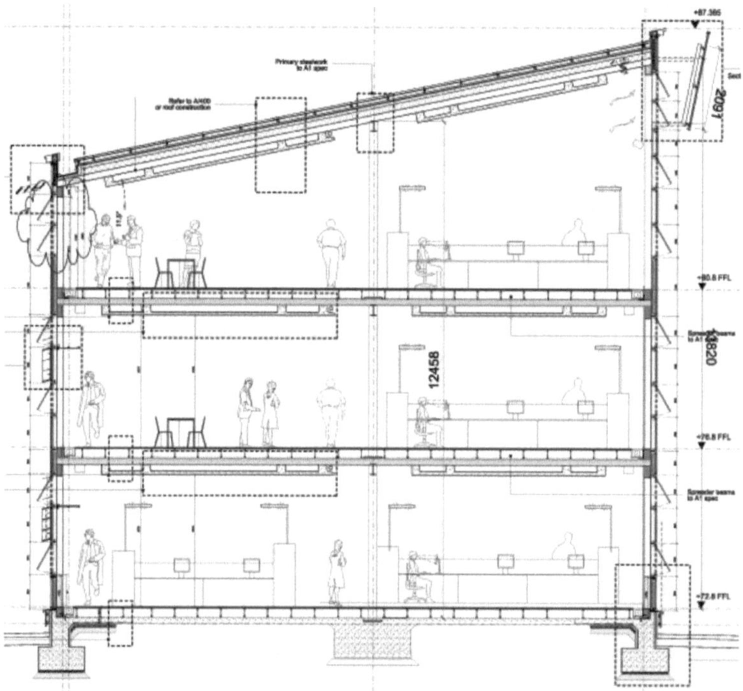

Fig. 6.13 Building section showing the high-level automated windows, low-level manual windows, the concrete radiators and the baffled air outlet at the top of the pitched roof. Note also the internal solar control louvres and light shelves to the south façade. (FCBStudios)

Fig. 6.14 The void allows air to rise to roof level and helps daylight to penetrate deeper into the floorplates. Wall-washing lights up the end walls. (Peter Cook)

6.19 Materials and Embodied Carbon

Again, the design team wanted to deliver naturally ventilated building with low running costs and low carbon emissions. However, there was also desire to try to reduce the amount of embodied carbon in the structure, primarily by reducing the volume of concrete.

This posed an interesting question, how can one reduce the amount of concrete involved, but still provide enough thermal mass to enable a night-cooling strategy to work? We focussed on a predominantly timber structure, incorporating the minimum volume of concrete necessary to keep the building cool to the required CIBSE Design Summer Year data for Nottingham. This generated the idea of constructing walls, floor and roof of the building in cross-laminated timber panels and fixing concrete panels to the soffits.

The concrete panels work as a structural composite with the timber; the reinforcement bars are bolted through the timber floor panels and modelling showed that a depth of 80 mm of concrete would facilitate night cooling under most conditions. What resulted was an array of panels referred to as 'Concrete Radiators'. As a novel solution, a sample panel was tested in the factory to check for stress cracks and movement over a 6-week period prior to construction.

An analysis of carbon dioxide generated by the structure was undertaken and an assessment made of any savings generated compared to a conventional concrete solution. We estimated that there would be a reduction of 400 tons of embodied carbon when compared with an entirely concrete structure, and the timber fabric of the building would sequester over 500 tonnes. The sum of this is equivalent to approximately 10 years of carbon dioxide emissions from the occupation of the building.

Throughout the design process, the pros and cons of the timber structure were compared against standard steel and concrete alternatives to ensure that the most cost-effective solution was still being chosen. This demonstrated that in addition to the reduced embodied carbon, there were also advantages of reduced time on site, simple detailing and reduced substructure. Once the decision was taken to use cross-laminated timber for the main structure, it very soon became logical to use it elsewhere in the building, including the lift shaft, stairs and balustrades.

As a result of the cross-laminated timber, the building achieved an air tightness of 2.4 $m^3/m^2/h$ at 50 Pa—less than 25% of the level required by Building Regulations. All external vertical joints are taped, and all interfaces with the cross-laminated structure are sealed with EPDM flashings. The air-tightness and thermal continuity contributed to very low heating loads.

6.20 Low- and Zero-Carbon Technologies and Renewables

During the design stage, the team assessed the potential for the generation of various renewable energy systems and calculated the payback periods. The result was that rather than renewable technologies, the most beneficial energy investment would be to focus on the reduction of the energy use of the server room, and to incorporate efficient servers and free cooling techniques. Nevertheless, the roof pitch of the building is designed to allow the potential for future installation of solar water heating and photovoltaics.

6.21 IT Strategy

Following Heelis, the significant contribution that ITC can make to the overall energy consumption was much more apparent during the design stage.

The Trust decided to install a thin-client system which had the advantage of moving the majority of the computing power and therefore the heat gains, from the office floorplate into the server room. This benefits the natural ventilation and comfort conditions in the office.

Higher-than-normal chilled water temperatures (flow 10 °C) mean that the system can operate in free cooling mode (i.e. the compressors do not run) and save energy for more of the year than would be the case had a more traditional approach been adopted. The initial design of independent chilled water and free-cooling systems was value-engineered out and replaced with a less-expensive packaged chiller with a free-cooling coil. Unfortunately, the chiller failed every few months bringing down the client's entire ICT, thin client and telephone systems. The manufacturer was remote and local support proved difficult to obtain, so a basic DX system had to be added as backup before the problems were eventually fixed.

6.22 Commissioning and Handover

The BSRIA Soft Landings process (BSRIA 2014) was informally adopted by the designers, but not formally incorporated into the procurement process. This was largely successful in embedding the lessons learnt from Heelis into the design and procurement and also managing expectations.

A facilities manager was appointed as the building works started on site; she had an important role in reviewing the design and preparing for the move-in. The handover was reasonably smooth but the occupier would have better understood the technical systems and control, had this appointment been even earlier. It would have also allowed the design to be reviewed more comprehensively. We produced a draft logbook for completion by the contractor, but unfortunately they did not develop this as hoped.

The design team had a more frequent presence on site than normal; however, without a stand-alone budget it was difficult to get the contractor to focus enough on submetering, BMS data logging and optimising controls algorithms.

6.23 Post-Occupancy Evaluation

We were able to secure funding from the Technology Strategy Board for a Building Performance Evaluation to review the performance in use of the building (Bordass and Burgon 2014a). This was carried out by the design team, assisted again by Bill Bordass of William Bordass Associates.

During this study, further funding was secured for a detailed investigation into the operation of the night cooling—monitoring temperatures and heat flows with infra-red thermography (Bordass and Burgon 2014a). The study found that over the first 2 years of operation (2011/2012), despite the relatively cool summers there was a tendency for the offices to overheat. The initial underperformance was found to be due to a number of factors:

- Restrictions on window openings due to overly tight health and safety and insurance requirements which were in excess of normal office practice.
- The outdoor sensor was incorrectly located and affected by solar gains and the internal temperature was referencing some of the wrong internal sensors.
- The building could feel too cold in the morning; furniture and other light-weight items were at a much lower temperature than the concrete. Windows are now closed 2 h before occupancy to allow internal temperatures to come into balance and be more comfortable.
- The controls strategy needed to be simplified.

The report also highlights key points to note for future designs:

- The concrete radiators achieved their objective: increasing thermal capacity and lowering peak temperatures. However, the effect is relatively small and needs to be complemented by a strategy that also minimises unwanted internal and solar gains.
- Care needs to be taken in designing, commissioning and fine-tuning of night ventilation systems. Simple and straightforward proposals may prove to be best, with some user intervention useful.
- Take care in positioning outdoor temperature sensors. Good locations can be difficult to find.
- Concrete is slow to cool, so the night cooling process can make lightweight elements including furniture too chilly the next morning. To maximise heat removal while avoiding comfort complaints, a rest period between night ventilation and initial occupancy will often be beneficial.
- Ensure that night ventilation openings are secure from break-in and daytime openings arranged to prevent any risk of occupants falling out.

6.24 Energy Consumption

The building's BMS and metering system unfortunately proved much less effective at collecting useful and detailed energy data than anticipated. However, a comparison with Heelis indicated that the base building energy use was similar to predictions and much lower than Heelis for all building-related end-uses except IT (Bordass et al. 2014).

At the Woodland Trust, with its low energy use for heating and lighting, the ICT systems—predominantly the server room including its air conditioning—were responsible for nearly 80% of total annual electricity use and 72% of annual CO_2 emissions. This highlights the need to pay increasing attention to these loads (Small 2015).

6.24.1 Comfort and User Experience

The Building Use Studies (BUS) occupant questionnaire survey was undertaken in March 2012, and the results were relatively good. Overall satisfaction levels were high with air quality, lighting, design and perceived productivity also scoring well. The only indicator that was significantly below average was for noise and unwanted interruptions. This may have been partly due to the move from cellular offices to open plan. There was also a large flexible meeting area open to the rest of the office which has now been dealt with using moveable acoustic screens to reduce noise levels at source.

There were some initial problems with ventilation and stuffiness in summer and winter, which were considered closely in the detailed concrete radiator study. As a result, there have been improvements in summertime temperatures and air quality.

It was slightly surprising how many aftercare issues arose in what is essentially a relatively small, simple and well-executed building, and one that performed much closer to its design intent than most. This reveals the need for buildings to receive much more fine-tuning and for there to be suitable resources and budgets to do so. As the facilities manager said:

> The Woodland Trust is lucky to have less complication than most. It is difficult enough to cope with the complication we have got.

6.24.2 Awards

Architects' Journal AJ100 Building of the Year—Shortlist 2011
Green Organisation Apple Award 2011
Sustain Magazine Architecture and Design 2011

Building Magazine Building Services Awards: Project of the Year—Shortlist 2011
British Council for Offices Corporate Workplace—shortlist 2011
RIBA East Midlands Award 2011
RICS Regional Sustainability Award 2011

6.25 The Hive, Worcester Library and History Centre

The Hive in Worcester came out of a collaboration between Worcestershire County Council and the University of Worcester. It is the UK's first joint-use library. It also incorporates a Multi-Agency Service Centre 'the Hub' for the city council, a local History Centre, the county archive and accommodation for the archaeologists of the Historic Environment Service. It replaces a range of provisions previously located across the city, uniting them in a single location to create an innovative new typology. The Hive exploits the potential for synergies in the provision of a range of public services in terms of operational efficiency and user experience. The project partners created the ambitious vision for this new public amenity with holistic sustainability at its heart.

The Hive is located just outside the line of the medieval city wall, immediately to the south of the prominent nineteenth-century railway viaduct and some 100 m to the east of the River Severn, the floodplain of which it abuts. The development presented the opportunity to heal the eroded fringe of the city with the creation of a new permeable and accessible city block that incorporates a significant public realm and connects the heart of the city to the recently developed University of Worcester Castle Street Campus. As a significant new public building, the Hive marks the northern boundary of the city centre much as the historic cathedral marks its southern limits; both are prominent from the river valley. As the cathedral is a beacon for faith, the Hive celebrates knowledge, from the repository of the county's rich history to life-long learning for the whole community.

The brief required a variety of study settings ranging from 'active' to 'reflective' in order to provide for the diversity of users anticipated. Early analysis of the site area and the likely footprint suggested that public areas should be arranged over several levels, while the desire for physical and intellectual access implied that these should be well connected. This led to the concept of a 'social landscape' with excellent connectivity both horizontally and vertically. A series of atria provide visual connection and allow daylight deep into the plan providing a strong sense of coherence from much of the public zone.

This connectivity, inspired by the aspiration for a truly inclusive building, presented the design team with the challenge of providing an appropriate acoustic environment for each activity. Great care has been taken to provide acoustic absorption and, where necessary, separation such that users with very different requirements can happily coexist. The meeting 'pods' that hover over the children's library are acoustically isolated from the noise below while sharing the fabulous views out to the river and the Malvern Hills.

The active/reflective gradient runs broadly from west to east rising up through the building. At lower ground floor overlooking the flood basins, there is a 'social learning' zone where paper coexists with digital media in a range of settings for group and individual study. Directly above is the children's library where zones are tailored to the needs of different age groups from pre-school to older teens. This shares the entrance level with the quick access library, the hub, the café and a studio theatre to create a dynamic, all-age environment. At first floor, the History Centre includes both a secure study area for examination of archive material and an open access collection. At this level, there is also a suite of bookable public meeting rooms.

The principal library is at second floor where the six-perimeter roof cones imply separate 'reading rooms', each of which provides a subtly different environment with contrasting views and light, accentuated by contrasts in colours used for floors and furniture. Book stack heights are varied to provide differing degrees of enclosure and acoustic environment. Even within this floor, plate users can choose a study setting to suit their preferences.

At third floor, a single cone accommodates a silent study area that is completely separate from the rest of the library volume. This 'eyrie' sits within the laminated timber cone itself and enjoys fabulous views out between the roof forms to the river valley.

The basement archive, which houses over 260,000 records, is the one area of the building serviced independently from the rest of the library. This thermally massive concrete box comprises 250 mm-thick waterproof concrete walls surrounded by 50 mm of insulation. The archive is designed to meet the strict requirements of BS5454 to obtain national archive certification. The air-handling plant maintains strict environmental conditions and includes a full backup system; however, in the event of a total failure of plant the well-insulated heavyweight enclosure will maintain stable conditions for a number of days (Figs. 6.15, 6.16, 6.17).

Fig. 6.15 Aerial shot of the Hive in context with the River Severn. (Simon Kirwan)

Fig. 6.16 Brise-soleils to the south elevation provide shading and help with control of road noise. The west elevation has solar control glass. (Hufton and Crow)

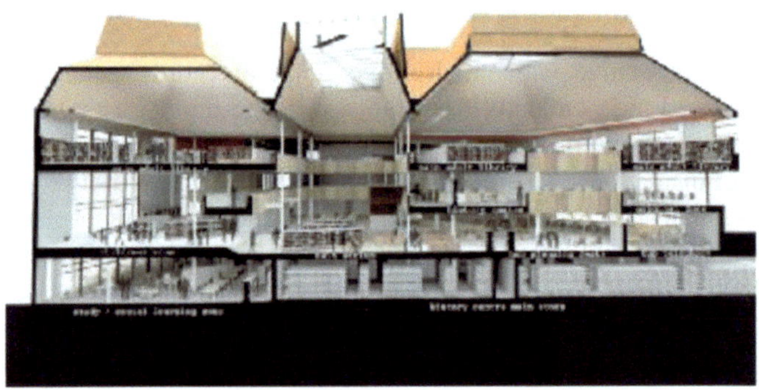

Fig. 6.17 Section through the building showing the arrangement of spaces. (FCBStudios)

6.26 Procurement

Procurement through the Private Finance Initiative entailed an extended period of competitive dialogue which fostered a positive teamwork and a spirit of partnership between the bid team and the end-user representatives. The regular day-long dialogue sessions were as much focused on the evolution of the new organisation as on the emerging design proposals. As the design team crystallised proposals for the building, the end-users began to envisage how they might work together in their new home.

6.27 Briefing

The project partners set pioneering targets for environmental sustainability. They recognised that this prominent new public building has the capacity to educate users by being a sustainable exemplar. The brief required BREEAM 'Excellent' as a minimum. It also required the building to have CO_2 emissions less than 50% of those allowed under 2006 Building Regulations and to be designed to adapt to the predicted impact of climate change to 2050 (as defined by the UK Climate Impact Programme, UK-CIP). The Hive exceeded its brief requirements and achieved an 'Outstanding' rating under BREEAM 2008.

6.28 Natural Ventilation and Thermal Mass

The challenging CO_2 reduction targets required the building to exploit natural ventilation and thermal mass to temper internal conditions; the iconic and carefully engineered roof forms drive natural ventilation regardless of prevailing wind. Air enters through the external elevations and a large below-ground duct delivers tempered air directly to the central atrium.

During winter, the relatively smaller amounts of fresh air required are brought in through low-level vents in the same plane as the floor voids, through perimeter trench heaters which heat up the incoming air and avoid cold draughts. In summer, windows open up to increase the airflow and provide more cooling and air movement. The windows that overlook the main street (to the south) have horizontal shelves below them, which reduce the amount of street noise as well as shelves above providing solar shading.

Plastic pipes, for chilled water, have been embedded in the in situ concrete frame further 'activating' the thermal mass to increase its capacity to act as a heat sink and maintaining user comfort during warmer weather.

The complex roof-form meant that there would be a very turbulent airflow at roof level, such that wind direction and pressure coefficients would be very variable and unpredictable. The opening windows were located in the vertical upstand at the top of the roof-cone and shielded by a baffle in front to create a 'trough'. The trough ensures a negative pressure at roof-level regardless of wind direction and so encourages air to exit via the roof vents at all times. Internal shading fins below the rooflights are also structural elements and are designed to capture a large proportion of the solar gains; they both prevent overheating below and heat up themselves to help drive stack effect ventilation. A scale model was tested in the wind tunnel at Cardiff University; this confirmed the turbulent nature of the airflow and the need for the baffles. The test results also better defined the pressure coefficients around the building and were used to improve the accuracy of the thermal model and tighten up the provision of window openings (Fig. 6.18).

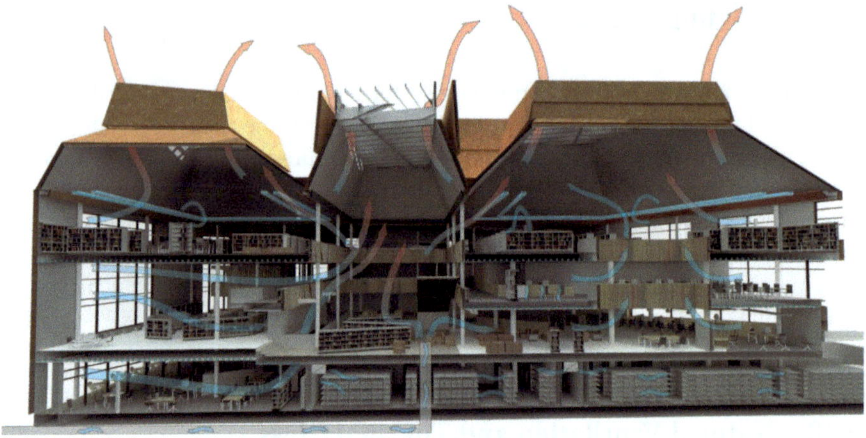

Fig. 6.18 Building section showing the natural ventilation strategy. Air enters via the perimeter and the ground duct and exits the roof cones under negative wind pressure

6.29 Building and Energy Modelling

As part of the PFI procurement process, a total annual energy consumption target was written into the contract documents with penalties to be applied if this was exceeded. Consequently, the predictions were carefully calculated, allowing for both regulated and unregulated energy use.

The summertime thermal target was slightly more relaxed than the CIBSE standard, allowing for some degree of adaptation under future climate projections without predicating the use of air conditioning. It stated that the internal temperature should generally be maintained below 25 °C, but where it exceeds this it should always be 2 °C less than the external temperature.

In addition, the scheme had to be sufficiently robust to be able to maintain comfort conditions under the impact of climate change in 2020, and the building had to be capable of being adapted to cope with the more extreme environmental conditions expected in 2050. The building was modelled using the UK Climate Impact Programme projections for these years and demonstrated the increasing need for supplemental cooling to the natural ventilation to ensure the building remains comfortable, hence the embedded pipework.

6.30 Daylight and Sunlight

Delivering good quality daylight to the heart of the deep-plan, three- and four-storey building was a key design generator. The highly articulated section with multiple atria connecting the 'social landscape' of the principal floor levels introduces daylight deep into the plan via the glazed tops of the seven irregular roof cones.

Fig. 6.19 Daylight is controlled by the structural timber baffles. Acoustic panels to the ballustrades and bottom of the cones provide a good acoustic environment. Lights are hidden behind the acoustic panels. (Hufton and Crow)

The perimeter glazing primarily provides the views out, particularly to the River Severn, and supplements the roof glazing to provide an average daylight factor of 3 % to the library.

Substantial modelling was carried out to define the detail of the internal shading to the rooflights. It needed to deliver very good levels of daylight and allow some direct sunlight in, without it affecting the occupants working below (Figs. 6.19, 6.20).

Fig. 6.20 The roof cones provide daylight and ventilation. External baffles ensure a negative wind pressure (Hufton and Crow)

6.31 Artificial Lighting

The natural light is complimented by a low-energy electric lighting scheme. The strategy was to generally provide lower than usual background levels of light (around 100 lx) sufficient to enable people to find their way around, supplemented by task lighting for detailed work.

Fluorescent tubes integrated behind the acoustic absorption panels light up the internal surface of the cones and circulation routes below providing much of the background lighting. This is supplemented by occupant-controlled task lights in the reading areas and supplementary lighting fixed to bookshelves under daylight and PIR control. Daylight and PIR lighting controls also control lighting in the separate administration and office areas.

6.32 Materials and Embodied Carbon

The concrete frame incorporates ground-granulated blast-furnace slag (GGBFS), a by-product of iron and steel-making, in lieu of 50 % of the cement content, thus reducing embodied energy. The original design of standard cast in situ concrete was developed by the concrete subcontractor Northfield into a post-tensioned solution that reduced the required thicknesses of floor slabs and overall volume of concrete.

The seven irregular roof cones inspired by the undulating ridge line of the Malvern Hills to the west and by the iconic brick kilns of the Royal Worcester pottery were initially designed in steel and concrete and later reconceived in solid laminated timber which generated a carbon saving in excess of 2000 tons compared to the original approach. This was achieved using award-winning, bespoke parametric modelling software developed by our team. It incorporated structural and environmental criteria and allowed us to manipulate the forms of the roof cones to achieve the desired internal environment while minimising material use.

Wherever possible, materials were sourced locally; Forrest of Dean Pennant forms the plinth and surface to the public realm, and locally sourced bricks are used in the adjacent structures, echoing the materials of the historic city fabric. The carapace of copper alloy shingles, though not locally manufactured, is durable and entirely recyclable and makes reference to Worcester's heritage of boiler making.

6.33 Low- and Zero-Carbon Technologies and Renewables

The brief required the carbon emissions to be 50 % of that allowed by building regulations. Half of this reduction is met by the environmental design strategy for the building. The remainder is achieved by offsetting consumption against carbon savings from the use of renewable technologies: river cooling and biomass.

The River Severn, the flood plain of which extends into The Hive's site, is exploited as a heat sink, providing 'free' cooling without disrupting the river's habitat value and with a significant reduction in CO_2 generation compared to standard air-cooled chillers.

Water is drawn from the Severn and passed through a heat exchanger before being returned back to the river. For much of the year, when the Severn's temperature is below 14 °C, direct heat exchange with the chilled water circuit provides sufficient cooling for the building. The water is pumped through pipework embedded in the lower part of the pre-stressed concrete floor plates, helping to cool the underside of the structure and increasing the radiant cooling output to the room below. The large expanse of concrete soffits mean that the water needs only to be slightly cooler than ambient temperatures to have a significant impact.

On the top floor where there are no concrete soffits, passive chilled beams fed from the same water circuit provide additional cooling with control to prevent condensation.

The river system also serves water-cooled chillers delivering high-grade cooling at a flow temperature of 6 °C. This system is primarily used to cool the IT server room and meeting rooms as well as the archive. In summer, when the river temperature rises, cooling from the high grade circuit is 'blended' with the river circuit to lower its temperature to ensure it can still cool the main body of the building via the embedded pipework.

River abstraction was not without its challenges. We had to ensure the system could cope with water levels in both flood and drought conditions and satisfy British Waterways that abstraction and return of the water would not affect the river traffic. The system also had to meet the Environment Agency's requirements, particularly that return water should not be more than 3 °C above ambient to comply with the Freshwater Fish Directive. In practice, the system is allowed to raise the temperature of the return water by up to 5 °C as mixing at the outlet quickly reduces the warming to acceptable levels. We estimate that the river cooling solution uses less than a third of the energy of conventional air-cooled chillers to run and takes up much less space in the building.

A 550-kW biomass boiler provides the primary heat source and is served by a well-developed local woodchip supply chain. Heat loads are minimised by the building's airtight construction; the building has an air leakage rate measured at 4.3 m^3/m^2 façade at 50 Pa. Nevertheless, the biomass boiler is supported by three 250-kW gas boilers to meet peak heating load and to provide backup.

Heat from the boiler is circulated through the embedded pipework in the concrete soffits, and the chilled beams. It also serves the perimeter trench heaters which temper incoming air in winter and combat downdrafts from any large expanses of glazing.

6.34 Post-Occupancy Evaluation

The building's energy consumption in use was initially more than 50 % over the design prediction. There followed a detailed review over two years by Max Fordham and the facilities management team. The main reasons for excessive energy consumption were generally due to incomplete commissioning.

The building has an automated lighting system involving presence detectors, a roof-mounted directional daylight sensor and individually addressable lights that need to be set up to respond to varying daylight conditions. The system was not fully commissioned before handover, with the result that most lights were on continuously at full brightness.

Completing the commissioning so that lights dim down when there is sufficient daylight and turn off when they are not needed has made the greatest contribution to reducing electricity consumption. Further reductions were made by changing the bookshelf lighting over to LEDs.

In addition, a number of the pumps and fans serving systems such as the river cooling, archive air conditioning, meeting room fan coil units and the main heating system were running continuously at a fixed speed. These have now been properly set up to run as variable speed and turn off when not required.

Figure 6.21 shows how the monthly electricity usage was gradually brought down to below target over the first 2 years after opening to the public in July 2012.

The building continues to be monitored, and energy consumption and thermal comfort conditions are reported monthly and displayed on a screen in the entrance foyer. The archive environmental conditions are reported weekly for the ongoing National Archive Certification.

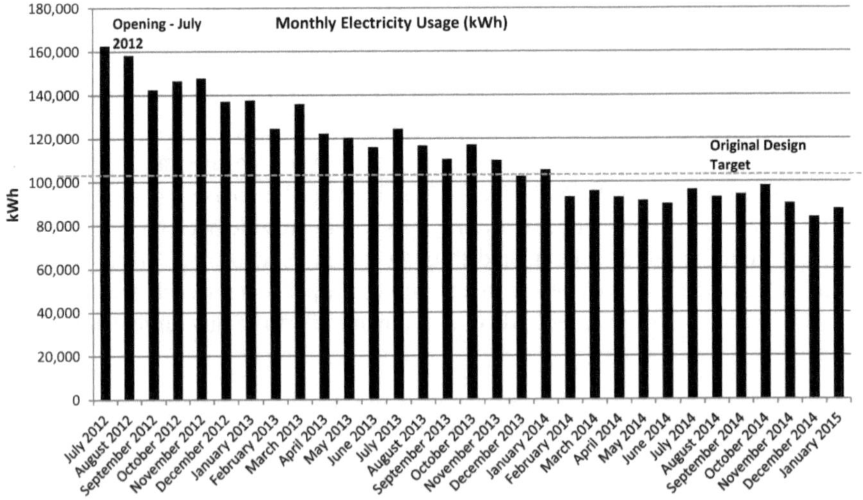

Fig. 6.21 Electricity consumption over first 2 years reducing to below the design target

6.34.1 Comfort and User Experience

While a detailed building user survey has not been carried out, The Hive obtained initial feedback on the opening of the building from visitors. There were positive comments from both visitors and members of staff on the light, airy and open feel of the building and the success of the acoustics in providing a good environment for conversations.

6.35 Other Sustainability Features

In addition to its low energy credentials, the scheme also includes a rainwater harvesting tank for water collected from the building's golden roof. Rainwater is used by the low-flush WCs and by the archaeology department for washing archaeological finds. It provides around 40 % of the total water used. Local area shut-off valves and mains supply leak detection help reduce wastage from leaking taps or damaged pipes.

> We believe that the development of Worcester Library & History Centre, now known as the Hive, sets a new standard for the procurement of significant buildings in the public sector. The project partners, Worcester County Council and the University of Westminster set a challenging operational brief and very demanding standard for design quality and environmental performance. These demands had to be met within the constraints of the PFI Competitive Dialogue Process. There was a seamless transfer of the detailed design and construction phase to Galliford Try Construction. In our view, this phase represented the construction industry at its best, being characterised in particular by closely integrated working between the constructor, supply chain, design team and client. The project completed on time and within budget. This is an exceptional project in every way—**Peter Parkes, Head of Property Services, Worcestershire County Council.**

6.35.1 Awards

2013 RICS National Award
 2013 RIBA Regional Sustainability Award
 2013 Building Awards—Sustainable Project of the Year
 2013 Civic Trust Award
 2013 CIBSE Building Performance Awards—New Build Project of the Year
 2012 PFI Partnership Award—Best Sustainability in a Project
 2009 Bentley Awards—Innovation in Generative Design

6.36 The A.G. Leventis Gallery, Nicosia, CYPRUS

The A.G. Leventis Gallery represents the fulfilment of Anastasios Leventis' vision to create a public gallery in his homeland where the extensive art collection acquired during his lifetime could be shared with his fellow Cypriots and visitors to Nicosia. Established over 30 years ago, the A.G. Leventis Foundation has expanded the collection since his death and now finally realised his vision. The building has already become a destination for residents of and visitors to Nicosia and a valuable educational resource for all generations with a regular programme of lectures and activities. The collection of more than 800 works of art dating from the Renaissance to the twentieth century is the biggest collection of European art in Cyprus.

The A.G. Leventis Gallery houses a formerly private collection of Cypriot, Greek and European art on a site just outside Nicosia's sixteenth-century Venetian walls. The building, which includes a restaurant and ten apartments, is conceived as a stone sculpture, eroded to create courtyards, balconies and roof gardens with far-reaching views. The building mediates between the scale of historic Nicosia to the north and the twenty-first century city rising to the south. The three-storey gallery, embracing an entrance courtyard which opens to the street, forms a plinth to the sculptural apartment tower which is topped by a roof garden providing spectacular mountain views.

Our design process started from a careful analysis of the context and evolution of this part of Nicosia; the site sits just outside the sixteenth-century Venetian walls which encircle the old (and currently divided) city centre. The site is part of a zone first urbanised by Lord Kitchener in the late nineteenth century—remnants of this remain, comprising finely detailed two- and three-storey stone villas. To the south, twentieth-century Nicosia sprawls: Heavily trafficked highways weave between mid- and high-rise 'modernist' buildings which pay little heed to the fierce climate, their highly glazed elevations and lightweight skins, reliant on energy-hungry mechanical cooling. Our inspiration for the gallery was the traditional buildings of the historic core, stone-built with carefully shuttered bays and balconies which open to allow mid-season and winter sun while excluding the searing summer heat.

The gallery is organised around a central atrium with a monolithic timber-lined, bronze-finished steel stair giving access to the three public floors of galleries, auditorium and café. Judiciously orientated openings provide views to the city, connecting the collection of art and artefacts to a broader cultural context. The upper two-gallery floors are connected by light shafts which allow the collection to be viewed in carefully controlled daylight from the roof lights which puncture the stepped, planted roof plains (Fig. 6.22).

Fig. 6.22 A.G. Leventis art gallery and apartment tower

6.37 Briefing

The brief required a highly sustainable building which minimises carbon footprint and running costs. The gallery is designed to accommodate three permanent collections. Additionally, it houses a temporary gallery designed to meet the requirements of the Government Indemnity Scheme in terms of environmental control and security, allowing work of international significance to be displayed. Fine art, decorative art, furniture and sculpture are displayed in domestic scale settings with controlled daylight throughout. Internal conditions are allowed to reflect fluctuations externally through the year and range from 22 to 26 °C, while relative humidity is carefully controlled. The gallery also incorporates a restaurant, a shop, an education suite and offices for the directorate.

In addition to the gallery, the development includes ten luxury apartments with far-reaching views across the city and the mountains beyond.

6.38 Daylight and Sunlight

The ability to introduce controlled daylight to the galleries was central to our design approach, both to reduce the use of artificial lighting and to exhibit the paintings and artefacts in conditions somewhat similar to those in which they were created. The combination of roof lights, light-shafts and inter-pane louvres were carefully mod-

Fig. 6.23 The galleries are day-lit from the rooflights; sunlight is controlled with mid-pane louvres and retractable blinds

elled to make sure that no direct sunlight falls on any of the artwork while providing naturally day-lit spaces. Large north-facing windows provide further daylight and views over the city to the mountains beyond.

Both gallery and apartment windows are screened with laser-cut metal panels perforated to a pattern derived from William Morris's olive leaf design and reminiscent of traditional slatted timber shutters. The combination of sliding and fixed panels to windows and balconies provides flexible shading, filtering intense sunlight when necessary and then retracting to open the apartments up to the city. The pattern has a free area of 20% which allows airflow when the sliding shutters are closed. The intensity of sunlight is such that this permeability provides useful internal daylight levels while eliminating solar gain and casting delightful, dappled shadows (Figs. 6.23, 6.24).

6.39 Natural Ventilation and Thermal Mass

Both the gallery and apartments are intended to demonstrate replicable strategies which enhance the building's sustainability while responding to the challenges presented by the Cypriot climate. To withstand the potential for seismic activity, the building has a heavy, reinforced concrete frame which also provides thermal mass to temper fluctuations in internal conditions. The high levels of insulation and airtightness together with the exposed thermal mass allow the gallery conditioning to be turned off at night without the temperature or humidity drifting outside of the required conditions.

Fig. 6.24 Dappled light and shadows from the sliding shutters

The flats have opening windows so that they can be naturally ventilated for as much of the year as possible with mechanical ventilation and cooling/heating coming on when necessary.

6.40 Low- and Zero-Carbon Technologies and Renewables

An array of piles drilled below the footprint of the building provides a closed-loop ground source heat exchange system which takes advantage of the consistent ground temperature fed by an underground aquifer. This is connected to reversible heat pumps to meet the base heating and cooling loads by either rejecting or taking heat from the ground. Peak loads are met by bringing on either additional boilers or chillers to supplement the ground-source system. Water extracted from a borehole is used for irrigation of the landscape, in particular the green roof of the gallery which helps to limit the gallery temperature due to evaporative cooling. A large roof canopy which sails over the top floor roof terrace is made up of an array of evacuated solar-thermal tubes which provide hot water to the whole building and shading to the roof terrace (Fig. 6.25).

Fig. 6.25 Solar-thermal panels provide hot water to the building and shading to the roof terrace

6.41 Delivery

FCB Studios and Max Fordham worked on the early-stage design and then handed over to local engineers and architects who delivered the majority of the detailed design. FCBS Studios were retained to oversee construction quality via regular site visits.

With the split design responsibility and non-UK construction market, it was particularly important to communicate the design principles with the local team for a smooth handover. There were a number of key areas that we felt important to deliver this:

- Understand the local construction market and what could be delivered
- Agree the key performance criteria
- Discuss and agree the design approach in detail with the local consultants so that the design intent was both achievable and maintained.
- Set out commissioning and handover procedures that reflect best practice and involve the necessary people to oversee their implementation.

While the design development generally progressed well, unfortunately the commissioning and handover suffered from a lack of input from the UK-based team.

6.42 Post-Occupancy Evaluation

A year or so on from completion, the building manager has been measuring the energy consumption and found it to be well over the design projections. At the time of writing, Max Fordham had just visited the site and uncovered some of the key issues. In general, the building and its systems were meeting the brief for environmental comfort conditions but not doing this in the most efficient manner. Following the visit, the UK and local design team, client and contractor have discussed a series of actions to address the energy consumption.

> In our visitors' book, guests wrote that the people of the gallery work together like a close knit family—with a devotion for a higher cause. They called our spaces welcoming, noting that in fact they felt so comfortable that they would take their shoes off and step barefoot on the soft wooden floors. That everything happens quietly, humbly and effectively. I cannot think of a better compliment. Loukia L Hadjigraviel, Director, A.G. Leventis Gallery

6.43 Awards

2015 Building Awards—International Project of the year—Shortlist

6.43.1 Keynsham Town Hall

- Keynsham Town Hall does not strictly form part of this series as the architects were AHR (formerly AEDAS) rather than FCBS, so it is only briefly addressed. However, it allowed Max Fordham to contribute many of the lessons from the other case studies in this chapter and so is worth touching on. The brief was to deliver a DEC A building with a very low energy use in operation and contractually binding energy targets were set.
- Natural ventilation, thermal mass and daylighting have been key drivers in the passive environmental design strategy which is similar in nature to the Woodland Trust, although secure acoustic ventilation openings were necessary to deal with the noisy town centre site.
- A steel frame and timber structure limits the embodied carbon; half the timber decks are replaced with concrete to provide thermal mass. Pipework is embedded in the concrete for future connection, thus future-proofing against addition cooling requirements due to climate change.
- There was a concerted effort to try to reduce the complexity and quantity of the M&E systems and associated maintenance, and the approach of background lighting, wall-washing and task lighting has been followed again. In addition, out of hours working is restricted to a small area of the building so that the remaining areas can be shut down via 'kill switches'.

- A thin client system has been used with water-cooled chillers to the server room; the recovered heat offsets 20 % of the building heating load. The south facing roof accommodates a PV array which offsets much of the electrical demand and the annual income from the Feed-in-Tariff is in excess of the anticipated annual utility bills.
- The Soft Landings process was followed with a Soft Landings Champion, and an energy risk register and operational risk register were developed. These identified potential areas during design, construction, handover and operation where either energy consumption could increase or there might be other problems with operation. Responsibility was then allocated to the party most able to deal with it and ring-fenced budgets provided for fine-tuning. The process has formed the basis for the BSRIA Guide: how to procure soft landings (Bunn 2014).
- The contractor produced an illustrated guide on the operation of the building for the occupants, and induction sessions have taken place to ensure everyone understands the building. Both the contractor and design team are retained for 2 years after practical completion to optimise energy performance and comfort with seasonal commissioning and occupant surveys.
- So far, an EPC of 5 (2.85 kg CO_2/m^2) has been achieved; suggesting an almost zero carbon base build (EPCs do not include non-regulated energy such as IT). At the time of writing, the building is not yet fully commissioned and we wait to see how it performs in practice.

6.43.2 Summary

We have only been able to touch on some of the key moves and findings within these case studies. There are references at the end of the chapter for further reading should you be interested in finding out more detail.

Aside from the general design approaches explained, there are some general lessons that we have learnt along the way that can be usefully applied to future projects.

- We have been lucky to work with forward-thinking clients who have given us the opportunity to be adventurous in our designs. However, finding a common language to draw out their aspirations during the briefing stage (e.g. using the sustainability matrix (Max Fordham LLP 2010)) has been very helpful.
- A robust base design using passive design principles can help provide reasonably comfortable conditions even when the commissioning of the systems has not been completed. It can also contribute significantly to the health and well-being of the occupants. Provide a well-considered secure natural ventilation with controlled daylighting and appropriate thermal targets that consider future climate. Review whether sufficient draught-free wintertime ventilation be achieved?
- Keep it simple and user-friendly—in particular, the usability and clarity of controls. Avoid unmanageable complication of controls to deliver further energy

savings; it is quite possible they will not work that well in practice. Think about the experience of using a space.

- As the energy performance of buildings improves and computing power increases, the energy consumed by the ICT and fit-outs such as cafés can become even more significant than the base building systems. These need thinking about carefully and integrating within the design approach.
- Adopt a soft landings approach right from the start of the project and identify champions to drive it all the way through. It is important that all of the key people involved engage with this process. Allocate responsibility for dealing with energy consumption and other operational issues.
- Where possible, involve the facilities management staff early, ideally during design or construction. Make sure that they are well trained in the use of the systems and particularly the BMS. There should be a clear description of the operational principles and expected energy consumption and induction of the building occupants.
- Identify specific allowances for dealing with commissioning and fine-tuning during and after practical completion. Fine-tuning is likely to take up to 2 years and it is helpful to have a contingency budget to deal with the unexpected and any minor alterations quickly and effectively.
- POE can be contractually difficult, particularly during the defects liability period owing to the contractual situation and the grey area between defects and enhancements. Potentially better advances are made with direct relationships with BMS and electrical contractors.
- Contribute to a constant stream of feedback as we can all learn from every project. Building on previous experience should help to get as much as possible right first time; however, this may not be enough. Some degree of review, fine-tuning and feedback will undoubtedly be necessary to deal with the unexpected!

References and Further Reading

Heelis

Bunn R (2007) Trust in construction. BSRIA J, pp 10–13

Max Fordham LLP Sustainability Matrix (2010) Online http://www.maxfordham.com/news/publications/. Accessed 20 Feb 2015

Nevill G (2007) So, how are you doing? Build Serv J 32–37

Usable Buildings (Online) Probe Studies http://www.usablebuildings.co.uk/. Accessed 20 Feb 2015

Woodland Trust

Bordass B, Burgon P (2014a) TSB building performance evaluation, woodland trust, July 2014. https://connect.innovateuk.org/web/building-performance-evaluation/woodland-trust. Accessed 20 Feb 2015
Bordass B, Burgon P (2014b) Early findings from non-domestic projects. Innovate UK—building performance evaluation programme (Online). https://connect.innovateuk.org/documents/3270542/19792080/BPE%20Early%20Findings%20Report_Nov%202014. p 11. Accessed 20 Feb 2015
Bordass B, Burgon P, Brough H, Vaudin M (2014) Trees of knowledge. CIBSE J 6(10):20–26
BSRIA (2014) Soft Landings https://www.bsria.co.uk/services/design/soft-landings/. Accessed 20 Feb 2015
Small T (2015) Chip & sin. CIBSE J, 7(1): 42–43

Hive

Pearson A (2013) The Hive—the UK's first fully integrated public & University library CIBSE J (Online) http://www.cibse.org/Knowledge/Case-Studies/The-Hive
The Hive (Online) Archdaily (Online) http://www.archdaily.com/225962. Accessed 13 April 2012
The Hive by Feilden Clegg Bradley Studios (Online) DeZeen http://www.dezeen.com/2012/04/12/the-hive-by-feilden-clegg-bradley-studios. Accessed 12 April 2012

Keynsham

Bunn R (2014) How to procure soft landings—specification and supporting guidance for clients, consultants and contractors (BG45/2014) *BSRIA*

Chapter 7
Building Low Carbon High Rise Buildings in a Sub-Tropical Climate: Case Study

Conrad T. C. Wong, Albert T. P. So and Andrew Platten

7.1 Introduction

As the planet faces the issues of climate change, it is now accepted that global warming will have a significant and threatening impact upon climate, which will have a significant impact upon nearly every aspect of economy, environment and society. It is also widely accepted that buildings contribute to as much as one third of total global greenhouse gas emissions (UNEP 2009), and this is primarily through the burning of fossil fuels in their operation, consuming up to 40 % of all energy. It is therefore a necessity that each nation seeks to make the reduction of green house gas emissions from buildings as part of their climate change strategy.

The United Nations Environmental Programme (UNEP) notes the objective that by 2050, greenhouse gases need to be cut by 50 % and at least by 25 % in by 2020. The Kyoto Protocol (United Nations 1998) set objectives for each signatory nation to reduce its greenhouse emissions. This contribution target is now subject to negotiation, but is still to be finalised at the time of writing.

It is also widely acknowledged that the building sector continues to grow through the development of the world's economies and in response to population growth. Without further actions, the greenhouse gas emissions from buildings are set to double in the next 20 years.

In response to the commitment targets set by the Kyoto Protocol, we have seen the development of a range of building control processes, each seeking to develop systems by which to control design so as to limit the impact of new and existing

C. T. C. Wong (✉)
Yau Lee Holdings Limited, Hong Kong
e-mail: conradwong@yaulee.com

A. So
City University of Hong Kong, Hong Kong

A. Platten
Faculty of Arts, Environment and Technology, Leeds Beckett University,
Broadcasting Place, Leeds LS2 9EN, UK

© Springer International Publishing Switzerland 2016

M. Dastbaz et al. (eds.), *Building Sustainable Futures*,
DOI 10.1007/978-3-319-19348-9_7

construction upon the environment. For example, in the European Union the Energy of Buildings Performance Directive (EBPD) of 2002/2003 came into effect in member states over 2006–2009 so as to improve the energy efficiency of buildings and ensure the convergence of building standards across Europe. With 70 % of domestic energy demands arising from the need to provide hot water and heating, the directive took high-level importance. In the directive, energy consumption was to be reduced by 25–40 % by 2020, with savings of 80 % predicted by 2050. The implementation of the EBPD within the EU has influenced step changes in national building standards and design codes so as to achieve this aim.

Likewise, the Building Research Establishment Energy Assessment Methodology (BREEAM) was first published by the Building Research Establishment (BRE) in the UK in 1990 (2012); BREEAM is the world's longest established and most widely used design tool, which is used to assess, rate and certify the sustainability of buildings. The methodology now comprises 12 different guides including residential and commercial buildings. Globally, other assessment tools have been developed as a result of the lead taken by the BRE. They include in the USA the Leadership in Environmental and Energy Design (LEED), HK BEAM (founded as Hong Kong Building Environmental Assessment Method), the China Three Star code, Green Star in Australia and the Estidama Pearl Rating System, which is used in the Gulf. The range of standards may be confusing to the sector. However, they reflect the growing importance across the globe and meeting the differing needs of buildings according to their location, local legislation and planning needs and of course climatic conditions. A comparative study of BREEAM, HK BEAM and LEED (Lee and Burnett 2008) sought to develop a basis for benchmarking the standards.

The common theme throughout the noted methodologies and others all seek to focus upon limiting the energy use per unit area of floor space by the consideration of the thermal properties of the building structure, consideration of the relative heat gains and losses by occupation, use, climatic conditions and air permeability. It may be said that such tools all seek to limit the energy and thus green house gas emissions by improving performance by incremental changes in design, building fabric and performance. Of particular note and contrast is the Passivhaus approach (Cottera and Dadeby 2012), which takes a particular focus to radical reductions in the energy demand of the building by employing very low thermal transmittance values and low air permeability setting a limit of 15 kWh/m^2 year This is achieved by optimising solar heat gain, removing thermal bridges and providing good indoor air quality by ventilating the habitable space and employing highly efficient heat recovery technology. These approaches have now been applied to domestic buildings in temperate European climates, but extending this approach to sub-tropical climates and to multi-occupancy and commercial premises is an imperative.

In this study, we focus upon the sub-tropical environment of Hong Kong. The thermal demands are normally converse to the European context. Normally, the winter is cool and dry, whereas the spring is changeable and wet. The summer is hot and very humid with the highest rain fall occurring over May to September. Autumn is the most pleasant time of the year. These climatic conditions therefore lead the design team to focus upon the thermal comfort of occupants by the consideration of the avoidance of excessive solar gain through both the structure and fenestration, the reduction of internal heat gains and attempting to make the best use of passive ventilation, whilst providing adequate protection against rainfall over the summer months. This task is not easy with the summer weather, which is normally over the range of 24–32 °C and 78–83 % relative humidity (Hong Kong Observatory, accessed 2013).

A further dynamic of the Hong Kong scenario is the extensive building development in the Special Administrative Region (SAR) of China. The small region has a land mass of 1104 km^2 and a population of 7,234,800, making it one of the densely populated areas of the globe. This coupled with its property market makes for a very densely developed area (where buildings can be safely built in-between its hilly terrain) so promoting large building developments, which often employ mixed use transportation stations, shopping malls, schools and residential apartments that often rise to 40, 60 or more storeys.

In consideration of air quality, reducing the heat load from the mass of air conditioners is critical. From an energy perspective, Hong Kong uses annually just under 600,000 TJ (Hong Kong Electrical and Mechanical Services Department (EMSD 2013)). With no indigenous energy sources in the region, energy conservation is an economic necessity.

The region's energy mix comprises 23 % nuclear power generated from the mainland, 54 % coal and 23 % natural gas. Of the region's carbon emissions, electricity generation makes the biggest contribution at over 25,000 kt of CO_2 (76 % of the region's total figure) compared to less than 5000 kt from transport. The density of the city and its integrated public transport makes the region's transportation policy a model for others to follow. The carbon dioxide emissions per capita have seen a small reduction from 2007 to 2011 at 6.0 t of CO_2 per person, less than that of mainland China and two thirds that of Japan (Hong Kong Environmental Protection Department (EPD) 2015). As a coastal city region, which is exposed to extreme climate conditions, the mitigation of carbon emissions is a real policy driver. The Hong Kong government is fully aware of the impact of climate change and thus the increase in local temperature, higher night-time temperatures, increasing extreme rainfall and thunderstorms and sea level.

The EPD has focused on a range of measures to encourage reduction in the region's emissions targets, and they focus upon reducing energy usage, the use of clean fuels and reducing the reliance upon fossil fuels. The Hong Kong government has highlighted the adoption of electric vehicles and the Building Energy efficiency scheme as policy goals in 2007.

Hong Kong's construction industry is one of the major drivers of the Hong Kong economy, particularly by nature its buoyant property market. The Hong Kong government reports for instance the index sales prices for apartments rising from circa 70 in 2009 to 140–150 by 2014. With increasing demand, the government policy is to further develop land for building. Hong Kong is presently therefore in a time of growth in its building stock. The current decade is characterised by the development and extension of the Mass Transit Railway (MTR), the building of the Hong Kong/Macao bridge, the new Hong Kong cultural district, extending its utilities infrastructure and the revitalisation of its building heritage among others. The Hong Kong Development Bureau reports that of the $ 480 billion approved by the Legislative Council's (LegCo's) Finance Committee over the last 5 year period, 60 % is devoted to the Ten Major Infrastructure Projects and the remaining 40 % to other projects of varying scales. It is expected that the annual capital works expenditure will exceed $ 70 billion in meeting this commitment. These projects are all set in the context of a developing tourist market that raises continually the demand for hotel accommodation and all the ancillary services that are demanded by visitors whether this is related to trade or leisure (Hong Kong Development Bureau 2014).

As a consequence of this setting, it is timely to consider one of the landmark projects within the Hong Kong Special Administrative Region. This focuses upon the construction of the Holiday Inn Express located in the busy Soho area on Hong Kong Island. The following section reports the needs of such a building and how carbon reduction technology can be employed in a 36-storey building.

7.1.1 Purpose and Design of the Hotel

The hotel, Holiday Inn Express Soho Hong Kong (the Hotel hereinafter), shown in Fig. 7.1, is located at 83, Jervois Street, Sheung Wan on the Hong Kong Island, which is within the central business district of Hong Kong, a 5-min walking distance from the Sheung Wan MTR (name of the subway in Hong Kong) station. Construction commenced on 10 September 2010 and was finished on 15 March 2012. Total cost of construction was around HK$ 350 millions, i.e. US $45 millions or £ 28 millions. The land lot is about 612 m^2 in size while the overall gross floor area is 9173 m^2. This high-rise hotel is 36-storey tall, consisting 274 guest rooms.

Fig. 7.1 Holiday Inn Express Soho Hong Kong

The project has won praise for its design and performance receiving numerous international prizes and achieving three key levels of certification, including US LEED Platinum (first of this nature in the whole world), HK BEAM-Plus Platinum, Singapore BGA Green Mark Platinum, Three Star of China Green Building Council, Green Building Award 2012, Intelligent Hotel (Distinction) of Asian Institute of Intelligent Buildings 2012, Hotel Investment Conference Asia Pacific Sustainable Project Design 2013, and Quality Building Award 2014. The description on various systems and features of the Hotel in this chapter is based on an article published by the leading author of this chapter (Wong et al. 2014).

7.1.2 Construction, Orientation and Passive Features

The building is oriented towards the NEN—SWS direction, path of sunshine, which is shown in Fig. 7.2.

The climatic conditions of Hong Kong are noted previously in this chapter, wherein it is seen as a sub-tropical city, but it is acknowledged that the city is gradually moving towards a tropical environment due to the global climate change. Based on the official Web site of Hong Kong Observatory (www.hko.gov.hk), the average temperature rise of Hong Kong from 1885 to 2014 is 0.12 °C/decade, 0.15 °C/decade from 1947 to 2014, and 0.16 °C/decade from 1985 to 2014, indicating that the increase in average temperature of Hong Kong is accelerating (Hong Kong Observatory 2015).

The summer in Hong Kong is hot and humid. As a result of these conditions, substantial solar load comes from the west, in particular, between 1 and 6 pm, as shown in Fig. 7.3. Because of this condition, the western side of the Hotel is designed with open staircases, electrical/mechanical plant rooms and 200-mm-thick structural wall for good thermal insulation. The western stair case is designed with projecting platforms,

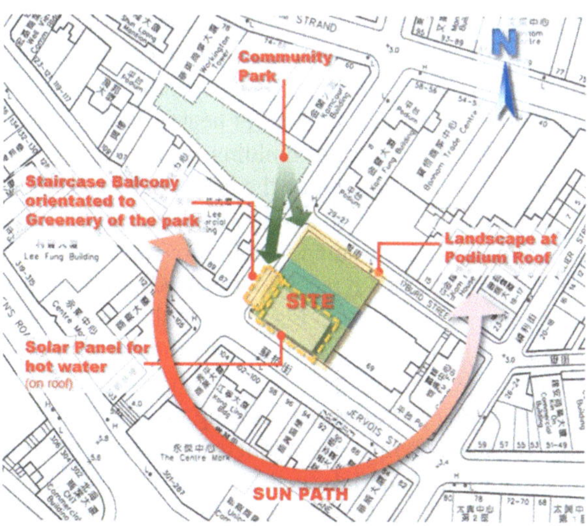

Fig. 7.2 Location and orientation of the hotel

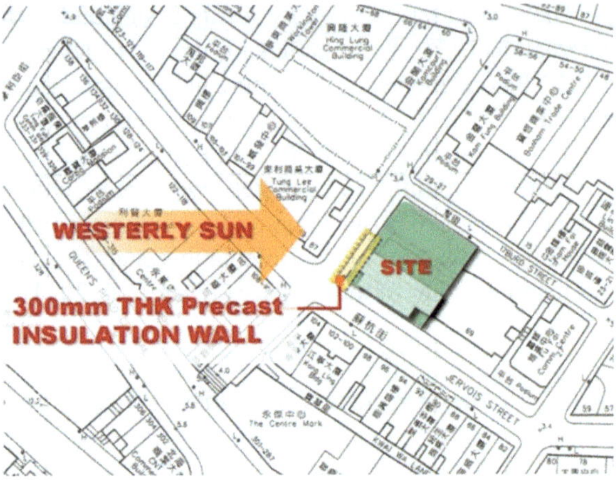

Fig. 7.3 Significant solar load from the west

serving as a "breakout" area for hotel guests. Low-E coating and argon gas-filled insulated glass units are used for all of the windows to the guest rooms which face north and south so as to optimise the use of daylight, while minimising solar heat gain.

With the use of Low-E glass and a shading coefficient of 0.42, the thermal transmittance or U-value for windows is 1.3 $W/m^2 °C$, versus 2.7 $W/m^2 °C$ without the use of IGU (insulated glass unit) Low-E coating. Low-E coatings can help to minimise the amount of ultraviolet and infrared light that can pass through glass without compromising the amount of visible light that is to be transmitted. In Hong Kong climatic conditions, the U-value is representative of the amount of heat penetrating into the interior from outside in summer, thus the lower the better. The spandrel panels are made up of layers which include a spandrel mono glass, air space, aluminium back frame, insulation, concrete and gypsum plaster, achieving an overall U-value of 0.35 $W/m^2 °C$. The roof of the Hotel is composed of a layered construction, which includes concrete tiles, cement sand, protection board, membrane, cement sand and reinforced concrete slab, achieving an overall U-value of 2.3 $W/m^2 °C$ (Lee 2010). In terms of Hong Kong design code, the Hong Kong Government Buildings Department specifies an overall thermal transfer value (OTTV), which is defined as the average amount of heat in Watts transmitted from outdoor to indoor through a square metre of building façade during summer when the outdoor temperature is much higher than the indoor (Hong Kong Buildings Department 1995).

Modular and standard design was implemented throughout the construction project so that construction and demolition waste was minimised. This accounted for more than 50 % for a typical floor plan. Off-site fabrication (precast) building materials were used, shown in Fig. 7.4a, so that on-site waste could be further reduced. The erection of the lift shafts was by the use of metal formwork to enhance buildability, shown in Fig. 7.4b.

A new low carbon green building material, Starfon™, was widely applied. It is an eco-friendly building material using a cementitious composite extrusion technology, made with materials such as cementitious binders, fibre, recycled glass and other inorganic fillers. During the curing of the substrate, only a small amount of heat

Fig. 7.4 **a** Precast staircase. **b** Metal formwork

is required to support the curing process. By applying ultraviolet (UV) curable inks and coatings to the products, no volatile organic compounds (VOCs)were produced while energy consumption was reduced. It should be noted that low VOC products were extensively used throughout the whole building, including adhesives and sealants, paint and coatings, composite wood, and carpet, etc. In addition, 100 % of the timbers used for construction were either from recycled sources or timber sourced from sustainably managed forests (Forest Stewardship Council).

With regard to project design and management, the application of building information modelling (BIM) was extensively implemented for the construction of subterrainian utilities and the construction of floors and the external façades. The 5D BIM strategy which was applied involved (i) 3-dimensional REVIT (one popular BIM software used by the construction industry) modelling, (ii) clash detection, (iii) construction programming, (iv) standard method of measurement for the take off of quantities, and the development of accurate construction cost estimation and verification. The combined process enables errors to be identified prior to construction, so reducing wastage on site, reducing the time of the construction phase and importantly minimising construction waste.

The project also makes a significant contribution to the bio-diversity of what is a central downtown area, which would have previously made very little contribution to the local environment. The Hotel provides 47.5 % greenery to the density developed land lot, and this was achieved by providing 165 m^2 of horizontal green roof and podium and 126 m^2 of vertical green wall as shown in Fig. 7.5. The plant species selected were adaptable to the local climate of Hong Kong, resistant to drought, requiring limited fertilisation and less maintenance. This involved the introduction of tuberous sword fern and springers asparagus which are used for the green wall, while carpet grass, China-berry and Persian lilac are for the green roof and podium, respectively.

The green roof and green wall can further reduce the U-value of the façade, so that less heat can penetrate into the interior during hot summer. An extensive implementation of the green roof policy by a city can mitigate the "Heat Island Effect". This process refers to the abnormal increase in mean air temperature of a city over its natural surroundings (Environmental Protection Agency of the United States of America 2015), and heat islands can affect communities by increasing summertime

Fig. 7.5 Vertical green wall of the hotel

peak energy demand, air conditioning costs, air pollution and greenhouse gas emissions, heat-related illness and mortality, and water quality. The use of green roofing and walls infers a degree of transpiration and expiration of water from the foliage and natural evaporative cooling and absorption from its sedum. The impact being that the process of water evaporation induces a cooling effect upon both the building and the surrounding environment.

With regard to water supply, two recycling storage tanks, 5000 L each, are installed in the Hotel to collect both rain water and condensed water from the air-conditioning system for irrigation. A filtration system with UV sterilisation is used to treat the collected water before its reuse.

7.1.3 Building Services Systems and Energy Conservation

The Hotel received full marks for "Optimize Energy Performance" under the category "Energy and Atmosphere" of LEED 2009 for New Construction. Obviously, most building services systems, besides their normal functions to serve the occupants, were designed to save energy. The major energy consumer of a hotel in Hong Kong is of course the central air-conditioning (AC) system.

Two water-cooled screw-typed variable speed twin-compressor chillers are installed, each with a capacity of around 190 refrigerant tons (668 kW), the coefficient of performance (COP) is designed to be 5.48 (0.64 kW/ton in US units). It should be noted that the chiller cannot operate alone; there are other accessories including cooling towers, condensing water pumps, chilled water pumps; therefore, supervisory control is necessary. Therefore, the COP of the entire chilled water system is lower than 5.48. However, the COP of a chiller is still the most important parameter addressing the energy performance coefficient of the unit. This is defined as the ratio of the amount of heat extracted from the chilled water (kW) by the chiller to the amount of electrical energy consumed by the unit (kW), thus the higher value the better. The COP is rather sensitive to the condensing water temperature, thus the lower the better. When cooling towers are used, the condensing water temperature is closely related to the dew point temperature of the outdoor environment. According to the *Code of Practice for Energy Efficiency of Building Services Installation,* Electrical & Mechanical Services Department, HKSAR Government, 2012 (www.emsd.gov.hk), the nominal COP of a screw-type chiller over the range 50–1000 kW is about 4.7, indicating that the two chillers installed at the Hotel perform beyond normal design. The advantage of using variable speed compressor is that it improves performance during part-loaded operation. There are generally two types of compressors for chillers, constant speed drive (CSD) and variable speed drive (VSD). For CSD operation, the valve controlling the injection of refrigerant into the compressor is the only device under control. For VSD operation, the speed of the compressor is under control and therefore less energy is consumed under a part-loaded condition when the speed is reduced. Similar to the chillers, cooling towers and chilled water pumps at the Hotel are also driven by variable speed drives for optimal energy performance.

An Energy Optimization Solution (EOS™) was developed to optimise the overall energy consumption performance of the AC system based on a continuous monitoring the overall cooling load of the building, 24/7, as well as weather conditions using parameters such as CO_2 density, indoor temperature, electromagnetic chilled water flow rate, differential pressure, outdoor temperature and humidity, etc. All these values are fed to the EOS™ software for processing and further to the building automation system of the Hotel for optimal control including optimal chiller operation, limited maximum chilled water flow rate of any individual branch and supervised chilled water return temperature under part-loaded condition with the aid of zonal and terminal thermal energy control valves. For example, if the setpoint of the chilled water return temperature is at 12 °C, the valve opening is reduced when the temperature is lower than 12 °C, and vice versa. At the same time, a PowerBox™ online energy management and monitoring system was developed to report energy performance, 24/7. It is an online energy data logging software with analytical functions to reduce peak demand and initiate appropriate alarms. The components of it are shown in Fig. 7.6a. The PowerBox™ reporter, shown in Fig. 7.6b, provides benchmarking reports and gives advice on energy improvement.

Chilled water flows into the fan coil units (FCUs) in the guestrooms, corridors and function halls, where return air is cooled down before recycled to the supply air. In the guestrooms, FCUs are operating 24/7, even when the guests are out of the room because of the consideration of sanitation. The key card of the guest is the

Fig. 7.6 **a** Components of PowerBox^{TM}. **b** The dashboard of PowerBox^{TM}

Fig. 7.7 Permanent magnet motor of iFCU^{TM}

control parameter, and when it is not inserted, this raises the in-room temperature setting to 27 °C and reduces the fan speed of the FCU to its low-setting so as to save energy with both the chiller and the FCU. Since hundreds of FCUs are operating in the Hotel simultaneously, energy saving with FCU operation is critical. The technology of iFCU^{TM}, shown in Fig. 7.7, was developed, where permanent magnet motors are used, versus the traditional alternating current permanent split capacitor motors.

Traditionally, the permanent split capacitor motor serving a FCU consists of two coils, one main and one auxiliary, in the stator. The rotor is the conventional squirrel cage design. Without any permanent magnet or commutator within, a single-phase supply from the city mains cannot rotate the rotor. Therefore, a capacitor is installed to split the single phase into two phases. It is well known that the current through a capacitor is, in principle, 90° leading the phase of the voltage supply. Hence, the current in one winding basically lags that in another winding, thus producing the rotating torque. The main winding has terminals so that the applied voltage can be applied to any one of them to fulfil speed selection, thus Hi-speed, Mid-speed and Lo-speed as determined by the switch position of a conventional thermostat with 3-speed settings. This approach, though effective in speed control, is not energy efficient. By placing permanent magnets on the rotor, we still need a rotating magnetic field produced by a power electronic drive. The current applied to various stator windings comes sequentially based on the current position of the rotor.

To drive a permanent magnet motor, a direct current voltage is applied under a pulse-width-modulation mode so that energy is saved because no current is needed to provide the magnetising current from the stator. Also, the intrinsic magnetic field produced by the permanent magnets on the rotor strengthens the interaction with the rotating magnetic field, thus a much higher torque to size or weight ratio. This technology is called brushless direct current (BLDC) or electronically commutated motor (ECM) approach. Table 7.1 shows the energy performance of the iFCU™ compared with the use of a conventional permanent split capacitor motor for the FCU of a typical guest room. The total power factor of the iFCU™ is always higher than 0.85 as statutorily required and energy saving could be up to 80 % under low-speed operation as shown in Table 7.1. It should be noted that in a hotel guest room, the FCU is operating at low speed for most of the time, i.e. when the guests are absent during the day and are sleeping during the night. This technology also provides advantages including higher static pressure provision, lower motor heat dissipation and noise level, and more accurate temperature control up to ±0.5 °C.

The Hotel also offers a distinct opportunity to generate its own power using renewable technologies, because of its large surface area and its location. In this respect, solar power is the most compelling renewable energy source in Hong Kong and it is utilised on the rooftop of the Hotel. For this project, the solar thermal option has been adopted. There are 24 aluminium solar-collecting panels, each 3 m^2 in size, as shown in Fig. 7.8, together with a schematic representation of the hot water

Table 7.1 Energy performance of iFCUTM versus conventional for a typical 600 cfm FCU

Operation Mode	Conventional AC Motor	PM Motor with Intelligent Control Driver	Saving	
	Power consumption (W)		(W)	%
Low	85.0	17.0	68.0	80%
Med	98.2	33.4	64.8	66%
High	112.0	67.2	44.8	40%

Fig. 7.8 Solar hot water collecting panels and the water circuit

circuit diagram. It is generally accepted that central hot water supply is one critical service of a hotel as guests need showering, hand and face washing, and restaurants need a continuous supply for washing and cooking. The solar panels therefore become the first heat source to pre-heat cold potable water.

The second heat source comes from the returning chilled water of the AC system. Return chilled water is directed by variable speed pumps to four heat pumps, where the heat is extracted from the chilled water and injected to the pre-heated potable water. At the same time, the temperature of the return chilled water could be lowered accordingly to improve the efficiency of the chillers. According to the *Heat Pump Water Heaters Pamphlet* of the Energy Efficiency Office of Electrical & Mechanical Services Department of the HKSAR Government (www.emsd.gov.hk/emsd/e_download/pee/HeatPumpPamphlet.pdf), the amount of heat and cool energy produced at the condenser and evaporator, respectively, of a heat pump depends on factors including the temperature of the media surrounding the condenser and evaporator of the heat pump, the type of refrigerant used, and working pressure of the refrigerant vapour entering the condenser. For electrically driven heat pumps that serve both heating and cooling, the coefficient of performance (COP) is defined as the ratio of total heat and cool energy output to the electrical input. Generally, a heat pump may achieve a $COP_{Heating}$ of 3 and the corresponding $COP_{Cooling}$ could be 2, thus a total COP of 5. Finally, as standby, a boiler or calorifier is used to ensure the temperature of final hot water delivered to the guest rooms is at 65 °C.

In high-rise buildings, a key draw upon energy usage is that for the provision of vertical transportation. In this respect, the project sought to change the approach to lift design and focus upon energy conservation. This approach was therefore a further opportunity to gain additional points for innovation. The Hotel is served by four high-speed lifts, one of which is mainly used by staff as a freight lift. Three lifts, L2, L3 and L4 of rated speed 3 m/s and contract capacity of 1275 kg carry passengers from the ground floor to the top floor, serve all 33 levels of the building. All four lifts are equipped with a re-generative braking device so that electrical energy can be produced and fed back into the city mains whenever the heavier side, either car or the counterweight, is moving down or even when the heavier side is moving upwards under deceleration. It was reported that up to 25 % of energy could be saved by using this re-generative braking approach (ASME 2015).

Each lift controller is continuously monitored by a networked power meter with an accuracy up to 1 Wh (=3.6 kJ). A new benchmarking parameter, *J/kg-m*, was implemented onto the system for checking the energy performance of every lift on an hour-by-hour basis. This parameter has been incorporated in the *Technical Guidelines on Code of Practice for Energy Efficiency of Building Services Installation* 2012 of Electrical & Mechanical Services Department of HKSAR Government (downloadable from the Web site: www.beeo.emsd.gov.hk/en/pee/TG-BEC_2012%20(Rev.%201).pdf) as a good engineering practice recommended to the industry. A period of time, *T* seconds, is determined and fixed, say 30 min, 1 or 2 h, etc. Once *T* is determined, it should be permanent for the purpose of benchmarking. During such period of time, a lift car makes *N* number of brake-to-brake journeys. A brake-to-brake journey is defined from the moment the brake is released until the moment the brake is re-applied. Within the *i*th journey, *i* running from 1 to *N*, the car carries w_i kg of passengers and travels d_i m of distance, irrespective of direction. And during such period of time, E_T Joules of electrical energy is consumed as recorded by the networked power meter. Then, *J/kg-m* is defined as:

$$J/kg-m \overset{\Delta}{=} \frac{E_T}{\sum_{i=1}^{N} w_i d_i}$$

It is generally known that when the lift car is moving upwards, the counterweight should be as heavy as possible to save energy and when the car is moving downwards, the counterweight should be as light as possible. However, it is impossible to change the counterweight frequently and there is a rule of safety that the counterweight has to be within a range, usually equal to the empty car weight plus 40–55 % of car capacity due to the frictional requirement as shown by the formula:

$$\frac{T_1}{T_2} \leq e^{fa}$$

Here, T_1 and T_2 represent the tension on either side of the traction sheave, one suspending the lift car and the other suspending the counterweight; f is the equivalent coefficient of friction between the hoisting ropes and the grooves of the sheave and α is the angle of wrap which is related to the contact perimeter that the ropes make on the sheave. Subject to this inequality, the two tensions cannot be significantly different. The counterweight setting is usually determined by the lift manufacturer, based on the overall design. With a fixed setting, energy efficiency very much depends on the traffic patterns. In general, if a lift is always lightly loaded, the counterweight should be adjusted to the minimum allowable value and the opposite case is applicable if the lift is always heavily loaded. The fact is however that traffic patterns change continuously. The method adopted at the Hotel records all journeys of each of the three lifts, L2, L3 and L4, and over a period of 2 weeks, an optimal counterweight setting is estimated and suggested for the maintenance contractor to implement, of course, within the safety margins, based on the *J/kg-m* parameter. Such optimal setting, though sometimes more energy efficient, while sometimes it is not, however can still achieve the minimum energy consumption for a period of two previous weeks. If the history repeats in the subsequent 2 weeks, energy consumption will certainly be reduced. After the system has been operating for a year, the system keeps an annual history of traffic patterns and the best counterweight setting over time can be suggested accordingly. It was reported that as much as 13 % of energy saving could be achieved by using this method (Lam et al. 2006).

7.1.4 Customer Needs, Comfort and Energy Saving

Up to now, this chapter has discussed those building services which are mainly inaccessible by guests, except the Low-E glass windows. There are various features however that the guests can personally interact with.

If we consider water use, opportunities to conserve its use are to be found, particularly in a large hotel building. For instance, low-flow sanitary fittings, as shown in Fig. 7.9, are installed throughout the Hotel. Potable water flows through the faucets (which can achieve a 51.4 % reduction in water flow) and shower heads (which achieve a 47.5 % reduction in water flow) by the use of aerators. When water passes through the aerators, air bubbles are injected into the water stream, creating the

Fig. 7.9 Low flow sanitary fittings

same feeling of volume flow rate to the user, but using less water. A further saving of potable water is the use of seawater for wc flushing, which is a common practice in Hong Kong. Therefore, seawater is used to flush the Hotel's water closets and urinals. Low-flow urinals achieve 16 % reduction in water flow while dual flush water closets (20 % reduction in water flow) provide both low-flush and high-flush options. These low-flow sanitary fittings could reduce the loading upon the building and municipal drainage systems.

As we have discussed previously, the windows offer scope to limit solar gain and heat transfer from the warm external ambient air into the building. Naturally, the use of blinds or curtains can support this function. In consideration of fenestration to the hotel rooms, roller blinds inside all guest rooms are motorised and interlocked with the keycard control. This enables shading and energy retention to be initiated. Whenever a guest leaves the room and removes the keycard out of the holder, which is close to the door, the curtain is automatically closed, so serving two functions. The blind is highly reflective towards the exterior and can significantly reduce the amount of solar radiation injected into the room, thus lowering the solar heat gain and the resultant cooling load. Furthermore, the motorised blinds are programmed to be closed at 11 pm in every evening, thus preventing light pollution to the surrounding environment due to the in-room lamps. Once the guest returns and inserts the keycard back to the holder, the curtain is automatically opened to present the guest with a lovely external view.

At the same time, when the keycard is removed, the fresh air damper is closed, thus further reducing the in-room cooling load, bearing in mind that the in-room temperature setting is also elevated to 27 °C while the fan speed of the iFCU™ is set to low. The absence of fresh air, when the room is unoccupied can also reduce the burden on the primary air units, which cools down the outdoor air. When the room is occupied, a sensor continues to monitor the CO_2 density to regulate the fresh air damper so as to admit the right amount of fresh air.

Indoor lighting also offers the potential to reduce energy usage. Therefore, all lamps in the guest rooms are LED based with high efficacy, while some are dimmable. Efficacy, (lm/W), is a measurement of the energy efficiency of lamps, which is defined as the ratio of the light output in lumens to the electrical energy consumed in Watts. The normal efficacy of traditional incandescent lamp bulbs is around 10 lm/W, while that of the CFLs (compact fluorescent lamps) is around 55 lm/W. The efficacy of state-of-the-art LED lamp bulbs could be over 100 lm/W, while there is still room for improvement. In the staircases and offices away from windows, T5 fluorescent tubes with electronic ballasts of high efficacy are installed.

In public circulation areas, lighting control is also a potential energy-saving measure, and there is also scope to modulate air conditioning according to the transient occupancy of these areas. Along the corridors outside the guest rooms, a Pattern Recognition Energy Saving Solution (PRESS™) was developed by making use of the security CCTV cameras installed there. Therefore, in addition to the normal function of detecting unauthorised intruders inside the Hotel, the PRESS™ software helps to check the presence of occupants along a corridor or at a particular zone. A prolonged absence of occupants initiates the turning off of the lighting and raises the temperature setting and lowers the speed of the corresponding iFCU™. There are two merits of this system, first an additional sensor being unnecessary and second further energy-saving being accomplished.

One novel innovation was put on trial in selected guest rooms. This development helped the Hotel obtain one credit in the area of "Innovation" of LEED. Conventional air-conditioning design is known to waste energy because the whole indoor environment is cooled down irrespective of the position of the occupant. In the guest room, when the guest lies on the bed and sleeps during midnight, it is unnecessary to keep the whole room under a low temperature, wherein 22 °C is normally applicable in Hong Kong. The idea is therefore to provide a localised cool space about 200 mm above the bed surface. The principle of operation is shown in Fig. 7.10.

Cool air, called "cold air" in Fig. 7.10, at around 22 °C, is supplied from a diffuser on the headboard above the head of the guest with a very low speed, less than 0.1 m/s. Relatively warm room air, called "hot air" in Fig. 7.10, is extracted back above the headboard and re-cooled. In this way, the localised space above the bed is appropriately cool enough, while the room could be relatively warmer than usual. The control logic is that when the average illumination level of the room is below 5 lx and within the time period from 11 pm in the evening to 7 am in the morning, the room temperature setting of the iFCU™ is adjusted to 25 °C instead of the normal 22 °C used by the guest. At the same time, the fan speed is adjusted to low. The "cold air" coming out from the diffuser at the headboard is able to lower the air temperature above the bed, in particular around the head of the guest, by 2–3 °C, and at the same time pushes the "hot room air" upwards. Once the illumination level is above 5 lx or time is after 7 am in the morning or the guest manually adjusts the thermostat, room temperature setting will be restored to the value before the operation of this mode but the iFCU™ is still running under a low speed.

The cool air from the headboard is not produced by using chilled water because, firstly the installation cost is huge and secondly, it is difficult to make sure there

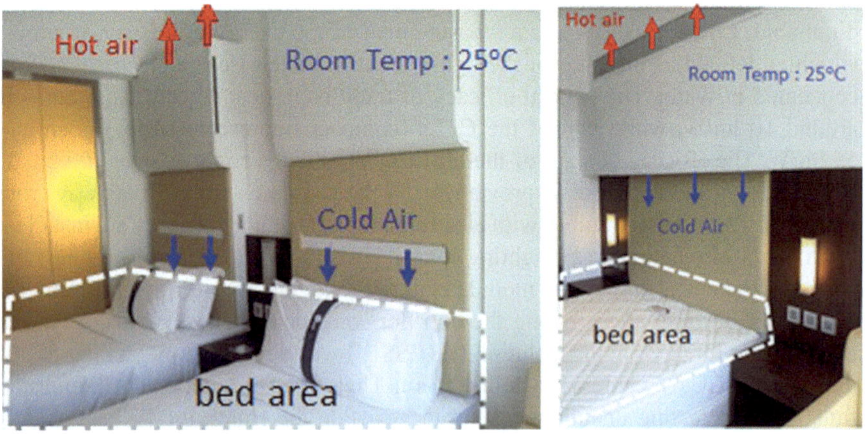

Fig. 7.10 Principle of operation of the Peltier headboard

is no condensation above the head of the guest. The design therefore involves the use of electronic cooling, and this is a thermoelectric effect, which is the direct conversion of temperature differences to electric voltage and vice versa. Here, the Peltier effect is utilised, which is the presence of heating or cooling at an electrified junction of two different conductors discovered by a French physicist Jean Charles Athanase Peltier (Pollock 1985) in 1834. When a direct current flows through a Peltier module, one side becomes cool and the other side warm; this is effectively similar to the evaporator and the condenser of a typical air conditioner. Heat drawn on the cool side, added to the electric power consumed, has to be dissipated on the warm side. It is this effect that is adopted within the headboard. However, both the cooling capacity and COP of a Peltier module are not high and therefore the effect is limited to only small applications, and this is ideal for the case used for the headboard for a single hotel guest. The temperature reduction is also limited to approximately 3 °C, but it is significant in terms of the comfort of the occupant. Room air is used to carry heat away from the warm side of the module. In this way, the temperature setting of the whole guest room is elevated throughout the night, saving substantial energy consumed by both chillers and chilled water pumps though some energy has to be consumed to energise the Peltier modules and the small fans blowing air out of the diffusers on the headboard. The whole energy exchange is still a net saving on traditional air conditioning.

7.1.5 Achieving LEED/BEAM Plus/AIIB IBI/Green Mark

At the beginning of this chapter, it was mentioned that the Hotel has received several awards, including LEED Platinum, HK BEAM-Plus Platinum, Distinction Intelligent Hotel of AIIB and the Singapore BGA Green Mark Platinum.

Leadership in Energy and Environmental Design (LEED) is a verification rating system developed by US Green Building Council (USGBC) in the mid-1990s, focusing on environmental performance designed for new and existing buildings. The idea of the creation of LEED was due to an intention to produce a US version of the UK's Building Research Establishment Environmental Assessment Method (BREEAM) as well as Canada's Building Environmental Performance Assessment Criteria (BEPAC). LEED version 1.0 was released in August 1998 and then version 2.0 by 2000. By 2007, version 3.0 was released with the ongoing modifications.

The Hotel achieved LEED version 3.0 2009 in the category of "New Construction", while "Major Renovation" was considered for another category. The updated version of LEED Rating System consists of the following categories as shown in Table 7.2.

Within each category, there are prerequisites that must be satisfied and credit points that can be given to a project. There are four levels of certification, namely "certified" (40–49 points), "silver" (50–59 points), "gold" (60–79 points) and "platinum" (80+ points).

Table 7.2 LEED rating system 2009

Green building design and construction	New construction and major renovation Core and shell development Schools Retail Data centres Warehouses and distribution centres Hospitality Health care
Green interior design and construction	Commercial interiors Retail Hospitality
Green building operations and maintenance	Existing buildings Schools Retail Hospitality Data centres Warehouses and distribution centres
Green neighbourhood development	Plan Built project
Green home design and construction	Homes and multi-family low-rise Multi-family mid-rise

Table 7.3 shows the achievement of the Hotel. All credit points related to "Optimize Energy Performance" were awarded indicating that the improvement in energy saving was more than 48%, a world record for a high-rise hotel at the time of assessment. As you can see from the table, the project achieved 28 credits out of 35 in the area of energy and atmosphere. As a matter of fact, the project achieved maximum points in the consideration of energy saving. For the submission to LEED, the annual energy consumption of the project was presented versus a baseline annual energy consumption. Table 7.4 provides a summary of the energy data for reference (Wong and Mo 2011; Leung et al. 2012).

Building Environmental Assessment Method (BEAM) was developed by the BEAM Society in Hong Kong by 1996. Version 1/96 was published for new office designs and version 2/96 was published for the existing office premises. By 1999, version 3/99 was published for new residential buildings. By 2004, version 4/04 was published as a revised standard for new buildings while version 5/04 was published as a revised standard for the existing buildings. By 2009, BEAM Society worked with other organisations in Hong Kong to establish Hong Kong Green Building Council (HKGBC) and BEAM Plus for new and existing buildings, Version 1.1 was launched in 2010 with an enhanced version, and version 1.2 was published for green buildings in 2012. By 2013, a new standard called BEAM Plus Interiors was launched for occupied spaces within buildings.

The Hotel also achieved BEAM Plus New Buildings version 1.2, which shares a similar structure to LEED. The required and credit points are assessed in six categories or aspects, namely site aspects (SA), materials aspects (MA), energy use (EU), water use (WU), indoor environmental quality (IEQ) and innovations and additions (IA). The award classifications of BEAM Plus 2013 are listed as Table 7.5. It can

Table 7.3 LEED new construction scorecard of the Hotel—platinum

Category	Achieved scores	Max points	Percentage achieved (%)
Sustainable site	23	26	88.5
Water efficiency	13	14	92.9
Energy and atmosphere	28	35	80.0
Material and resources	1	14	7.1
Indoor environment quality	13	15	86.7
Innovation and design process	4	6	66.7
Overall score	82	110	74.5

Table 7.4 Energy summary for LEED submission of the Hotel

System	Baseline annual energy consumption (kWh)	SoHo annual energy consumption (kWh)
Mechanical ventilating and air conditioning	1,286,042	619,193
Lighting	302,038	224,598
Receptacle	392,477	392,477
Hot water	889,631	221,951
Total	2,870,188	1,458,219

Table 7.5 Award classifications of BEAM Plus 2013

Class	Overall (%)	SA (%)	EU (%)	IEQ (%)	IA	Remark
Platinum	75	70	70	70	3 credits	Excellent
Gold	65	60	60	60	2 credits	Very good
Silver	55	50	50	50	1 credit	Good
Bronze	40	40	40	40		Above average

be seen that some minimum requirements with respect to SA, EU, IEQ and IA are stipulated for awards.

The overall percentage is based on a weighted sum of the five categories, 25% for SA, 8% for MA, 35% for EU, 12% for WU and 20% for IEQ. The percentage of every category depends on the number of credit points achieved out of the total. Table 7.6 shows the achievement of the Hotel in terms of BEAM Plus.

The Asian Institute of Intelligent Buildings (AIIB) was established in 2000 in Hong Kong as Asia's first learnt association promoting the research, promotion, construction and maintenance of intelligent buildings. AIIB defines intelligent buildings as "An Intelligent Building is designed and constructed based on an appropriate selection of *QUALITY ENVIRONMENT MODULES* (QEMs) to meet the user's requirements by mapping with the appropriate building facilities, termed Elements in the IBI Manual, to achieve a long-term building value".

Here, IBI refers to "Intelligent Building Index" which is a real number between "0" and "100", but not "0", indicating how intelligent a building is based on the IBI Manual. There are altogether ten QEMs, namely green, space, comfort, working

Table 7.6 BEAM Plus new building scorecard of the Hotel—platinum

Category	Achieved credits	Applicable credits	Achieved percentage (%)	Weighting factor (%)	Weighted achieved score	Category grade
Site aspects (SA)	14	19	73.7	25	18.4	Platinum
Material aspects (MA)	6	20	30.0	8	2.4	–
Energy use (EU)	31	37	83.8	35	29.3	Platinum
Water use (WU)	6	8	75.0	12	9.1	–
Indoor Environmental quality (IEQ)	22	28	78.0	20	15.6	Platinum
Innovations and additions (IA)	5	1	500	–	5	Platinum
Overall rating					79.8	Platinum

efficiency, culture, e-services and technology, safety and structure, management practice and security, cost effectiveness, and health and sanitation. Each QEM consists of a number of elements with formulae aiding the auditor to give scores, which are also real numbers between "0" and "100" but not "0". A combination of scores of all elements within a QEM gives the sub-index of that QEM and a combination of scores of all ten sub-indices gives the overall IBI. Besides the formula, each element is given a weight, a real number between "1" and "9". The most distinctive feature of IBI versus other assessment schemes is the way of combining scores together. Instead of using the linear addition of credit points, with or without weighting factors, the Cobb–Douglas function is employed, as shown by the following equation, where X is the combined score and x_i is the score of individual element with a weight w_i, for $i=1$ to N.

$$X = \prod_{i=1}^{N} x_i^{\frac{w_i}{W}} \quad \text{where } W = \sum_{i=1}^{N} w_i$$

It can be seen that it is similar to the geometric mean, where multiplication instead of addition is manipulated. There are advantages with this method: (i) the approach is closer to human thinking and judgment as this method has been widely used by the economists, (ii) a clear distinction between failure with and non-applicability of an element could be ensured, (iii) extremely poor performance of one element could significantly pull down the overall IBI. By the conventional linear addition approach of most assessment tools around the world, no credit is given irrespective of failure of non-applicability. In IBI, a failure by an element is penalised by a low score while a weight="0" is given to an element which is not applicable. It is known that any real number to the power zero returns the value of "1" and multiplying "1" to anything does not change anything. A "distinction IB" is awarded if the final normalised IBI is 90 or above, "credit IB" if between 70 and 89.9, "fair" if between 60 and 69.9, "to be improved" if between 1 and 59.9. The Hotel was awarded an Intelligent Hotel (Distinction) in the year 2012.

The BCA Green Mark of Singapore, developed and administered by the Building and Construction Authority, is a green building rating system to evaluate a building based on its environmental impact and performance, details are referred to at (www.bca.gov.sg/greenmark/green_mark_buildings.html). It is endorsed and supported by the Singaporean National Environment Agency, providing a comprehensive framework for assessing the overall environmental performance of new and existing buildings to promote sustainable design, construction and operations practices in buildings to promote energy savings, water savings, healthier indoor environments as well as the adoption of more extensive greenery. For existing buildings, the building owners and operators shall meet their sustainable operations goals and to reduce adverse impacts of their buildings on the environment and occupant health over the entire building life cycle. The rating system is similar to LEED in structure, wherein the assessment criteria covers five key areas, namely energy efficiency, water efficiency, environmental protection, indoor environmental quality and other green features and innovation. A "platinum" is awarded if the total green

Table 7.7 BCA green mark scorecard of the hotel—platinum

Category	Achieved scores	Max points	Percentage achieved (%)
Energy efficiency	63.5	72	88.2
Water efficiency	9.93	14	70.9
Environmental protection	17.00	27	63.0
Indoor environmental quality	4.00	4	100.0
Other green features	1.50	7	21.4
Overall score	95.93	124	77.4

mark score is 90 and above, a "gold plus" if between 85 and 90, a "gold" if between 75 and 85, and a "certified" if between 50 and 75. Table 7.7 shows the achievement of the Hotel in terms of BCA Green Mark.

7.1.6 Summary—What Went Right and Lessons Learned

The Hotel was awarded Triple Platinum certification, which was a unique achievement at the time of its opening.

Table 7.8 (Mak et al. 2014) shows the top three categories achieved by the project, indicating that the strengths are mainly with indoor environmental quality, energy and water use. Obviously, energy, air and water are the three most important criteria relating to both the environment and human needs. A sustainable supply of clean water, fresh air and energy can guarantee our future sustainability. Buildings have less effect on the outdoor air quality and therefore indoor air quality becomes the critical concern.

Innovations within the project should be the key and signature to success. Usually, developers tend to ensure their construction projects comply with local statutory regulations and be executed at a minimum cost.

For this project, the developer, from the inception of the project, has determined to construct a green and sustainable hotel and therefore a slightly higher construction cost was expected and prepared. As part of the project, there was an objective to develop and implement a range of research and development sub-projects, which would go beyond the practical limits suggested by standard design approaches, building code and trade practices.

As a result of the innovation in reducing the cooling load and adopting energy-efficient cooling plant, some startling outcomes are evident. The use of two 190-ton chillers to cool the building was originally considered inadequate for a high-rise

Table 7.8 Top 3 categories of percentage achieved

LEED		BEAM plus		Green mark	
Water efficiency	92.9%	Energy use	83.8%	IEQ	100.0%
Sustainable site	88.5%	IEQ	78.0%	Energy efficiency	88.2%
IEQ	86.7%	Water use	75.0%	Water efficiency	70.9%

hotel of 33 storeys subject to the sub-tropical climate of Hong Kong. However, the Hotel has been operating for almost one and a half years, where the adequacy of this provision has been validated. Very often, even in summer, it is possible to run only one chiller, when the Hotel is over 75 % full. This is due to the provision of so many novel energy-saving measures adopted in the Hotel, such as the optimal chiller plant control, the iFCU^{TM}-based FCUs, in-room fresh air control, Low-E glass, automatic blind operation, energy-saving lifts and the heat pumps/exchangers to lower the temperature of return chilled water. Most of these solutions were not readily available in the market at the initiation of the project; but the research and development conducted by the project team, which was strongly supported and driven by the developer, realised the goal of a low energy building. This is the primary reason why in all three green building assessment certifications, i.e. LEED, BEAM Plus and Green Mark, innovations and additions are emphasised.

The benchmarking parameter to measure the overall energy efficiency of a building is called energy use intensity (EUI) measured in (kWh or MJ)/m^2/annum in SI unit or kBtu/ft^2/annum in Imperial units.

For a typical hotel in Hong Kong during the design stage of this Hotel in 2007, the EUI was around 437.5 kWh (1575 MJ)/m^2/annum while that in the USA was around 401 kWh (1443.6 MJ)/m^2/annum.

The EUI of this Hotel, at the design stage, was estimated to be around 181 kWh (651.6 MJ)/m^2/annum, i.e. a saving close to 60 %, based on an occupancy rate of 80 %. Now, it is confirmed that the current EUI of the Hotel is 209 kWh (752.4 MJ)/m^2/annum based on a continuous occupancy rate of over 95 % throughout the previous year, thus validating the design objectives.

This Hotel project is however not an end goal, but just the beginning of a more challenging mission. The concept of "Zero Net Energy Building" (ZNEB) or even a net positive building (i.e. a building that contributes energy to the grid) is becoming more popular and more essential to the future energy equation.

The US Congress passed EISA 2007 where the objective is the realisation of 100 % newly constructed commercial buildings be ZNE by the year 2030 and all commercial buildings, including those existing, be ZNE by the year 2050. It is currently impractical to supplement the whole EUI of over 400 kWh/m^2/annum for high-rise buildings merely by renewable energy sources. So, the first step is to significantly lower the EUI, say by two thirds (So et al 2014), while the remaining consumption is supported by renewable energy sources, either on-site or off-site.

7.2 Conclusion

This project offers a view of the demands of building structures in tropical and subtropical environments. The need for thermal comfort is a key driver, particularly in the hotel sector. Therefore, for all buildings, the design imperative is to limit solar heat gain, ambient and occupancy thermal gains and therefore limit the overall cooling load of the building.

In a densely populated city, such as Hong Kong, difficulty of this task is increased with the need to build vertically, with added impact of other buildings (the heat island effect) as well as the housing of electrical plant for lifts.

In respect of energy conservation and carbon reduction, this task becomes particularly onerous. However, the carbon reduction objective in Hong Kong is now clearly central to its new buildings programme.

The sustainability agenda does not only focus upon the energy demand of the building. It also relates to water conservation, the introduction of green infrastructure but the adoption of responsibly sourced timber, green walls, roofing and other landscaping where possible. It is also important to note the careful selection of building materials that do not introduce volatile organic compounds (VOCs) into the environment. Furthermore, the efficient use of materials so as to minimise site waste and negate landfill provides the challenge for the design team and the building contractor.

Such objectives are not achieved simply by the imposition of building code or by the adoption of design methodologies. Project leadership by the client and developer is critical in wishing to progress these changes. The building team also needs to adopt new approaches to project management from inception so as to achieve these aims.

In this project, we have reported the adoption of a distinct objective by the developer to maximise the sustainable credentials of the building. This is emphasised by the adoption of building information modelling from the onset, integrating design, electrical and mechanical and structural engineering in the 3-dimensional representation of the building. The management process was enhanced by 4D BIM (time management) and clash detection so as to simulate the building process, eliminate errors and avoid wasteful site activities before committing to the on-site activity, this in itself being a huge aid in limiting waste and limiting costly design changes on site. We also note the adoption of modular off-site production of standard components as a means of limiting on-site practices and opportunities for further materials wastage. Where on-site construction was necessary such as the construction of the lift shafts, the adoption of extruded concrete technologies allowed for energy savings in the construction process.

The building structure focused upon maximising orientation to limit solar gain and using the fabric to limit thermal gains. Particular attention was paid to window design to further aid this objective. The installation of a green roof and green walling also aided the overall building performance, whilst enhancing the local ecology of the surrounding area.

This project is distinctively characterised by its detailed and innovation approach to the building services installation to the building, seeking to reduce energy usage and improve the efficiency of operation. This included such innovations as the chiller efficiency and their controls, the adoption of solar thermal panels to preheat hot water, heat exchangers used to remove heat from the chilled water circuit to further reduce the energy required for hot water supply.

With respect to vertical transportation, the lift operation also provided scope for energy conservation, using a regenerative braking device and using a smart approach to lift movements and learning from occupant usage.

In the hotel rooms, low-flow technology was used to reduce the demand on potable water and sea water was used for wc flushing. The use of automated roller blinds allowed for additional reductions in solar heat gain in rooms when the occupant left the room. Similar savings were made by adjusting the room temperature, when the guest rooms were unoccupied.

The lighting design in the building was such that opportunities to reduce the energy demand were made with respect to luminaire selection and the control of corridor lighting levels according to occupancy. This technology was also used to control air conditioning in corridors.

A novel approach to guest room air conditioning was also observed by the introduction of a localised chiller over the bed.

The lesson shown is that a multi-storey building can achieve an excellent environmental building standard. By learning from this example, strategies to achieve a zero-carbon building can be explored, but focusing further upon controls technology, developing the performance of the building fabric and utilising more passive technologies in respect of environmental control.

Acknowledgement The authors would like to acknowledge the assistance of Miss Carmen Wong of Asian Institute of Built Environment and REC Green Technologies Company Ltd. throughout the compilation of this chapter.

References

ASME (2015) Energy efficient elevator technologies. https://www.asme.org/engineering-topics/articles/elevators/energy-efficient-elevator-technologies. Accessed 8 Feb 2015

BRE (2012) Building research establishment energy assessment methodology. BRE Global

Cotteral J, Dadeby A (2012) The passivhaus handbook. Green Books, Devon

Environmental Protection Agency of the United States of America (2015) http://www.epa.gov/heatisland/index.htm. Accessed 8 Feb 2015

Hong Kong Buildings Department (1995) Code of practice for overall thermal transfer value in buildings

Hong Kong Development Bureau (2014) Policy address initiatives of development bureau. HKSAR Government

Hong Kong Electrical and Mechanical Services Department (2013) Energy report. Accessed 1 Feb 2015

Hong Kong Environmental Protection Department (2015) Climate change report. Accessed 21 May 2015

Hong Kong Observatory (2015) Monthly metrological normals for Hong Kong. Accessed 1 Feb 2015

Lam D, So A, Ng T (2006) Energy conservation solutions for lifts and escalators at Hong Kong housing authority. In Proceedings of 16th international congress on elevator technology, Elevcon, Helsinki, IAEE

Lee WL (2010) Energy simulation run for Jervois Street Hotel, Hong Kong Polytechnic University Consultancy Report, Nov, 2010

Lee WL, Burnett J (2008). Benchmarking energy use of HK BEAM, BREEAM and LEED. Build Environ 43(11):1785–1990

Leung E, Wong R, Mo E (2012) Energy saving summary report—Holiday Inn Express Hong Kong SoHo, Telemax EEML, Apr, 2012

Mak MY, Wong CTC, So ATP, Chan ACM (2014) Beyond platinum: a hotel project in Hong Kong. In Proceedings of International Conference on intelligent systems, structures and facilities, AIIB, January, 2014

Pollock D (1985) Thermoelectricity: theory, thermometry, tool. American Society for Testing and Materials, Ann Arbor

So A, Katz D, Wacks K (2014) Toward zero net energy (zne) super high-rise commercial buildings. Continental automated buildings association white paper, June, 2014

United Nations (1998) Kyoto protocol to the United Nations framework convention on climate change. United Nations

United Nations Environment Programme (2009) Buildings and climate change. UNEP

Wong R, Mo E (2011) HKBEAM plus—Holiday Inn Express Hong Kong SoHo Sheung Wan energy simulation report for EUP1, EUc1, EUc2 and EUc6, Telemax EEML, Oct, 2011

Wong TC, Chan ACM, Wong CYS, Wong KW (2014) Holiday Inn Express Hong Kong SoHo: a hotel with multiple green awards, triple platinum and three stars. Zero Carbon Building Journal 2, pp 50–63

Chapter 8
Achieving Sustainability in New Build and Retrofit: Building Performance and Life Cycle Analysis

Christopher Gorse, Felix Thomas, David Glew and Dominic Miles Shenton

8.1 What is a Sustainable Building?

The definition of a sustainable building is not a straightforward one. There are many criteria upon which the sustainability of a building can be judged, including but not limited to energy performance, financial viability and environmental and social impact (Berardi 2013). Any determination of the sustainability of a building will be dependent upon the criteria used to assess it. Much of the work undertaken by the Leeds Sustainability Institute on building sustainability focusses on energy performance in buildings.

8.2 Attributes of Successful and Sustainable Buildings

What makes a building sustainable can be difficult to define, as there are a number of philosophies and approaches to sustainable construction, ranging from very simple low-impact buildings using natural materials isolated from the national grid, to complex high-technology, high-performance energy-plus buildings, feeding excess power, which is generated by renewable technologies such as photovoltaic panels, back into the national grid (Fig. 8.1).

C. Gorse (✉) · F. Thomas · D. Glew · D. M. Shenton
Leeds Beckett University, Broadcasting Place—Leeds LS2 9 EN, UK
e-mail: C.gorse@leedsbeckett.ac.uk

F. Thomas
e-mail: F.Thomas@leedsbeckett.ac.uk

D. Glew
e-mail: D.W.Glew@leedsbeckett.ac.uk

D. M. Shenton
e-mail: D.Miles-Shenton@leedsbeckett.ac.uk

© Springer International Publishing Switzerland 2016

M. Dastbaz et al. (eds.), *Building Sustainable Futures*,
DOI 10.1007/978-3-319-19348-9_8

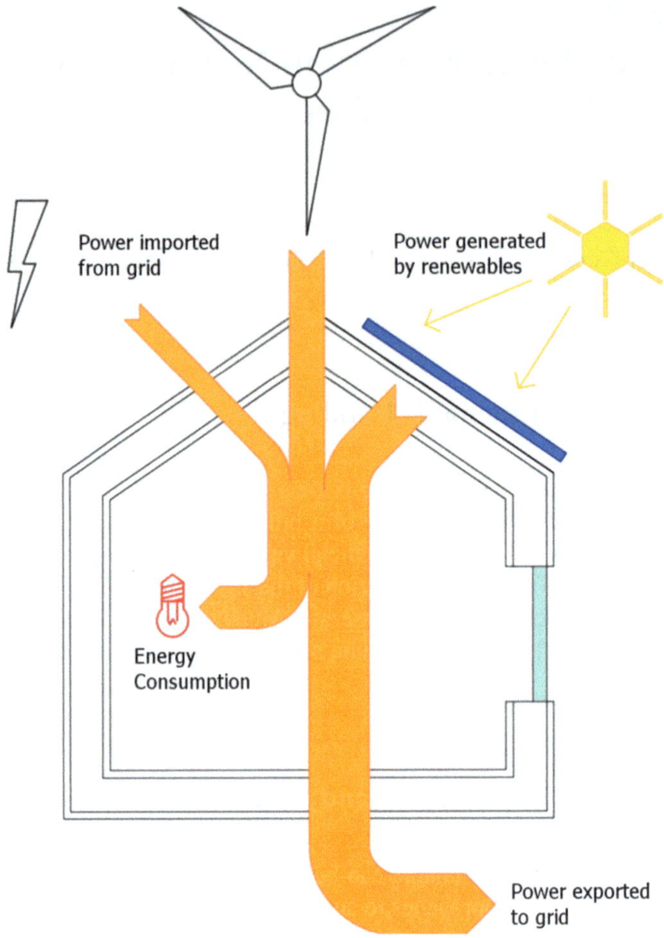

Fig. 8.1 Illustration of an energy-plus building that exports more power to the grid than it imports. (Thomas and Gorse 2015)

Despite the variations in the approach taken to sustainability, there are a number of attributes that many sustainable buildings have in common, and these can be used as indicators of sustainability. A successful and sustainable building will address a number of issues of sustainability, in the building design, construction and use.

The 'triple-bottom line' of sustainability consists of three categories: environmental/ecological, social and economic (Slaper and Hall 2011); and the concept initially devised by John Elkington as a way of measuring the sustainability of businesses is also applicable to construction and encompasses a number of attributes that make for sustainable buildings (Table 8.1).

Table 8.1 Attributes of sustainable buildings with relation to the triple-bottom line. (Adapted from Alwaer and Clements-Croome 2010)

Environmental	Social	Economic
Energy and natural resources	Usability and function	Flexibility and adaptability
Water conservation	Indoor environmental conditions—health and wellbeing	Economic performance and affordability
Material use, durability and waste	Architectural—cultural and aesthetic	Building manageability
Land use	Innovation and design	Whole life function and value
Transport and accessibility		
Greenhouse gases and pollution		

Environmental Environmental sustainability of a building is perhaps the most immediately obvious aspect of sustainability and is arguably the primary goal of sustainable buildings.

Energy and Natural Resources Consumption of energy and natural resources in the form of fuel within buildings is a major factor in the overall sustainability of a building. A sustainable building will minimise the amount of energy required in its operation, through thermally efficient fabric to reduce heating demands, use passive heating, lighting and cooling where possible and make use of efficient services to reduce energy requirements.

Water Conservation Minimising water consumption, particularly freshwater, contributes to the sustainability of a building, reducing strain on freshwater supplies as well as the energy and resource consumption associated with moving and treating water. Freshwater consumption can be further offset through the use of rainwater harvesting and the storage and reuse of 'grey water' (water used for washing, etc.) for non-potable uses, such as washing or toilet flushing.

Material Use, Durability and Waste This category applies primarily to the construction phase of a building, though it also applies to refurbishments and repairs. Materials should be used in a way that maximises the sustainability of a building, by reducing consumption of non-renewable resources and diverting waste away from landfill: where possible, the building using low-impact renewable materials, reused or recycled materials or materials that have high reuse or recycling potential. Materials should be selected and used for durability, to maximise the service life of the building and reduce the need for repair and replacement.

Land Use The site for a sustainable building should be chosen to minimise the impact upon the environment and ecosystems, avoiding development of previously undeveloped greenfield sites or productive land in preference to empty brownfield sites that have had prior development. Where a building has an impact on the surrounding environment and ecosystems, measures should be taken to offset detrimental effects.

Transport and Accessibility Building design and location should be conducive to sustainable methods of transport, providing bicycle storage, being in close proximity to public transport routes and providing a charging point for electric vehicles. The building should be designed to allow easy access to people with a wide range of physical capability.

Greenhouse Gases and Pollution Greenhouse gas emissions and pollution output caused by a building should be minimised, in the selection of materials, transport and construction methods used and the day-to-day operation of a building.

Social Social sustainability concerns how a building interacts with its users and the affect it has on the wider society in which it is located. To be sustainable, the building must be accepted, appreciated and used. A building that is rejected or not used appropriately will not be sustainable.

Usability and Function A sustainable building and the systems within it should be simple for the occupier to use, and the functionality of the building systems should be understandable to the occupant, coupled with simple controls to allow the building to be operated to meet the user's needs.

Indoor Environmental Conditions—Health and Wellbeing Sustainable buildings should be able to meet and maintain internal environmental conditions that meet the occupant's needs for comfort and their health. Maintaining a comfortable internal temperature and sufficient ventilation to control moisture build-up and an adequate supply of fresh air, these should be attained in an energy-efficient and sustainable way.

Architectural Considerations—Cultural and Aesthetic Though energy performance and resource consumption are of some of high importance, sustainable buildings should also find a place visually. If a building is considered to be visually unappealing or out of place, potential occupants may be driven away. It is possible to make sustainable buildings that respect the architectural merits of their local context with minimal impact to their design and overall sustainability. Some buildings are so environmentally appealing that they stimulate use and can evoke a culture of their own, encouraging wider sustainable practice.

Innovation and Design Innovation in design is at the core of many sustainable buildings, as reaching energy performance and CO_2 emission targets is no easy task, requiring significant innovation in design, materials, construction methods and technology. Innovation within the design of sustainable buildings can lead to higher levels of performance and more easily achieved performance targets.

Economic In a society driven by the exchange of capital and the accretion of wealth, buildings need to be sustainable from an economic perspective. Financial incentives help to drive the demand for sustainable buildings.

Flexibility and Adaptability A sustainable building should be flexible. It should be able to adapt its behaviour and operation to meet changes both in use and in the demands of the building users. Over the years the use and occupant need change, a test of sustainability is the capability of the building to meet the future needs and respond to new demands.

Economic Performance and Affordability Sustainable buildings can cost more to design and construct than comparable conventional buildings. The additional costs vary, in some cases costing less than a comparable building. The day-to-day cost of running a sustainable building should be lower than that of running a comparable conventional building, due to energy efficiency of the building fabric and services. Renewable electricity generation technology such as photovoltaic panels and wind turbines reduces electricity imported from the national grid and may export excess electricity to the grid, generating income for the building. The benefits of energy savings and potential income from renewables should offset any additional costs incurred from the design and construction of a sustainable building.

Building Manageability The day-to-day operation of a sustainable building should be easy to manage, incorporating smart technology and efficient systems that allow the cost of operation to be controlled whilst still meeting internal environmental requirements for user comfort. A number of building management systems have failed to achieve significant benefit simply because the users find the systems too difficult to operate. More intelligent systems are being developed, but the key to effective operation of the building and services is simplicity. At all times, the building must be designed to be operated easily and meet the user requirements.

Whole Life Value A sustainable building should be able to retain its value for future sale, the use of durable materials will help to ensure that the fabric of a building does not degrade and reduce its value, and designing for ease of repairs and access to services will help reduce the cost of repair work during a building's service life. A building should also offer savings due to reduced energy and fuel consumption during its service lifetime.

The attributes outlined above are not an absolute set of requirements that sustainable buildings must fulfil, and what makes a building sustainable will vary from one project to another, depending on many factors such as the purpose of the building, where it located in the world and the environment in which it is located. Sustainability rating systems such as BREEAM and LEED vary the weightings of their scoring systems to suit the type of building, and a sports centre will have different sustainability priorities compared to an office or residential building.

In order to succeed as a sustainable building, it is important for the sustainability goals and targets to be determined at an early stage in the development of a project, so that a practical and achievable sustainability strategy can be adhered to throughout all of its stages, from design to construction. This increases the buildings' chances of delivering on the targets set out for it. Although it is normally considered effective for buildings to have a long operational life, they should also be designed for deconstruction, disassembly, demolition and reuse of the materials and components. The cradle to grave concept ensures that the components of the building are considered so that they can be reused at the end of the buildings life.

8.3 Sustainability in Buildings: Energy Performance

Much of the work undertaken by the Leeds Sustainability Institute is based on energy sustainability in buildings, thermal performance of building fabrics, air permeability and hygrothermal behaviour.

Building Fabric The energy performance of a building is in large part affected by its local climate and weather conditions. In the UK and other temperate regions, the heating of buildings accounts for a large portion of energy consumption, particularly in domestic buildings, where 73 % of CO_2 emissions result from space and water heating (UKGBC 2014). The thermal performance of a building's fabric is a major factor in how much energy will be lost in the form of heat through elements such as the walls, floors and roofs.

Heat is lost via conduction through building elements and by the exchange of warm internal air with cool external air, requiring additional energy input to replace lost warmth and maintain internal comfort conditions. Common measures used to reduce heat loss from buildings include the use of insulation material to increase the resistance to thermal conduction of building elements and using airtight construction techniques to prevent the uncontrolled exchange of internal and external air. In turn, this leads to lower energy usage for space heating, helping to improve the energy sustainability of buildings.

Though highly insulating and making buildings airtight has its benefits in terms of energy use there are some potential drawbacks. If the insulation layer is not continuous it can lead to areas of thermal bypass, where thermal resistance is significantly lower, resulting in increased heat loss compared to the surrounding fabric, and these cold spots can lead to the formation of condensation and eventually mould, especially in well-sealed but poorly ventilated buildings. Poor ventilation is a problem in buildings which have been sealed to reduce uncontrolled ventilation; insufficient rates of ventilation can cause excessive levels of humidity or stale air.

Figure 8.2 shows a basic illustration of the nearly zero carbon emissions concept or Passivhaus construction. By increasing the thermal insulation, to form a continuous thermal envelope, avoiding thermal bridges, whilst maintaining a continuous air barrier layer in contact with the insulation, throughout the fabric and junctions an airtight thermally isolated envelope is provided. If orientated correctly for optimum solar gain, avoiding overheating, the building maintains a habitable environment with minimal or zero energy.

Services The services within a building are another area in which energy consumption can be reduced, leading to more sustainable operation. The most basic energy reductions can be made by fitting efficient services such as boilers or LED lighting. Renewable and low-energy technology also offer possible methods to reduce energy imported from the national grid or gas supply.

Low-energy services such as mechanical ventilation with heat recovery (MVHR) and air/ground/water source heat pumps allow heat to be recovered/extracted and introduced to the internal environment of a building. Where these systems are operating optimally, a greater amount of heat energy is recovered than the energy expended to operate the system; however, systems used in non-optimal conditions can experience

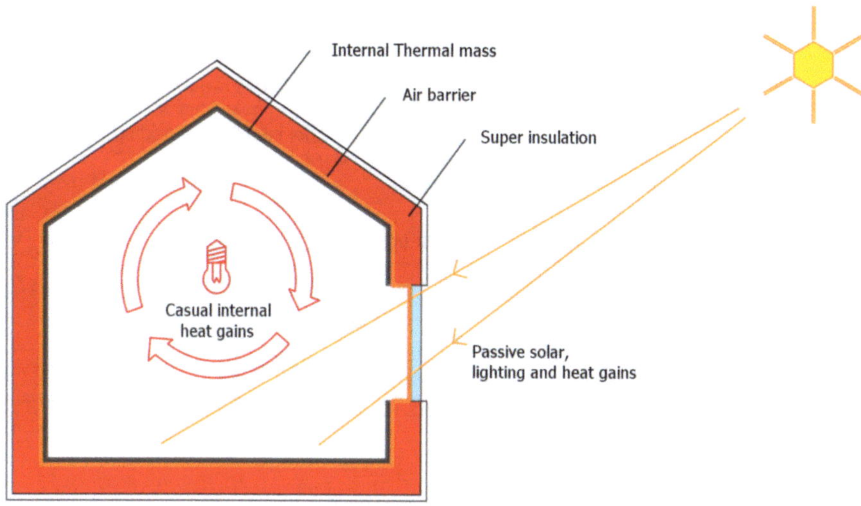

Fig. 8.2 Illustration of the principals of a nearly zero and Passivhaus (Thomas and Gorse 2015

efficiency losses. Renewable systems aim to provide energy or heating through the exploitation of renewable sources of energy, such as solar radiation or wind. The energy provided by renewable systems fluctuates due to seasonal variations and day-to-day weather, making such systems less than 100 % reliable. Nonetheless, the potential of free or low-cost energy still presents an appeal (Figs. 8.3 and 8.4).

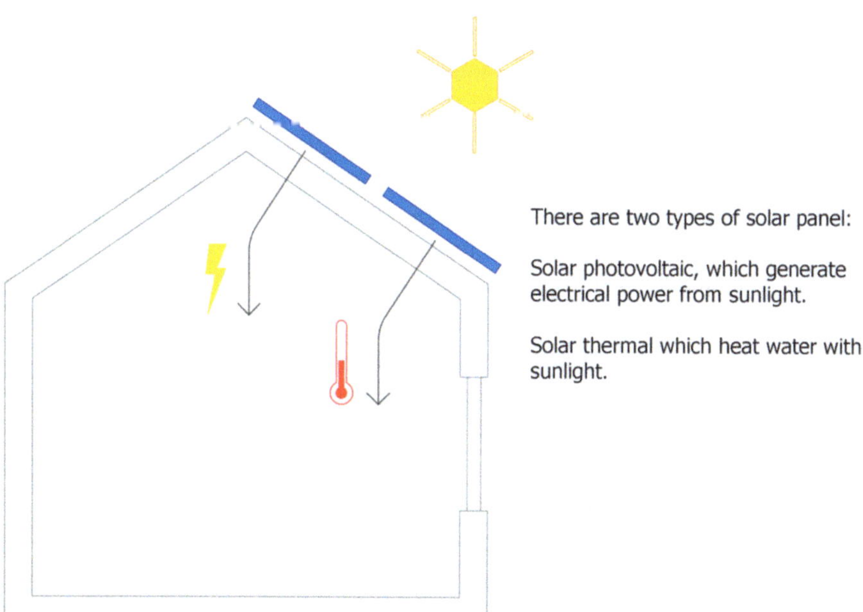

Fig. 8.3 Illustration of solar photovoltaic (PV) and solar thermal panels (Thomas and Gorse 2015)

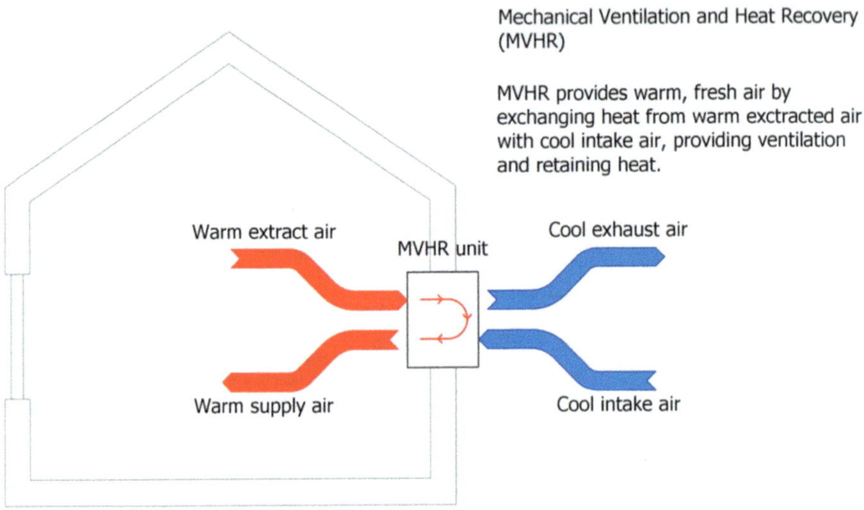

Fig. 8.4 Illustration of the operation of an MVHR unit (Thomas and Gorse 2015)

8.4 Embodied Impacts/Energy

The energy consumed in the operation of buildings is the most obvious factor affecting the energy sustainability of a building, though by no means the only one. The sustainability of the fabric that makes up the building itself is an important factor to consider. A significant amount of energy and resource consumption can be attributed to a building's fabric, known as embodied impacts.

Embodied impacts originate from a number of sources and can represent consumption of non-renewable resources, energy and associated CO_2 emissions at various stages.

Material Manufacture Energy and fuel is expended in the extraction of raw materials, and in the manufacturing processes undertaken to produce usable products from the extracted raw materials. CO_2 emissions may result as a by-product of processing and the generation of consumed energy. Extraction, processing and manufacture of materials and components account for up to 80 % of energy expended in construction of buildings.

Construction Energy and fuel is consumed during the process of constructing a building; transporting materials, machinery and personnel to site; powering equipment and machinery during construction activities; and providing facilities for the on-site personnel.

Maintenance As the fabric degrades over a building's service life, more energy and materials will be expended carrying out repair work.

Demolition Further energy will eventually be consumed when a building reaches the end of its serviceable life or is no longer required. Energy and fuel will be consumed operating machinery required to carry out demolition activities. There is the potential to recover materials and components intact for reuse, or recycle those that cannot be reused directly (Berge 2000).

The embodied impacts associated with a building can be reduced in a number of ways, some of which include using durable and long-lasting materials. This will reduce the need for repairs and maintenance activities, prolonging the building's serviceable lifetime (Akadiri 2011). Using low-energy or alternative methods of construction such as straw bale or rammed earth can reduce the amount of energy or non-renewable resources consumed for the building fabric (Halliday 2008).

Reuse of materials and the use of high recycled content materials in construction can also reduce embodied impacts of materials when compared to new and non-recycled materials. However, careful consideration should be given to these decisions, as recovery of materials for reuse and remanufacturing with recycled materials may offer little or no reduction in embodied energy (Denson and Halligan 2010).

8.5 Delivering Sustainable Buildings

Despite initial design intentions to build low-energy, thermally efficient buildings, this aspiration is not always realised at completion. Actual energy performance often falls short of the performance predicted at the design stage. This shortfall between as-designed and as-built energy performance is often referred to as 'the performance gap'. The vast majority of buildings do not have their energy performance put to the test post-construction, and without confirmation of their energy performance, there may be many buildings that do not reach their performance targets without their occupants ever realising.

The performance gap can potentially be addressed, in order to deliver buildings that live up to as-designed performance targets. Very large variations have been discovered when testing finished buildings, and in tests performed by Leeds Beckett University at the Leeds Sustainability Institute, shortfalls from 6 to 140% have been found with a median shortfall of 34.5% (Johnston et al. 2014). It should be noted that in buildings designed to be energy efficient, the percentages can make the shortfall look much more severe than the absolute figure.

Design During the design process, care should be taken to ensure that elements that affect the energy performance of a building will be effective, ensuring continuity of insulation layers and eliminating or reducing sources of thermal bridging, particularly at junctions and around openings to reduce the loss of heat through the fabric. Ensuring the integrity of the air barrier throughout the building is maintained, especially where openings and penetrations through the fabric occur. Carrying out accurate modelling of energy performance at design stages will help to indicate whether target energy performance is possible to achieve.

Communication Low-energy buildings are often constructed using unusual materials, products and techniques. This can lead to problems during construction as site operatives may be unfamiliar with their use. Producing clear, well-annotated detailed drawings, plans and briefing materials can help to clarify the make-up of complicated elements and avoid confusion on site, potentially avoiding mistakes.

Quality Control Faults and mistakes in the construction of a low-energy building can be detrimental in its ability to achieve performance targets upon completion. Strict quality control during the construction process can help discover flaws and mistakes, whilst it is still practical to remedy them. Carrying out interim tests at stages of completion can help identify whether the building is on track to perform as intended, thus allowing any remedial actions required to be carried out and ensure its performance targets are met (Miles-Shenton and Wingfield et al. 2010).

8.6 Construction

8.6.1 Modern Methods of Construction

Modern construction techniques often make greater use of insulation materials compared to more traditional construction methods. Even modern variants of traditional masonry construction, such as brick and block cavity walls, can be adapted and modernised to use large amounts of insulation. The increasing emphasis placed on thermal efficiency of buildings is making such high levels of insulation more commonplace.

With different combinations of cavity wall construction, it is relatively easy to achieve low U-values, and the combination of lightweight blocks, insulants with low thermal conductivity and airtight construction can achieve U-values of less than 0.2 W/m^2K.

For example, lightweight blocks are now available with thermal conductivities of 0.11 $W/m \cdot K$ (R value for 100 mm and 140 mm $= 0.1/0.11 = 0.909$ m^2K/W and $0.14/0.11 = 1.273$ m^2K/W), used in combination with PIR 0.022 $W/m \cdot K$ (R value for 100 mm $= 0.1/0.022 = 4.55$ m^2K/W), a regular brick thermal conductivity 0.6 $W/m \cdot K$ (R value 102.5 mm $= 0.102/0.6 = 0.17$ m^2K/W), internal surface resistance 0.12 m^2K/W and external surface resistance 0.06 m^2K/W can achieve significantly low U-values of 0.16 W/m^2K (the U-value for a wall built up with 140-mm lightweight block and 100 mm PIR). Such walls, with high thermal resistance, are now becoming commonplace.

Advanced, high-performance materials such as aerogels or vacuum-insulated panels offer much lower thermal conductivities compared to more common materials such as mineral wool or PIR foam boards. Silica aerogel has an extremely low thermal conductivity of 0.03 $W/m \cdot K$. Aerogels and other high-performance insulation can achieve the same result using far less insulation than more common materials, making them very useful for situations where space is limited. With the exception of a few situations, such insulants are very costly compared to more conventional materials and currently not commonly used across the industry. The price of these advanced thermal insulants is likely to reduce as their use increases. We are likely to see thinner materials with higher levels of thermal performance, becoming much more common in the future.

8.6.2 *Masonry Cavity Walls Built to Passivhaus Standards*

Most zero carbon or Passivhaus buildings are engineered or timber frame, few properties are built to Passivhaus standards using masonry walls, though this does not need to be the case. The Denby Dale passive house is a masonry building that did adopt Passivhaus principles (Green Building Store 2013). In this instance the outer skin of blockwork was constructed of stone, with the inner leaf blockwork, 300 mm of low conductivity Dri-therm ridged slabs of water repellent glasswool were used (0.037 W/m · K). The skins of the cavity were tied using extra-long wall ties with a low thermal conductivity (0.7 W/m · K). The very low thermal conductivity wall ties avoid cold bridging and risk of spot condensation. The ties prevent thermal bridges and cold spots that would have been experienced with steel ties. The Teplo Basalt fibre wall ties are 200–425 mm in length and can suit cavities up to 300 mm width. Each wall tie has thermal conductivity 0.7 W/mK and 7 mm diameter, with a sand rough textured finish to provide a mortar key. The tie penetrates across the three 100-mm boards of insulation to provide the structural tie between the two masonry walls.

8.6.3 *Exterior Wall: 0.113 W/m²K*

The depth of some cavity walls built to Passivhaus construction in this way has a built-up assembly of over 500 mm thickness of different materials. They combine to make a highly insulated cavity wall. The construction has a low air permeability and high thermal resistance. Although still referred to as cavity walls, the voids are now completely filled with insulation using 300 mm of mineral wool and lightweight blocks with plaster used to seal and create an airtight fabric. Highly insulated fully filled cavity walls can be constructed to achieve U-values of 0.1 W/m^{2}K, meeting nearly zero fabric standards (Fig. 8.5).

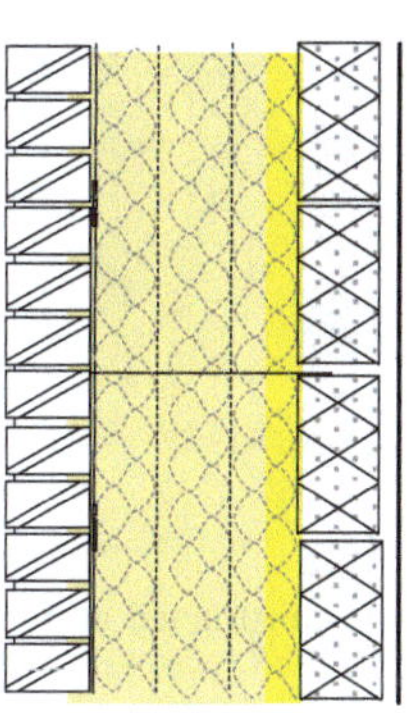

Fig. 8.5 Cavity wall built to Passivhaus standards. (Adapted from Green Building Store 2013).

8.7 Case Study: Temple Avenue-Prototype New Build Dwellings

The Temple Avenue project consisted of several prototype low-carbon dwellings by the Joseph Rowntree Housing Trust. Two of the dwellings were new build houses using modern methods of construction, one using structurally insulated panels (SIPs) and the other constructed with thin joint masonry (Fig. 8.6).

Both houses aimed for high standards of fabric performance, in terms of heat loss through the building fabric and air tightness. Though the finished buildings delivered high levels of fabric performance, they still fell short of the levels predicted using design information.

As these buildings were prototypes, setbacks and mistakes were expected to arise as part of the learning process for all parties involved, and they serve to illustrate some of the problems that may be encountered in the construction of low-energy, sustainable buildings.

Some problems occurred within the design of the dwellings. A number of potential areas of thermal bypass and paths of air infiltration were identified, though

Fig. 8.6 One of the two prototype dwellings constructed at Temple Avenue. (Image by Leeds Beckett University)

measures were taken to remedy these where possible. Services and M&E design in some cases disregarded air tightness and thermal performance, displacing insulation and adding penetrations to the air barrier.

Sequencing of the build process caused problems on site, due to unclear instructions or lack of planning. Poor sequencing leads to some processes being carried out incorrectly and preventing some from being carried out completely. A lack of clear specification and instruction in design documentation leads to some work being carried out unsuitably. This particularly affected the air barrier, where inappropriate tapes and sealants were used. There was also a lack of instruction for maintaining air barrier continuity around junctions, openings, etc.

For more information on this case study, see Miles-Shenton D. et al. (2010).

Retrofit The principles that apply to new build construction also apply to the retrofit of old buildings, though retrofit brings its own set of challenges. Dealing with existing building fabric imposes constraints on the kind of works that can be carried out. Features of the existing fabric may prevent or hamper access for certain retrofit measures.

The compromises required to work around existing fabric and features may lead to reduced energy performance as less than ideal solutions may be adopted that lead to thermal bridges or paths for uncontrolled air infiltration.

Establishing the energy performance of an existing dwelling prior to a retrofit is a non-trivial task, requiring detailed surveys to be undertaken and exploratory tests to establish the nature and performance of the existing fabric (Miles-Shenton et al. 2011). Inaccurate estimation of pre-retrofit energy performance can make the final energy performance of a retrofit appear more or less effective than it is in reality.

There are a number of retrofit measures that can be carried out on existing buildings, depending on the nature of their construction (Fig. 8.7).

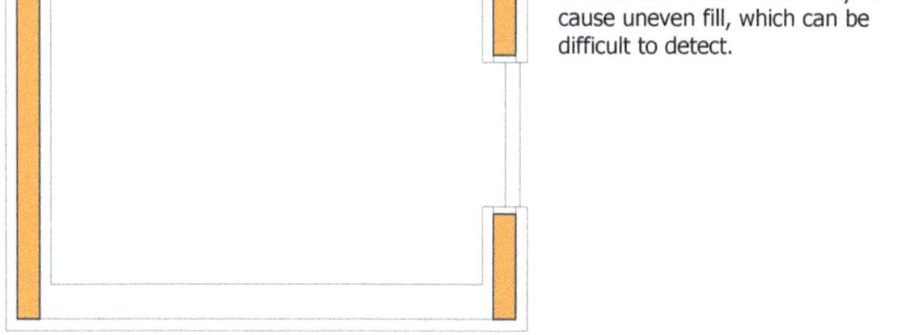

Fig. 8.7 Illustration of cavity wall insulation. (Thomas and Gorse 2015)

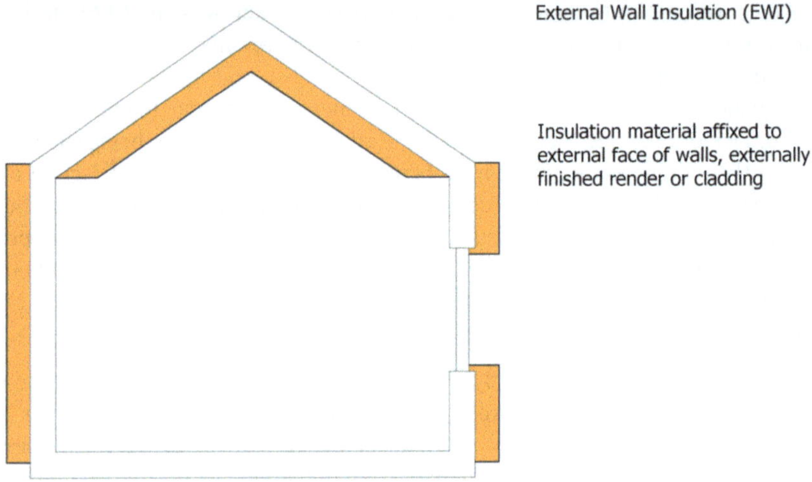

Fig. 8.8 Illustration of external wall insulation. (Thomas and Gorse 2015)

8.7.1 Cavity Wall Insulation

Where a cavity-walled building has a cavity that is suitable for filling, this method is relatively simple and quick to carry out. However, as it is difficult to see into the cavity without specialist equipment such as a borescope, it can be difficult to detect an inadequate fill.

8.7.2 External Wall Insulation

External wall insulation is used on hard-to-treat walls that lack cavities or have cavities unsuitable for blown insulation. Placing the insulation on the external walls can lead to thermal bridging where measures are not taken to seal the thermal envelope. Overhangs may need to be extended to accommodate EWI, and external fittings such as drainage pipes will need to be moved (Fig. 8.8).

8.7.3 Example of a Difficult-to-Treat Wall with External Insulation

Figure 8.9 gives an example of the construction of a hard-to-treat wall. Insulation cannot be introduced into the cavity, and it is not desirable to place it externally; the only option is to insulate on the internal face of the wall. Where walls are required to breath, it may be advisable to use a breathable insulation material such as mineral wool, sheep's wool, hemp or cellulose. Also, a breathable membrane is advisable to reduce the level of moisture entering from the internal environment whilst still allowing water vapour to escape from the wall.

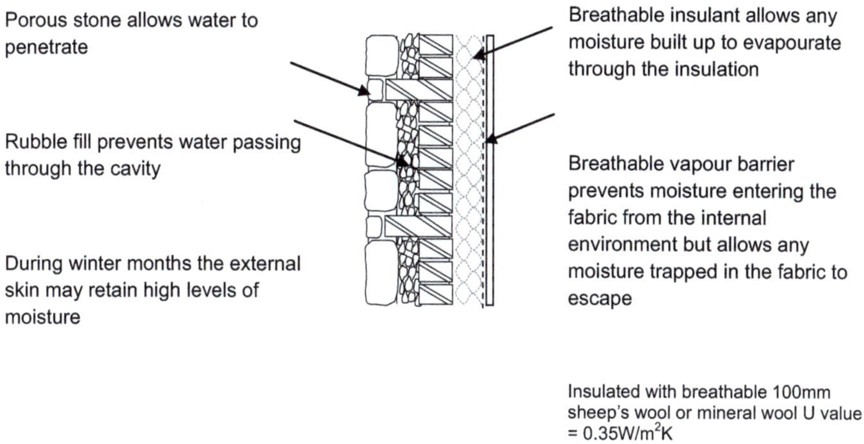

Fig. 8.9 Stone cavity wall, filled with rubble and insulated from the inside with cellulose or similar material (Thomas and Gorse 2013)

8.7.4 Internal Wall Insulation

Internal wall insulation can achieve a thermal envelope with less breaks than other methods if carried out well; however, fitting IWI is very disruptive to the buildings occupants. Internal floor space is also reduced, and any wall-mounted fixtures must be removed before fitting. Intermediate floor voids require additional attention and filling at perimeters to avoid the creation of a thermal bridge (Figs. 8.10, 8.11, and 8.12).

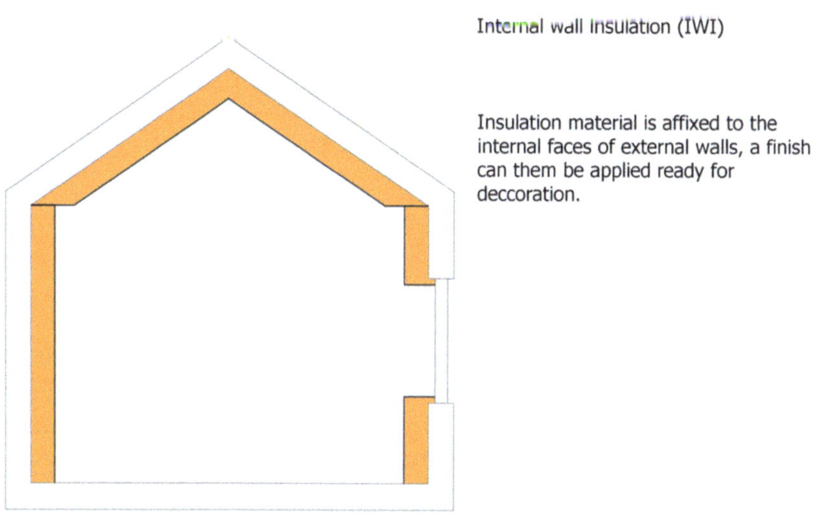

Fig. 8.10 Illustration of internal wall insulation (Thomas and Gorse 2015)

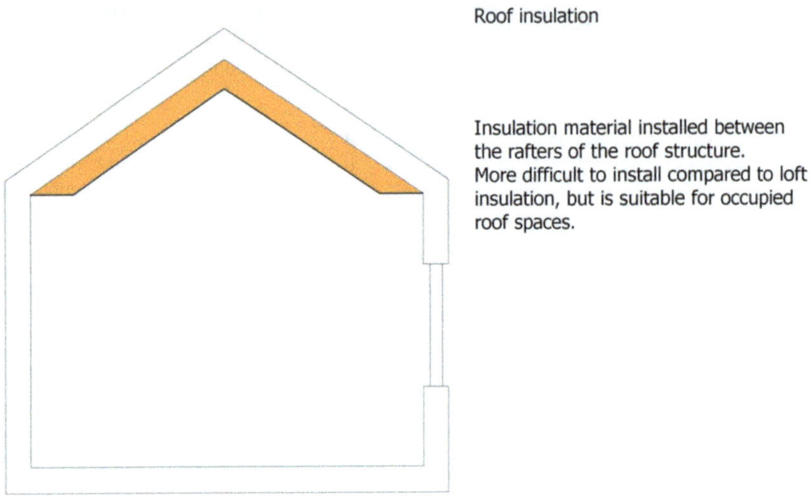

Fig. 8.11 Illustration of roof insulation. (Thomas and Gorse 2015)

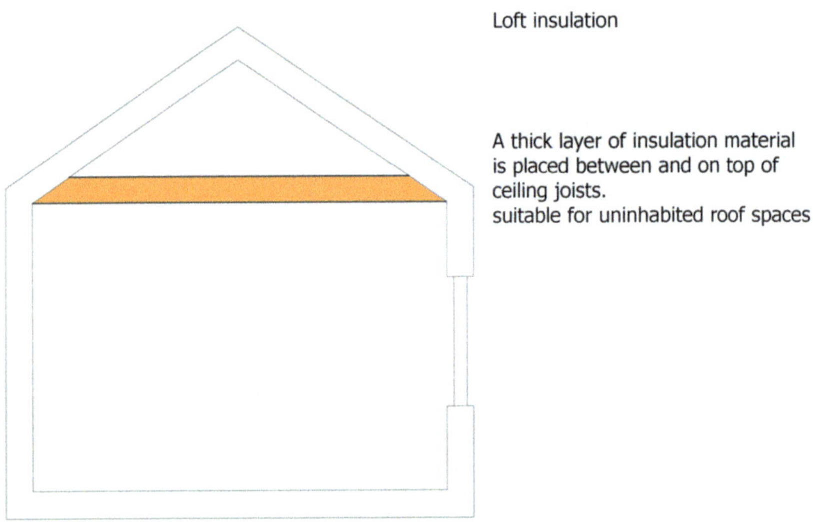

Fig. 8.12 Illustration of loft insulation. (Thomas and Gorse 2015)

Insulating the roof as well as the walls helps to create a continuous thermal envelope. The choice of loft or roof insulation will depend on the building undergoing retrofit, and particular attention should be given to ensure that insulation material is packed into the eaves, as failure to do so can lead to a thermal bridge.

8.8 Case Study: Temple Avenue-Prototype Retrofit Dwelling

This dwelling was part of the Joseph Rowntree Temple Avenue project. This 1930's semi-detached masonry cavity wall house underwent an extensive low-energy retrofit. The aim of this prototype was to investigate the potential performance improvements of retrofit on this typical UK house type (Fig. 8.13).

The retrofit achieved a measured level of fabric hat loss close to that achieved by the two new build prototypes in the project, though the as-build performance fell short of as-designed performance by a significant margin (37.6% more heat loss than predicted). This is partly due to limited survey data collected due to building occupation, and this lack of detail limited the accuracy of the estimated building performance modelling.

The design of the retrofit suffered from some problems, a lack of design detail in some areas necessitated solutions being developed on site, and this included repositioning of external drainage, the application of insulation around the bay win-

Fig. 8.13 Image of Temple Avenue retrofit property, post-retrofit. (Image by Leeds Beckett University)

dow and maintaining continuity of the air barrier around the windows. Insufficient specifications for tapes and sealants in the design documentation lead to the use of less suitable products on site.

The detailing of insulation around the ground floor left a thermal bridge around the ground floor exterior perimeter, which could have been reduced, had the external wall insulation continued below floor level. The fitting of loft insulation fully into the eaves proved to be problematic to achieve in practice, causing a thermal bridge.

The retrofit was carried out to a good standard, with much consideration given to thermal performance and air tightness; however, some of the work carried out proved to be problematic, and the blown fibre cavity wall insulation failed to provide even coverage and had to be remedied with further filling. It is unlikely that this would have been discovered if this not for the research team investigations. Additionally, the fitting of the MVHR system in the roof void was disruptive to loft insulation, mineral wool simply moved and displaced by duct work, requiring the contractors to make it good as best they could. Installation of the MVHR before loft insulation and better planning of duct placement could have reduced these problems.

For more information on this case study, see Miles-Shenton et al. (2011).

8.9 Life Cycle Assessment and buildings

8.9.1 What is Life Cycle Assessment (LCA)?

Life cycle assessment (LCA) is a term used to describe any attempt to measure the environmental impacts, commonly the greenhouse gas (GHG) emissions, of a product or a process. LCA is often used to articulate the impacts of construction projects, buildings and even building materials. The LCA methodology is defined by the ISO standard 14040-44 (ISO 2006). It has a relatively flexible framework made up of four essential steps, outlined in Fig. 8.14.

Life cycle costing (LCC) is a similar assessment method to LCA which considers only economic costs of products and processes, though it can include the monetisation of non-financial impacts and inputs. It is usually used in the construction industry to articulate the concept of payback time for an energy-efficient product or renewable technology; it does not describe the wider impact of these, whereas LCA considers *only* their environmental impacts. LCC is explored in great detail by Boussabaine and Kirkham (2004). Another alternative tool is *material* or *substance flow analysis* (MFA or SFA) which is expertly outlined in a *handbook of industrial ecology* (R. U. Ayres and Ayres 2002). These consider the quantities and types of materials used in a process or product's supply chain but stop short of converting these into environmental impacts, and LCA makes this final step.

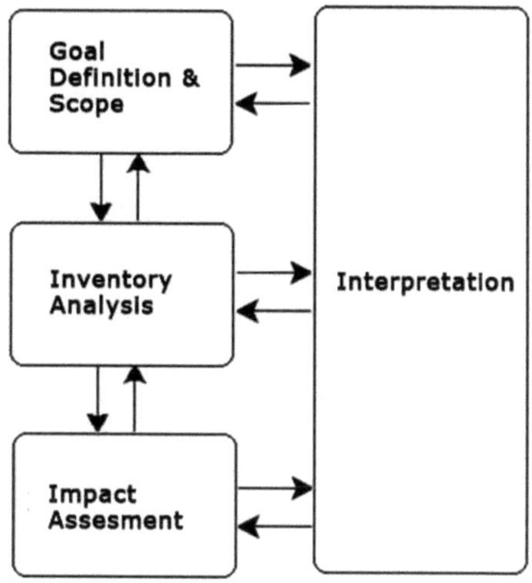

Fig. 8.14 LCA steps (ISO 14040 1997)

8.9.2 *Step 1: Scope and Goal*

The information an LCA provides should always be interpreted in the context of its scope and goal which sets out what will be looked at, what has been ignored and why. Often a schematic diagram or flow chart of the *system boundary* is used for clarity. This is important as it describes to the reader whether, for example, the LCA of a building includes only the inputs on the construction site (gate to gate), or if it includes the raw materials and processes required to make the components (cradle to gate), if it includes all the running costs until the building is demolished (cradle to grave) or finally if it also considers the reuse of the waste materials into new products (cradle to cradle). System boundaries can thus be simple or complex depending on the LCA goal, so long as all assumptions are explicitly identified.

For buildings, the dichotomy between embodied energy and energy use is noteworthy. Low-energy buildings are generally considered those which require little energy to support habitation (heat and power) and do not necessarily need account for the impact associated with the materials or processes consumed in its construction. This distinction is made in current UK legislation on energy-efficient buildings via the energy performance certificates (EPCs) and through the UK building regulations, neither of which require any consideration of embodied energy, only addressing energy consumption. LCA may be seen to be more comprehensive since it is concerned with impacts embodied in materials and processes as well as those associated with the use and disposal stages.

The final part of the scope is to set a *functional unit* against which all the data are reported. These may be simple, and more usefully, they are descriptive and incorporate specific conditions of time and space. Regarding time or *duration,* if building ‘A’ is reused beyond its expected or normal life, then this could be factored into a functional unit; for example, if building A is compared with a similar building 'B' which was demolished when it reached its recommended life, simply put, more durable buildings may appear to have lower impacts *ceteris paribus,* and this may be seen to justifiably or unfairly skew the intended LCA interpretation, depending on the scope and goal. Similarly, space or *location* should be factored into a succinct functional unit; a remote building will have more transport emissions to one located near raw materials *ceteris paribus,* yet this might also bias the LCA if this was not the intended subject to be investigated, for example if the assessment was only being done for the purposes of comparing two different building techniques. The functional unit is also an opportunity to identify the environmental variables of interest; for example, is use of water, GHG emissions or any other specific pollutants or impacts the main focus of the LCA?

It may in some instances be necessary to also include a component of utility or quality. For example, can one compare the carbon emissions of a high-achieving school which has lots of extracurricular activities, has a gymnasium, swimming pool, or breakfast club and is open most weekends, with that of a school that provides few additional facilities and which may be closed on weekends? Expanding on the granulation of what exactly is being assessed can reduce the chance of results being misinterpreted, but it is not always an easy task.

Thus, '*$KgCO_2eq$ per 1 m^2 of rural primary school classroom teaching space in the north of England over a 50-year life*' may be a more informative functional unit than '*the environmental impact of a school*' which leaves many unknowns and which could limit the ability to interpret the findings without a great deal of further investigation.

8.9.3 Step 2: Life Cycle Inventory (LCI)

Data on all inputs (e.g. kg concrete, kWh electricity, m^3 of water consumed) and outputs (e.g. kg waste or kWh electricity produced) that fall within the system boundary are recorded in the LCI and *normalised* against the functional unit, that is converted to the 'per m^2' or 'per building' being investigated. Primary data should be collected where practical; however, the use of some secondary data may be necessary. The LCI is a relatively straightforward task though it can be very time-consuming especially where primary data collection is undertaken.

8.9.4 Step 3: Life Cycle Impact Assessment (LCIA)

The LCI data are translated into environmental impacts in the LCIA using conversion factors generally sourced from other LCA databases such as Ecoinvent (www.ecoinvent.ch), which are licensed to practitioners or may be linked to LCA software such as openLCA[1], SimaPro[2] or GaBi[3]. Alternatively, freely available sources such as government reports and other published LCA may be used though these can be more time-consuming to collate, and in using multiple sources, one can introduce a greater number of assumptions regarding their data collection methodologies and therefore increase the degree of uncertainty in one's own LCA; therefore, there can often be no other realistic alternative. The source of these conversion factors is highly influential; thus, justifications should be made for using particular sources whilst excluding others. Alterations in the methodological choices and even the scope and goal may be necessary to accommodate the limitations of the LCIA data available.

8.9.5 Allocation

Within a construction project, there will be many inputs and impacts that are potentially shared across several buildings or projects. For example, timber used in construction may come from a sawmill that also sent its waste to the paper industry; groundwork undertaken in providing access roads for a new leisure centre may incidentally benefit a number of other buildings on or near the site; or a delivery van that is supplying bricks to site 'A' may also take a delivery to site 'B' on the same round trip. Deciding how to *allocate* the impact of processing the timber, laying the road or the emissions from the delivery van between the different products and processes is not necessarily straightforward; however, there are four accepted methodologies for doing so:

I. *Mass*: If the weight of bricks sent to site A is 70 % more than the bricks delivered to site B, site A should receive 70 % of the transport emissions.
II. *Economic: If the road to a new leisure centre which is valued at £ 900,000 also benefited two dwellings worth £ 50,000 each, the emissions resulting from the groundwork could be split into 90:5:5 between the leisure centre and the dwellings.*
III. *Energy: The calorific value of the timber going to site may be five times that of the waste wood going to produce books; thus, it should receive five times as much of the emissions.*
IV. *System expansion: If the road being constructed was part of a Sect. 106 planning requirement, it may be argued that its construction would have been required in the future anyway to provide adequate access to the dwellings and so the building of the leisure centre has offset the need to build the road later on; in this case, the emissions that would be expected by building an equivalent road to the dwellings can be subtracted from the total emissions of the leisure centre LCA (i.e. added as a credit).*

---

[1] www.openlca.org.

[2] www.simapro.co.uk.

[3] www.gabi-software.com.

8.9.6 Waste

If considering a 'cradle to grave' or 'cradle to cradle' LCA, impacts associated with waste management must be considered. Where building materials are sent to landfill, an emission or impact can be assigned to represent the transport to the landfill as well as the emissions from any biodegradation that may take place. Where recycling of a material has been removed from the building or construction site or if there are recycled materials in the building materials themselves, then it is necessary to incorporate these intricacies into the LCA. Two simple methods for doing this are outlined in the UK *PAS 2050 Specification for the assessment of the life cycle greenhouse gas emissions of goods and services* (BSI 2011).

- For **reusing** a product, the total lifetime and disposal emissions should be divided by the number of reuses, plus the emissions associated with each of its refurbishments, and uses should be added on top.
- For using a material with **recycled** content or for recycling a product, either i) the *cut-off method* is used where all the disposal emissions are given to the primary product, so the intermediary products only incur emissions associated with the recycling; or ii) the *closed loop method* may be selected where the overall disposal emissions are shared equally across all products and uses (similarly to the rules for reusing).

In addition, it is also possible to employ the *substitution method* (similar to the system expansion method of allocation) whereby in addition to the costs being distributed across all the products in the recycling chain, a credit may also be allocated for offsetting the use of virgin resources. If where the product being recycled loses a degree of quality (e.g. bricks used for hard core), then it is recommended to split emissions or credits between uses accordingly, that is the *loss of quality method,* perhaps using market value per kg as a proxy for quality. Conversely, the first and last products may be assigned all the burdens and credits irrespective of the number or quality of the intermediary uses, that is the *50:50 method.*

8.9.7 Step 4: Interpretation

The final stage of the LCA is to present the results relative to the original scope and goal of the LCA. This can be done in several ways; however, the conclusions should be made with reference to the aim of the LCA. For example, if the functional unit discusses only GHG emissions per m^2 of teaching space of a specific school, it should be highlighted that the LCA should not be used to compare schools with different facilities, for example schools with swimming pools.

Inevitably, there will be uncertainties within an LCA based on the data quality and comprehensiveness, methodological assumptions made when dealing with allocation and waste as well as limitations of the chosen system boundary and functional unit. It may therefore be necessary to undertake sensitivity analyses to explore the influence of major areas of uncertainty so that a degree of certainty or robustness can be identified.

8.10 Future of LCA

The four steps outlined leave much room for variability of method and many limitations in data collection requirements. This allows flexibility for the users to suit their own situations; however, such specificity can limit the ability to use LCA as a comparison tool. This can be problematic when setting emission standards for products, which is common in the form of legislation. Harmonising the rules and data sources is one means of gaining greater control in LCA, and such an approach has been adopted by the EU for setting emission limits for biofuels. Harmonisation may therefore play a role where LCA is required for certification and validation. Harmonisation is presently used in legislation affecting the construction via EPCs. However, harmonised LCA should not be used to answer specific questions for particular products and processes; the use of EPC thus for predicting specific carbon reduction values has already caused significant confusion and mistrust in the industry, and the same danger exists for LCA; thus, any future harmonisation must be used wisely for it to be a success.

As mentioned, the availability of data is often a limiting factor in LCA and it is not always clear where in a supply chain one can stop counting the impacts; for example, should the marketing and advertising impacts of a construction company be included in overall LCA of the building? The form of LCA discussed so far is termed *process LCA*. In process LCA, there inevitably is a point beyond which no more data can or will be collected; often, this is where additional inputs constitute less than 1% of the overall inputs. This is referred to as *truncation*. Truncation in LCA results in the omission of impacts from a process LCA which can range from the insignificant to tens of percentage increases; thus, a solution to address this has been proposed: *input–output* (IO) LCA (Suh 2009).

IO takes advantage of economy-wide reporting requirements on companies, such as those imposed on UK companies by the UK government and collated by the Office of National Statistics. Where a company's GHG emissions are reported alongside the amount of money (pounds) each sector spends on another, one can understand the emissions per pound spent on different sectors. Thus, if one knows how much money is spent on a company or sector when making a product, one can understand the average emissions that are incurred; in this way, LCA of entire products can be built up. Clearly, this only allows for average emissions to be investigated and so specific investigations cannot be undertaken, for example comparing different building materials or construction techniques. It also means that the LCA will be influenced by prices, which may be volatile and subject to fluctuations unrelated to the supply chain impacts. Merging *process* and *IO* LCA into a *hybrid* LCA is a relatively new development which has been proposed as a means of providing specificity and completeness in one assessment, though this is a relatively new form of LCA (Kagawa 2012); thus, its success will not be known until its uptake is more widespread.

8.11 Summary

LCA can be used to identify the impacts (GHG emissions, water use, etc.) of all types of construction projects and products. There is some standardisation in the rules of LCA to ensure that results are robust; however, these are flexible and so there are still many uncertainties within each LCA. The scope and goal and interpretation steps are therefore important to justify the method and provide context for the results, and also to identify whether any scenario or sensitivity analysis is needed. LCA is a mainstream tool with which to evaluate the impacts of buildings, and it is likely to become more so in the future. It is therefore important that practitioners in the construction industry are aware of and understand how to use and interpret LCA.

References

Akadiri PO (2011) Development of a multi-criteria approach for the selection of sustainable materials for building projects, Thesis, University of Wolverhampton

Alwaer H, Clements-Croome DJ (2010) Key performance indicators (KPIs) and priority setting in using the multi-attribute approach for assessing sustainable intelligent buildings. Build Environ 45(4):799–807

Ayres RU, Ayres LW (2002) A handbook of industrial ecology. Edward Elgar Publishing Limited, Cheltenham

Berardi U (2013) Clarifying the new interpretations of the concept of sustainable building. Sustain Cities Soc 8(0):72–78

Berge B (2000) The ecology of building materials—English edition. Architectural Press, Oxford

Boussabaine A, Kirkham R (2004) Whole life-cycle costing: risk and risk responses. Blackwell, Oxford

BSI (2011) PAS 2050:2011 Specification for the assessment of the life cycle greenhouse gas emissions of goods and services. British Standards Institute

Denson J, Halligan C (2010) Building materials and the environment: second edition, Stephen George & Partners. http://www.stephengeorge.co.uk/index.php?option=com_k2&view=item&id=36:sustainability&Itemid=38. Accessed June 2015

Green Building Store (2013) Denby Dale Passivhaus technical briefing. http://www.GreenBuildingStore.co.uk/page-denby-dale-passivhaus-technical-briefing.html. Accessed June 2015

Halliday S (2008) Sustainable construction. Butterworth-Heinemann, Oxford

ISO (2006) ISO 14040:2006 environmental management—life cycle assessment—principles and framework. ISO1400. International Organisation for Standardisation.

Johnston D., Famer D, Brooke-Peat M, Miles-Shenton D (2014) Bridging the domestic building fabric performance gap. Building Research and Information

Kagawa S (2012) Frontiers of environmental input-output analysis. Routledge, Oxon

Miles-Shenton D, Wingfield J, Sutton R, Bell M (2010). Temple avenue field trial—evaluation of design & construction process and measurement of fabric performance of new build dwellings. Temple avenue Project Part 1, Leeds Metropolitan University, Leeds

Miles-Shenton D, Wingfield J, Sutton R, Bell M (2011) Energy efficient renovation of an existing dwelling: evaluation of design & construction and measurement of fabric performance. Temple avenue project. Leeds Metropolitan University, Leeds

Slaper TF, Hall TJ (2011) The triple bottom line: what is it and how does it work? Indiana business review, Indiana University: Indiana Business Research Centre. http://www.ibrc.indiana.edu/ibr/2011/spring/article2.html. Accessed June 2015

Suh S (2009) Handbook of input–output economics in industrial ecology. Springer, New York

Thomas F, Gorse C (2013), Wall types and changes in construction; design and good practice, LSI technical paper 2013-TP-0005. Leeds Sustainability Institute

Thomas F, Gorse C (2015) An illustrated review of the factors contributing to differences between designed performance and in-use energy demand: LSI technical paper 2015-TP-0001. Leeds Sustainability Institute. http://www.leedsbeckett.ac.uk/as/cebe/resources.htm. Accessed June 2015

UKGBC (2014) Key statistics. Accessed 9 Sept 2014

Chapter 9
Building Simulation and Models: Closing the Performance Gap

Christopher Gorse, Matthew Brooke-Peat, James Parker and Felix Thomas

It is not practicable to test every aspect of all the buildings that are built. As we understand the behaviour of buildings from field and laboratory tests, the data can be used to produce generalised assumptions about the way a building and its component parts will behave. These models simulations are now an integral part of our understanding of the performance of buildings. While the assumptions made in models and simulations can be relatively imprecise when first developed, as their development is advanced the models become more detailed, reliable and intelligent. Researchers are constantly updating and calibrating the sensitivity of their models, using new data from the field and in-use studies to improve the reliability and accuracy with which the models can operate. The construction industry is heavily reliant on the use of models and simulations to perform a variety of design and analysis calculations, for predicting energy consumption and performance of finished buildings, and to demonstrate compliance with regulatory or voluntary performance standards (Fig. 9.1).

9.1 Building Performance Simulation and Models: Climate Change

Recently, scientific consensus has confirmed that man-induced climate change is inevitable and that emissions of greenhouse gases, particularly carbon dioxide (CO_2), need to be reduced to avoid the most extreme problems (Intergovernmental Panel on Climate Change 2013). Due to the size and scale of its impact, the built environment is considered to have the greatest scope for reducing CO_2. Built environment emissions accounts for approximately 35 % of the UK's total carbon footprint (Intergovernmental Panel on Climate Change 2007; Carbon Trust 2009). Thermal models and building

C. Gorse (✉) · M. B.-Peat · J. Parker · F. Thomas
Leeds Sustainability Institute, Leeds Beckett University, Broadcasting Place, Woodhouse Lane, LS2 9EN Leeds, UK
e-mail: C.Gorse@leedsbeckett.ac.uk

© Springer International Publishing Switzerland 2016

M. Dastbaz et al. (eds.), *Building Sustainable Futures*,
DOI 10.1007/978-3-319-19348-9_9

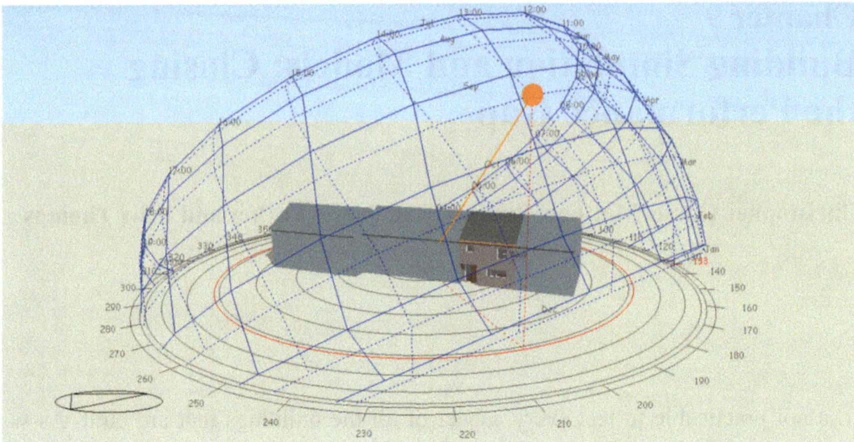

Fig. 9.1 Understanding and simulation building performance

performance simulation already play an important part in delivering more sustainable designs and buildings; furthermore, their use and development remains important in the rapidly growing field of building performance research. The models and simulations can be used at a micro- and macroscale. For example, hygrothermal models are used to look in detail at the movement and development of moisture through elements and thermal models to look at heat flow through junctions (Figs. 9.2, 9.3).

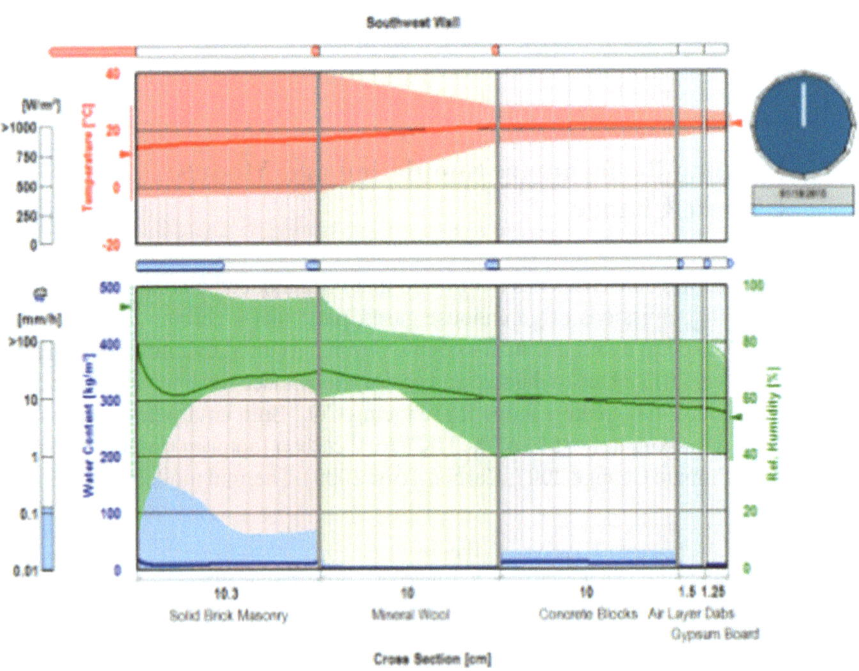

Fig. 9.2 Hygrothermal analysis: Temperature and water profiles within an insulated cavity wall

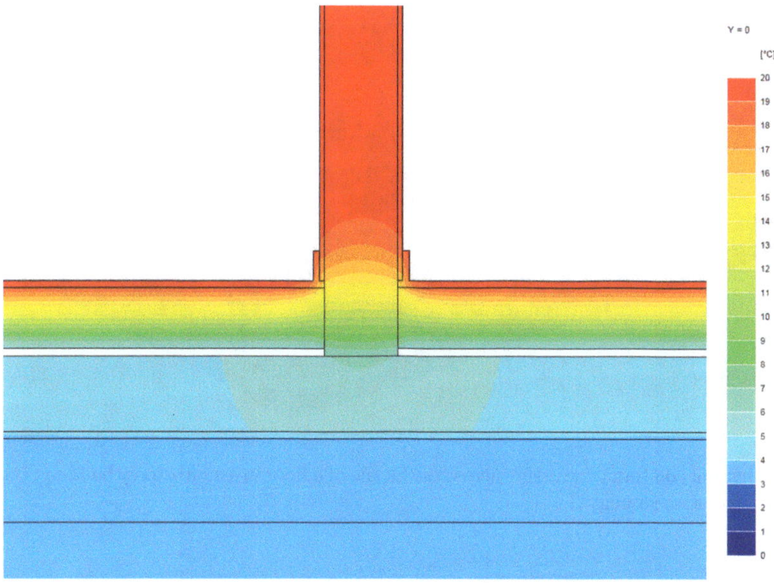

Q_1:	165.5655	w:	4.000	U_f:	0.159	T_{si}:	18.06
Q_2:	150.6437	t_{wall}:	0.238			**f_{min}:**	**0.903**
T_i:	20.00					**Ψ_c:**	**0.187**
T_e:	0.00					Ψ (W/m·K):	**0.224**
T_u:	4.99					**Ψ_{app}:**	**0.112**

Fig. 9.3 Thermal model of junction between party wall and ground floor

Building simulations are developing in both their power and flexibility. They play an important part in ensuring the detail and junctions perform as intended and the simulations ensure, as far as is practicable, the buildings are appropriately equipped for their use and exposure.

Building simulation is an important tool in the design and development of low-energy buildings. Simulations are at a microscale in the analysis of thermal comfort (Fig. 9.4) and can encompass surrounding buildings and features at what may be considered a macrolevel (Figs. 9.5, 9.6). District and city scale simulations are used to look at wind patterns, solar exposure, heat island effect and other interaction between building and the environment.

Architects, designers and engineers can now create highly detailed models of building performance due to cumulative and the ongoing advances in computational capacity and wide-ranging software innovations. Almost all elements of building performance can now be modelled from the structural integrity of an entire building to the fire escape strategy of a given facility. Building information modelling has the potential to revolutionise the design, construction and operation of facilities

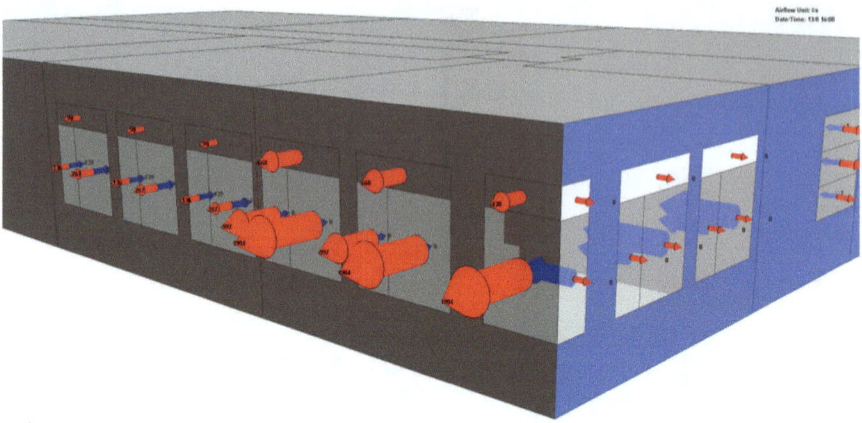

Fig. 9.4 Thermal comfort: air pressure across the façade of a floor in a multi-storey building. Used to inform ventilation strategy

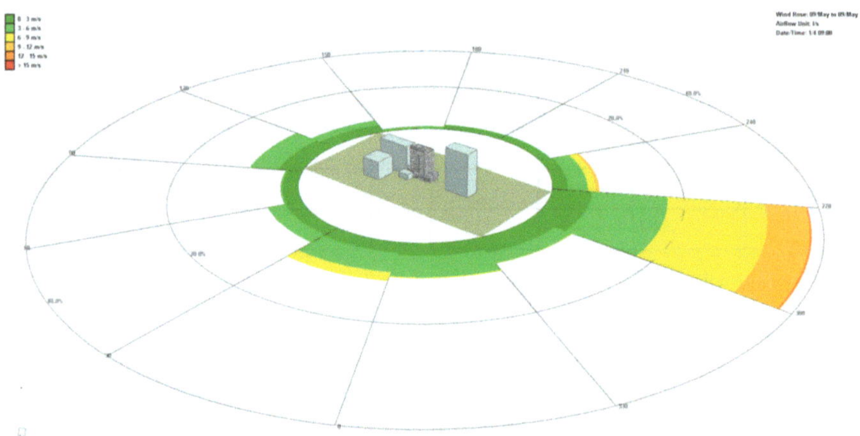

Fig. 9.5 Building simulation, used as part of a study on the influence of exposure to wind on buildings and affect on cross-ventilation

and has the inertia to result in performance simulation and computational analysis becoming common place for not just large complex non-domestic developments but for small scale domestic work as well. Engineering practice and research in the field of energy performance simulation has grown on a global scale and is represented by The International Building Performance Simulation Association (IBPSA 2015). IBPSA produces a number of resources and publications focused on building simulation.

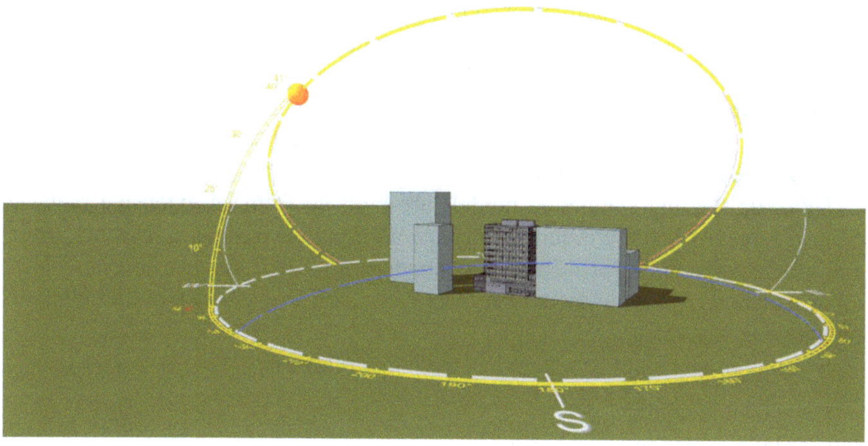

Fig. 9.6 Building simulation, solar exposure and heat gain, part of a study on thermal comfort and overheating

9.2 Dynamic Simulation Modelling

The following text specifically focuses on what is commonly referred to as dynamic simulation modelling (DSM) but is possibly more correctly described as dynamic thermal simulation. Due to the complexity of modern construction techniques and building systems, the challenges to reduce energy consumption are becoming more critical; traditional engineering design tools have a limited application; they do not sufficiently mimic the dynamic conditions encountered in reality, they assume static or steady-state parameters and create a simplistic version of what is an infinitely complex universe. Dynamic simulation modelling allows for a more sophisticated analysis of the conditions encountered in reality:

> Computational building performance modelling and simulation… is multi-disciplinary, problem-orientated and wide in scope. It assumes dynamic (and continuous in time) boundary conditions, and is normally based on numerical methods that aim to provide an approximate solution of a realistic model of complexity in the real world.
> Hensen and Lamberts 2011, p. 3

Despite this, it is important to understand that models are just that; models. It is not technically achievable or practical to exactly replicate the complexity of the real world; an effective and efficient model should only be as complex as is necessary. For example, the geometry of a thermal model used to analyse a fully air-conditioned building does not need every individual window to be added to each individual space; only the percentage of glazing for each given space needs to be represented. However, if the building in question is naturally ventilated, each individual window unit needs to be modelled and inputs controlling the window opening angle and schedule need to be added.

Dynamic simulation modelling can be used to predict many aspects of building performance including heating and cooling loads, energy consumption and CO2 emissions, ventilation requirements and effects, indoor thermal conditions, human comfort levels, daylight conditions and heating, ventilation and air conditioning (HVAC) system performance. Some (DSM) software applications are now also approved to produce regulatory compliance calculations and can be used to design too and provide evidence for a range of statutory requirements and sustainable architecture standards around the world.

Dynamic thermal simulation models are built up from an extensive range of input parameters that are required to create a virtual environment; it is this virtual environment that provides the boundary conditions for performance analysis. These input parameters provide the source data for a series of interactive and complex building physics calculations.

External model input parameters include the following:

- site global location
- detailed simulation weather files
  dry-bulb temperature
  wet-bulb temperature
  wind speed and direction
  direct, diffuse and global radiation
  atmospheric pressure and relative humidity
- building-specific parameters control values for building geometry
- construction materials
- heating, ventilation and cooling system
- air exchanges
- internal heat gains from equipment, lighting and occupants

Data outputs from the simulation files are extensive.

At a building level, thermal loads can be calculated and energy consumption/ CO2 estimates can be made for fuel types, broken down to end uses (for example boiler or chiller energy) if necessary. At a room level, a wide range of parameters can be analysed including thermal properties, ventilation rates, comfort levels and daylighting provision. At an elemental level, outputs can predict properties such as surface temperature or heat transfer. The full range of outputs will depend on the specific model/software that is being used.

Dynamic thermal simulation models have a wide range of applications from simple regulatory compliance calculations to advanced estimates of operational performance in large complex buildings. Advanced modelling techniques can be used to accurately predict thermal performance in both new build and retrofit projects; they can also be used to simulate day-lighting and complex air movement. Some DSM software is approved by the UK Government for the production of Building Regulation Part L compliance calculations and production of Energy Performance Certificates (DCLG 2011); more sophisticated models can be produced in the design and evaluation of energy efficient, low carbon buildings (Carbon Trust 2011). When used as part of an iterative design process, building simulation can provide initial energy performance estimates and these can be refined as the design is developed throughout the procurement process.

9.3 Regulatory Compliance and the Performance Gap

In the UK, the Department for Communities and Local Government (DCLG) legislates that non-domestic buildings that are newly built, being sold or being let have an Energy Performance Certificate (EPC) rating (DCLG 2008a); the requirement is in accordance with European Directive 2002/91/EC. It is also a mandatory requirement that all new or extensively refurbished buildings comply with Part L of UK Building Regulations (HM Government 2010a; HM Government 2010b). Compliance with these policy instruments requires the energy performance of a given facility to be estimated using the simplified building energy model (SBEM) which is underpinned by the National Calculation Method (DCLG 2011).

Public buildings with a floor area over 1000 m^2 are now legally obligated to display benchmarked and measured operational energy performance under the Display Energy Certificate (DEC) scheme (DCLG 2008b). Comparison of EPC performance estimates with DEC results and more detailed monitoring information has identified significant differences between the predicted and operational performance of non-domestic buildings in recent years (Burman et al. 2012; Carbon Trust 2011; Parker et al. 2012; Menezes et al. 2012); this difference has become known as the 'performance gap'. There are numerous reasons for this (partly perceived) gap including inaccuracies in modelling techniques, unregulated loads, out of standard hours operating, commissioning and management issues and the as-built performance of building elements (Carbon Trust 2011).

As the existence of a performance gap has become acknowledged, more research projects have been commissioned with the intention of identifying contributing factors (Menezes et al. 2012). Primarily, DSM software is used in the analysis of non-domestic buildings and is only approved for regulatory compliance calculations in this type of facility; the remainder of this section therefore focuses on the performance gap in non-domestic buildings. The phrase 'performance gap' in the built environment is thought to have been coined during early studies through the Post-occupancy Review of Buildings and their Engineering (PROBE) project which conducted detailed studies of 23 buildings; this study found that the majority of these used twice as much energy in practice as estimated at design stage. This early study of this phenomenon concluded that a lack of feedback between building operators and designers contributed towards the gap which is a common finding in this field of research to date (Bordass et al. 2001). The Low Carbon Buildings Accelerator and the Low Carbon Buildings Programme are more recent examples. Both of these projects are referenced by the Carbon Trust in a report aimed at construction clients interested in the procurement of low carbon buildings (Carbon Trust 2011). This report asserts that energy consumption can be up to five times higher than predicted in some of the worst cases. An industry led project, CarbonBuzz, is designed to provide an online platform for sharing of performance data. It has recorded an average gap in performance of 114 % (Kimpain and Chisholm 2011). Unfortunately, commercial pressures have restricted the amount of available data that is available in terms of predicted and actual performance. Although there seems to be a collective desire to produce more sustainable buildings and designs, many organisations are reluctant to share performance data that cast a bad light on their architectural designs, engineering or physical assets.

Some of the discrepancy between estimated and actual energy consumption has been described as a 'perception gap' (Austin 2013). This is due to the perceived idea that Building Regulations Part L and EPC calculations account for total energy consumption. Part L and EPC calculations only account for energy consumed by space conditioning and fixed lighting and do not account for 'unregulated' loads (Austin 2013; Carbon Trust 2011). These unregulated loads include plug loads (equipment), server rooms, security systems, vertical circulation and external lighting systems. This is particularly important in non-domestic buildings as unregulated equipment energy use can account for as much as 40 % of total CO2 emissions (Parker et al. 2012; Jenkins et al. 2009).

A summary report of the Technology Strategy Board's Building Performance Evaluation Programme (BPEP) describes preliminary results from 29 non-domestic buildings. In 26 cases, the Building Emission Rate (calculated for Part L compliance) accounts for less than 25 % of operational CO2 emissions (Cohen 2013). Quantitative data are being collected in all BPEP buildings using a network of sub-metres and building management systems (BMS). A paper from 2012 describes results from the analysis of a school building included in the BPEP (Burman et al. 2012). It notes that unregulated loads account for 41 % of total consumption. Sub-meters and data taken from the BMS allowed operational problems to be identified. In this example, the Ground Source Heat Pump system was not functioning correctly, demand control ventilation was not operating, and the inverters were not enabled (Burman et al. 2012). Certain assumptions made in the SBEM calculations relating to heating set points, auxiliary power and lighting density contributed to the performance gap in this example.

9.4 Reducing the Discrepancy in Modelling and Simulation

There has been criticism of the accuracy achievable by simulation software (Raslan and Davies 2010) but results from the case study noted above demonstrated that DSM models can be calibrated to closely simulate actual performance. It has further been proposed that '…the impact of modelling tools on the overall discrepancy between predicted and actual performance is constantly being diminished'. (Menezes et al. 2012, p. 356). It is mainly assumptions that made in regulatory compliance calculations account for the modelling-related gap in performance (Austin 2013; Burman et al. 2012; Cohen 2012). This provides further evidence that when used as a detailed design tool, DSM can produce accurate predictions of energy performance and aid the low-carbon design of buildings.

It has been proposed that a large proportion of the performance gap in non-domestic buildings is due to equipment loads not accounted for in regulatory calculations (Austin 2013; Burman et al. 2012; Cohen 2012; Menezes et al. 2012). This is exacerbated by out-of-hours operations (Menezes et al. 2012). A detailed case study of an office building shows actual electricity consumption to be three times higher than predicted by SBEM calculation. However, when using a predictive model to estimate electricity consumption based upon realistic equipment consumption and operating profiles, the estimate is within 3 % of the actual observed consumption (Menezes et al. 2012). Due to the significant role that these unregulated loads have in the non-domestic sector, the majority of research has focused on this and post-occupancy evaluation studies of buildings in use.

There is a consistent message in published research that examines the performance gap; a lack of effective feedback maintains the gap between actual and predicted performance. There is however scope to reduce the gap by feeding back performance data to designers and engineers. In turn, the data inputs and performance estimates from regulatory compliance calculations can be enhanced and improved by taking full advantage of the capabilities of DSM software.

9.5 Model Accuracy and Calibrated Whole Building Simulation

The very nature of a model is abstract, and they represent a simplified version of a complex reality. This has long been recognised by scientists in many fields of research and has been succinctly summarised in the past: 'Remember that all models are wrong; the practical question is how wrong do they have to be to not be useful?' (Box and Draper 1987, p. 74). Leading DSM software is validated for use against various global standards; the most prominent of these are the Chartered Institute of Building Services Engineers (CIBSE) 'TM33: Tests for Software Accreditation & Verification' in the UK and the American Society of Heating, Refrigeration and Air Conditioning Engineers (ASHRAE) 'Standard 140-2007: Standard method of test for the evaluation of building energy analysis computer programs' (CIBSE 2006; ASHRAE 2007). Using these benchmarks, software is validated against known parameters and compliance with these standards provides independent assurance that the software will produce accurate results when predicting the energy and thermal performance of buildings. For existing buildings, it is possible to calibrate the models against measured performance. Accuracy of calibrated models is most often checked by calculating the mean biased error (MBE) and coefficient of variation of the root mean squared error (CVRMSE) which quantify the differences between metered utility data and baseline simulation estimates. When using monthly metered utility data, ASHRAE Guideline 14 considers a model to be calibrated when the MBE is within 5 % and the CVRSME is within 15 % of metered data (ASHRAE 2002); when using hourly data, this extends to 10 % and 30 %, respectively.

Building simulation calibration has been described as the '...process of using an existing building simulation computer program and 'tuning' or calibrating the various inputs to the program so that observed energy use matches closely with that predicted by the simulation program' (Reddy 2006, p. 227). Calibrated models can be used to inform investment-grade decisions for deep retrofit projects; this a widely acknowledged function of calibrated models (Reddy 2006; Raftery et al. 2011a; Heo et al. 2012). Outputs from the simulations can be used in financial appraisal models, supplying the base data to complete sophisticated life-cycle assessment calculations that use metrics such as Net Present Value (NPV) or Internal Rate of Return (IRR). In scenarios that include multiple, interactive retrofits, a package of different retrofit technologies or long-term retrofit pathways, whole building simulation is a particularly useful technique.

Calibration of building energy performance models is the focus of a considerable body of academic research. In 2003, a research project was commissioned by ASHRAE with the aim of developing a '…coherent and systematic calibration methodology…' (Reddy 2006, p. 225). A broad literature review of the existing techniques was one of the main outputs of this work (Reddy 2006). This has more recently been developed by Coakley et al. (2014) to encompass an extensive range of calibration techniques and tools. Methodologies were originally divided by Reddy into four main groups, and despite advances in the field, Coakley et al. retain the same four categories: (a) manual, iterative and pragmatic interventions; (b) suites of informative graphical comparative displays; (c) special tests/analytical procedures; and (d) analytical and mathematical methods. However, they also provide a simplified categorisation of calibration approaches these being classified as either 'Manual' or 'Automated' techniques.

These alternative approaches to model calibration rely on a range of data collection and analytical tools. For example, hourly data sets are used in approach (b) and error ranges are visualised to identify aspects of the model/outputs that need refinement. Approach (c) uses intrusive 'blink-tests' where groups of end-use loads are turned on and off to determine consumption; this test data are then used to refine model inputs (Shonder et al. 1997; Soebarto 1997). Another approach in this category, short-term energy monitoring (or STEM tests) uses a similar process to refine model inputs (Subbarao 1988; Manke and Hittle 1996). Analytical/mathematical optimisation calibration techniques can produce inaccuracies due to the heterogeneous nature of buildings and limited number of parameters normally used in the process (Reddy et al. 2006; Raftery et al. 2011a). A robust mathematical approach based upon Bayesian regression for normative models has been shown to be as accurate as a calibrated transient simulation (Heo et al. 2012). This type of calibration can however be computationally expensive and time-consuming for larger models. Limitations can restrict the accuracy of the first iterative approach (a) although work has been done to refine this type of calibration process (Hunn et al. 1992; Yoon et al. 2003).

The manual iterative approach involves analysts updating models in stages by adjusting appropriate parameters until the required level of accuracy is reached. This approach is often based upon an analyst's own knowledge and understanding which can sometimes be applied in an ad hoc manner (Reddy et al. 2006). A recent improvement on the iterative approach uses the concepts of 'calibration signatures' and 'characteristic signatures' (Claridge 2011). Calibration signatures are calculated from the differences in simulated and measured consumption and characteristic signatures are calculated from the changes in consumption against a baseline simulation from a similar reference-building type. This approach is ideally suited to calibrating the measured and simulated heating and cooling system energy consumption. The signature values are plotted against a specified temperature range to allow errors to be visualised.

Contemporary research has sought to define standard systematic methodologies for calibrated simulation which use a combination of the four approaches previously defined plus a higher resolution and range of metered data from the subject building, most accurately using hourly end-use records (Reddy et al. 2007a; Reddy et al. 2007b; Raftery et al. 2011a; Raftery et al. 2011b). As mentioned previously, the work of Coakley et al. (2014) provides an extensive review of contemporary calibration techniques and tools. Despite there being a wide range of theoretically sound approaches, a number of issues still limit the use and accuracy of calibrated models, namely a lack of consensus on standard approaches; expenses associated with model development and data collection; simplification of model inputs; quality of model inputs; uncertainty relating to approximated and simplified inputs; identification of errors; and the lack of automated processes (Coakley et al. 2014). In practice the most common barrier to effective calibration is the sheer number of inputs required, which can be in the thousands for relatively simple buildings. Model calibration is described by Coakley et al. as a problem that is '…over-specified (i.e. too many data points) and under-determined (i.e. too few validation points)'. (Coakley et al. 2014, p. 135). Ultimately, the choice of calibration approach relies heavily on the data that are available for a specific building.

9.6 Elemental Modelling

9.6.1 Thermal Bridging

Revisions to the Building Regulations (Department of the Environment and the Welsh Office 1995; Office of the Deputy Prime Minister 2002, 2006; HM Government 2010a, 2010b, 2013) have successively increased the required thermal performance of the building envelope. As the thickness of insulation increases to the plane elements of buildings in an attempt to reduce heat loss, the effect of thermal bridging at the junctions between these elements can increase proportionately (Janssens et al. 2007; Berggren and Wall 2013). See Fig. 9.7 for an illustration of plane elements, thermal bridges and junctions.

The thermal bridges occur as a result of both building geometry and the way building materials are assembled. The thermal bridges that result from assemblies often occur due to the interface between different components, such as window and door lintels, jambs and typically elements used to add structural reinforcement or make structural connections, or where materials simply pass through the insulated

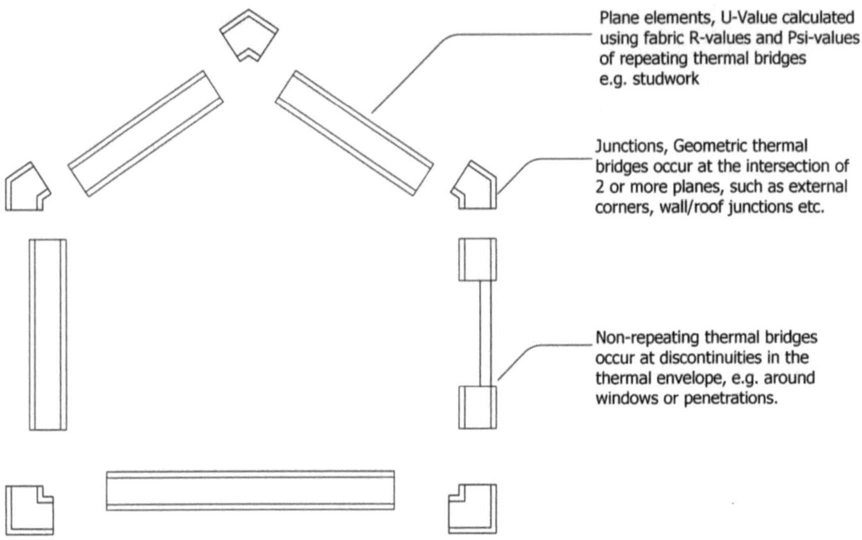

Fig. 9.7 Plane elements, junctions (geometrical thermal bridges) and thermal bridges. (Thomas and Gorse 2015)

fabric without having the benefit of the same insulation properties (Fig. 9.8). With careful detail design, the impact of such interfaces can be reduced.

Geometric thermal bridging is where the internal and external surface areas of the thermal envelope differ in size at the location of each junction (Janssens et al. 2007). A geometric thermal bridge with the potential for increased heat loss occurs at internal corners or curves, where the internal surface area is less than the external surface exposed. The increased exposure and material through which the heat transfer can take place lowers the thermal resistance, and provides additional paths for heat flow to the external environment, in cold climates.

The U-values of plane elements primarily consider heat transfer as unidirectional to the main envelope areas, whereas heat flow is multi-dimensional at thermal bridges (Ward and Sanders 2007). A geometric thermal bridge is shown in Fig. 9.9 that represents a temperature distribution of a normal corner in plan with the heat flow lines indicated. The effect of the thermal bridge in these instances can be positive or negative depending on whether the convention for measuring the heat loss area is based on internal (*ibid*) or external dimensions (Waltjen 2008), and also if the junction forms an internal or external corner in the building fabric (Ward and Sanders 2007). A constructional thermal bridge is where a material of high thermal conductivity penetrates a material of low thermal conductivity, thus creating discontinuity to insulating layers (Janssens et al. 2007). An example of a temperature distribution for a constructional thermal bridge is shown in Fig. 9.9 where a blockwork party wall extends into a cold roof void that is insulated at ceiling level. As a result of the higher rate of heat transfer at thermal bridges, a localised reduction in internal surface temperature occurs (Ward 2006). Thermal bridging must

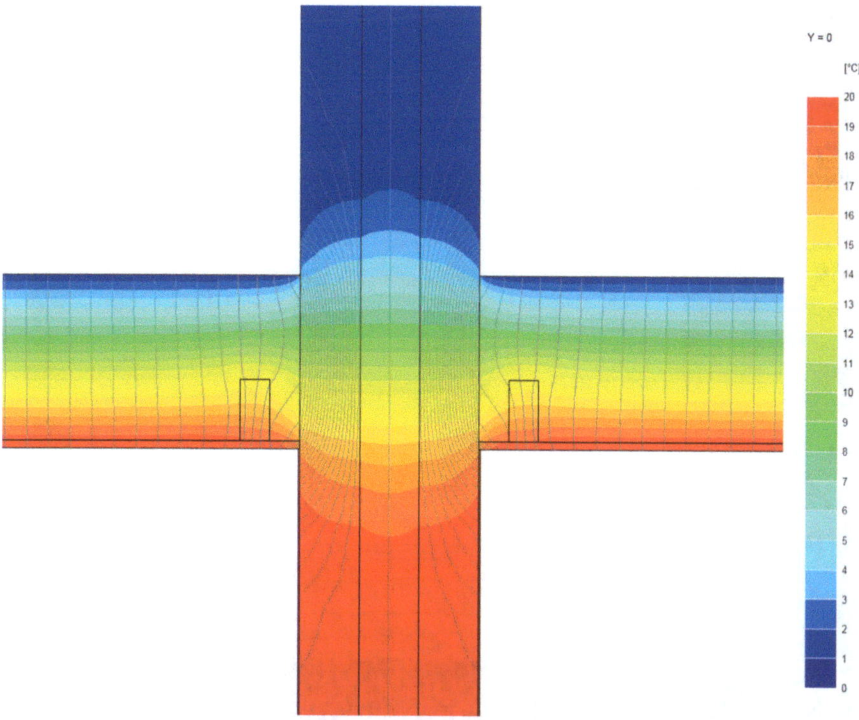

Fig. 9.8 Constructional thermal bridge. A blockwork party wall extends into a cold roof void that is insulated at ceiling level

be accounted for in order to make accurate predictions of thermal performance for design and regulatory purposes that seek to minimise energy consumption and limit the risk of surface condensation and mould growth.

Thermal bridges that occur at regular intervals within the main areas of building fabric (e.g. studwork) are known as repeating thermal bridges, and these are accounted for in the U-value calculations for plane elements (Ward 2006). Non-repeating thermal bridges require the undertaking of thermal bridging calculations to obtain the linear thermal transmittance (Ψ-value) and minimum temperature factor (f_{Rsi}). Point thermal transmittance (χ-value) is calculated for constructions with individual and isolated penetrations of insulation layers (*ibid*). The thermal transmittance of thermal bridges (Ψ-value or χ-value) is used for whole building heat loss accounting purposes (Ward 2006). The f_{Rsi} is a metric that is used to assess the potential risk of surface condensation and mould growth based on the predicted lowest internal surface temperature (*ibid*). Thermal bridging calculations are underpinned by a numerical modelling technique known as thermal modelling (Ward and Sanders 2007). This is undertaken using computer software to model multi-dimensional heat flow and predict surface temperatures. Thermal modelling software inputs include the geometry of the junction, the thermal conductivity of the materials incorporated in the construction and the boundary conditions to the environments

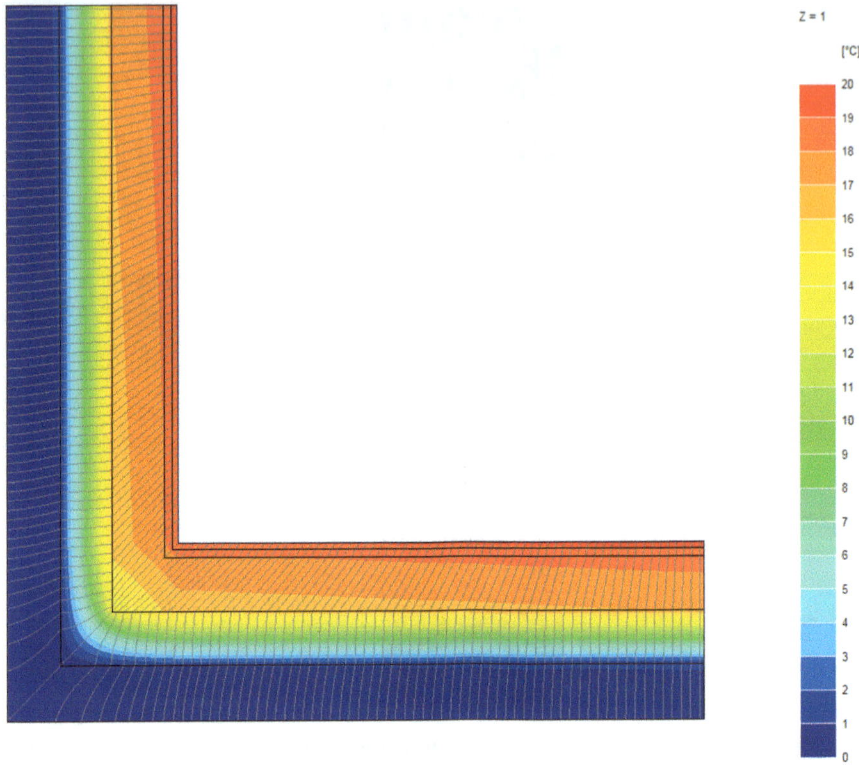

Fig. 9.9 A geometric thermal bridge showing a temperature distribution of a normal corner in plan with the heat flow lines

surrounding the subject of the model. Both two- and three-dimensional software programs are available, and these model either steady state or dynamic conditions. It is important that the correct software is selected for the type of analysis to be undertaken. The use of two-dimensional software is appropriate in most situations, but there are circumstances where three-dimensional effects must be taken into account (Ward and Sanders 2007). For example, where plane elements are non-uniform in the third dimension and the non-uniformity contributes to the thermal bridging, the construction must be analysed using a three-dimensional model. Steady-state models are used to determine the thermal bridging input parameters to whole building energy use and CO_2 emissions calculations that are based on one-dimensional heat transfer equations (Martin et al. 2011). The transient analysis of thermal bridges is undertaken using dynamic models to investigate specific issues such as the effects of thermal inertia, but this approach tends to be limited because of the limitations of whole building models (*ibid*). All thermal modelling software should be validated to the standard BS EN ISO 10211 (BSI 2007a). National application documents define the principles and procedures for thermal modelling (Ward and Sanders 2007) and outline the methodology for the treatment of thermal bridges (Ward 2006) under the Building Regulations (HM Government 2013).

Studies (Comiskey 2010; Peat 2013) have shown that careful design detailing that is assessed using thermal modelling can yield results that have the potential to minimise building fabric heat transfer and can prove advantageous for input to whole building energy use and CO_2 emissions calculations when compared to standard detailing. Therefore, it is essential when seeking to create low energy buildings that attention is paid to thermal bridging and that efforts are made at the design stage to minimise the impact of this phenomenon (Comiskey 2010).

9.6.2 Hygrothermal Behaviour

Increasing energy efficiency standards achieved by low rates of air permeability and high levels of insulation makes hygrothermal performance an important factor to consider in the design of new buildings and particularly when thermally upgrading existing buildings (May and Sanders 2014). Moisture transfer in buildings and the risk it presents to the building fabric and the health of occupants is a complex subject (*ibid*). Due to a failure to understand the contributing factors, reasons for the presence of moisture are sometimes misdiagnosed. Predicting hygrothermal behaviour and assessing the associated risks requires skills in the use of specialist software and knowledge of building pathology, materials science and understanding of building use. BS EN 15026 (BSI 2007b) defines the equations that form the basis for numerical simulations known as hygrothermal simulations. These use computer software to predict one-dimensional transient heat and moisture transfer in building fabric under dynamic conditions. Models require an extensive range of inputs that include the geometry of the construction, material properties (e.g. density, porosity, heat capacity, thermal conductivity, liquid transport coefficients,), internal and external climate data, orientation and boundary conditions. The outputs from hygrothermal simulations typically include temperature, heat flow, water content, moisture flow, relative humidity and vapour pressure. These conditions can be used to assess the risk of frost damage, moisture accumulation, timber rot, corrosion of metal components and mould growth. There are also some post-process software packages available to assist in the analysis of transient results from hygrothermal simulations. Hygrothermal bridges can be assessed using two-dimensional hygrothermal simulations, but these are outside the scope of the standard BS EN 15026 (BSI 2007b) and require an even higher degree of specialist capability.

9.7 Summary

The use of models and simulations to understand and predict building performance is increasing. While there remains some scepticism around models and their use, the sophistication and capability is rapidly increasing. The improvements are a direct result of the laboratory tests, field work and the development of advanced computer hardware with increased computational power, processing tools and software. The capability of the software and knowledge gained has significantly changed the way we understand building performance and improved the ability to design buildings that have resilience to the impact of more dramatic future climates.

References

ASHRAE (2002) ASHRAE guideline 14-2002, measurement of energy and demand savings. ASHRAE, Atlanta

ASHRAE (2007) Standard 140-2007: standard method of test for the evaluation of building energy analysis computer programs. ASHRAE, Atlanta

Austin B (2013) The performance gap: causes and solutions. Arup, London

Berggren B, Wall M (2013) Calculation of thermal bridges in (Nordic) building envelopes—risk of performance failure due to inconsistent use of methodology. Energy Build 65:331–339

Bordass B, Cohen R, Standeven M, Leaman A (2001) Assessing building performance in use 3: energy performance of probe buildings. Build Res Inform 29(2):114–128

Box GEP, Draper NR (1987) Empirical model building and response surfaces. Wiley, New York

British Standards Institution (2007a) BS EN ISO 10211: thermal bridges in building construction—heat flows and surface temperatures—detailed calculations. BSI, Milton Keynes

British Standards Institution (2007b) BS EN 15026: hygrothermal performance of building components and building elements—assessment of moisture transfer by numerical simulation. BSI, Milton Keynes

Burman E, Rigamonti D, Kimpain J, Mumovic D (2012) Performance gap and thermal modelling: a comparison of simulation results and actual energy performance for an academy in North West England. In: IBPSA England First Building Simulation and Optimization Conference, Loughborough, pp. 35–42, 10th–11th Sept 2012

Carbon Trust (2009) Building the future, today: Transforming the economic and carbon performance of the buildings we work in. HMSO, London. http://www.carbontrust.co.uk/Publications. Accessed 12th June 2013

Carbon Trust (2011) Closing the gap: lessons learned on realising the potential of low carbon building design. HMSO, London. http://www.carbontrust.co.uk/Publications. Accessed 12th June 2013

CIBSE (2006) TM33: tests for software accreditation & verification. CIBSE, London

Claridge DE (2011) Building simulation for practical operational optimization. In: Hensen J, Lamberts R (eds) Building performance simulation for design and operation. Spon Press, Abingdon

Coakley D, Raftery P, Keane M (2014) A review of methods to match building energy simulation models to measured data. Renew Sustain Energy Rev 37:123–141

Cohen R (2013) The performance gap in non-domestic buildings: evidence collected from the Technology Strategy Board's Building Performance Evaluation Programme. London, Verco Ltd. http://www.greenconstructionboard.org/images/stories/pdfs/performance-gap/CPG%20Evidence%20Base%20from%20BPE%20Final%20Report%2005Mar13_.pdf. Accessed 5th May 2013

Comiskey D (2010) Detailing to minimise heat loss at junctions of building elements. In: Akintoye A, Goulding J, Rahimian F, Brindley D, Emmitt S (eds) Proceedings of the ninth international detail design in architecture conference: innovative detailing: materials & construction methods for a low carbon future. University of Central Lancashire, Preston, UK, pp. 169–180, 4–5 Nov 2010

DCLG (2008a) Improving the energy efficiency of our buildings: a guide to energy performance certificates for the construction, sale and let of non-dwellings. Communities and Local Government Publications, Wetherby

DCLG (2008b) Improving the energy efficiency of our buildings: a guide to display energy certificates for the construction, sale and let of dwellings. Communities and Local Government Publications, Wetherby

DCLG (2011) National Calculation Method modelling guide (for buildings other than dwellings in England and Wales). Communities and Local Government Publications, Wetherby

Department of the Environment and the Welsh Office. (1995) Approved document L: conservation of fuel. The building regulations 1991. HMSO, London

Hensen JLM, Lamberts R (2011) Building performance for design and operation. Spon Press, Abingdon

Heo Y, Choudhary R, Augenbroe G (2012) Calibration of building energy models for retrofit analysis under uncertainty. Energy Build 47:550–560

HM Government (2010a) The building regulations 2000, approved document L2A: conservation of fuel and power in new buildings other than dwellings, 2010 edition. NBS, London

HM Government (2010b) The building regulations 2000, approved document L2B: conservation of fuel and power in existing buildings other than dwellings, 2010 edition. NBS, London

HM Government. (2013) Approved document L1A: conservation of fuel and power in new dwellings. The building regulations 2010. NBS, London

Hunn, B.D, (1992) Energy analysis of the Texas capitol restoration. Proceedings of the eighth symposium on improving building systems in hot and humid climates. Dallas, TX, pp. 165–173, 13–14 May 1992

IBPSA (2015) The international building performance simulation association. IBPSA. http://www.ibpsa.org. Accessed 2nd March 2015

Intergovernmental Panel on Climate Change (2007) Climate change 2007: synthesis report. http://www.ipcc.ch/pdf/assessmentreport/ar4/syr/ar4_syr.pdf. Accessed 27th Oct 2010

Intergovernmental Panel on Climate Change (2013) The physical science basis. http://www.ipcc.ch/. Accessed 2014

Janssens A, Londersele E, Vandermarcke B, Roels S, Standaert P, Wouters P (2007) Development of limits for the linear thermal transmittance of thermal bridges in buildings. In: Proceedings of tenth thermal performance of the exterior envelopes of whole buildings conference: 30 years of research. 2–7 Dec 2007, Clearwater Beach, USA. American Society of Heating, Refrigerating and Air-Conditioning Engineers, Atlanta

Jenkins DP, Banfill P, Pellegrini-Masini G (2009b) Non-domestic conclusions of the TARBASE project—reducing CO2 emissions of existing buildings. Urban Energy Research Group, School of Built Environment, Heriot-Watt University, Edinburgh

Kimpain J, Chisholm, S (2011) Tracking design and actual energy use: CarbonBuzz, a RIBA and CIBSE platform. In: International Conference on Passive and Low Energy Architecture Proceedings, 2011

Manke J, Hittle D (1996) Calibrating building energy analysis models using short term test data. Proceedings of the 1996 ASME International Solar Engineering Conference, ASME Solar Energy Division, pp 369–378

Martin K, Erkoreka A, Flores I, Odriozola M, Sala J (2011) Problems in the calculation of thermal bridges in dynamic conditions. Energy Build 43:529–535

May N, Sanders C (2014) Moisture risk assessment and guidance. Sustainable Traditional Buildings Alliance, London

Menezes AC, Cripps A, Bouchlanghem D, Buswell R (2012) Predicted vs. actual energy performance of non-domestic buildings: using post-occupancy evaluation data to reduce the performance gap. Appl Energy 97:355–364

Office of the Deputy Prime Minister (2002) Approved document L1: conservation of fuel and power in dwellings. The building regulations 2000. TSO, London

Office of the Deputy Prime Minister (2006) Approved document L1A: conservation of fuel and power in new dwellings. The building regulations 2000. NBS, London

Parker J, Oates M, Cropper P, Shao L (2012a) Comparison of EPC, DEC and dynamic thermal simulation results at Birmingham Airport. In: IBPSA England First Building Simulation and Optimization Conference, Loughborough, pp. 27–34. 10th–11th Sept 2012

Peat M (2013) Applying research in practice: developing a specialist service in the analysis of thermal bridging. In: Emmitt S (ed) Architectural technology. Wiley-Blackwell, Chichester

Raftery P, Keane M, O'Donnell J (2011a) Calibrating whole building energy models: an evidence-based methodology. Energy Build 43:2356–2364

Raftery P, Keane M, O'Donnell J (2011b) Calibrating whole building energy models: detailed case study using hourly measured data. Energy Build 43:3666–3679

Raslan R, Davies M (2010) Results viability in accredited building energy performance compliance demonstration software in the UK: an inter-model comparative study. J Build Perform Simul 3(1):63–85

Reddy T (2006) Literature review on calibration of building energy simulation programs: uses, problems, procedures, uncertainty and tools. ASHRAE Transac 211:226–240

Reddy T, Maor I, Panjapornpon C (2007a) Calibrating detailed building energy simulation programs with measured data—part I: general methodology. HVAC & R Res 13(2):221–241

Reddy T, Maor I, Panjapornpon C (2007b) Calibrating detailed building energy simulation programs with measured data—part II: application to three case study office buildings. HVAC & R Res 13(2):243–265

Shonder JA, Hughes P, Thornton JW (1998) Using calibrated engineering models to predict energy savings in large-scale geothermal heat pump projects. ASHRAE Transac 104(1):944–954

Soebarto VI (1997) Calibration of hourly energy simulations using hourly monitored data and monthly utility records for two case study buildings. IBPSA Conference Proceedings, Prague, Czech Republic, 13–15 Sept 1997

Subbarao K (1998) PSTAR- Primary and secondary terms analysis and renormalization: A unified approach to building energy simulations and short-term testing. TR-254-3175, National Renewable Energy Laboratory, Golden

Waltjen T (ed) (2008) Details for passive houses: a catalogue of ecology rated constructions. Springer, New York

Ward T (2006) Assessing the effects of thermal bridging at junctions and around openings. IP 1/06. Building Research Establishment, Watford

Ward T, Sanders C (2007) Conventions for calculating linear thermal transmittance and temperature factors. BR 497. Building Research Establishment, Watford

Yoon J, Lee EJ, Claridge DE (2003) Calibration procedures for energy performance simulation of a commercial building. J Sol Energy Eng 125:251–257

Part III
Sustainable Planning, Urbanism and Development

Chapter 10
Sustainable Urban Design

Lindsay Smales and Pam Warhurst

The role urban design and urban designers can play in helping to deliver high-quality, sustainable built environments is now well established, both in the UK and elsewhere. Previously known as 'Civic Design', with the 'urban' coming to the fore in the USA during the 1960s (Cowan 2005), this activity is multidisciplinary in its approach and holistic in its general outlook. Urban design practitioners pride themselves in filling the spatial gaps between the traditional functions of the architect, who usually concentrates on designing an individual building on a single site for a single client, and those of the planner, who works on policy making and delivery for whole neighbourhoods, towns, cities and even regions. Urban designers claim that, unlike their professional colleagues, they are uniquely able to focus on, amongst other things, the quality of the public realm, the street, the square and the spaces between buildings (Carmona 2004).

Even though urban design is now a recognised profession amongst those responsible for delivering our built environments (The Urban Task Force 1999), there is a tangible lack of confidence amongst those who aim to practise this profession. This manifests itself in an incessant need to define and continually restate the remit of urban design, and the role and function of urban designers (Sorkin 2011). It is also evident in their seemingly endless, but helpful, desire to identify and codify a series of rules and principles for designing good-quality, sustainable places.

This chapter examines and reviews the evolution of some of these keynote principles and tests their validity, usefulness and sensitivity by evaluating them against the physical and visual characteristics of an existing small town. This is a settlement which is relatively typical of many UK towns, not least as its physical fabric is mostly a product of the Victorian era. It is also one that is unique in that it has led something of a national, if not international, revolution in the way local communities can rethink and reuse public space greater level sustainability.

L. Smales (✉) · P. Warhurst
CBE, School of Built Environment and Engineering, Leeds Beckett University, The Northern Terrace, Leeds, LS2 8AG UK
e-mail: l.smales@leedsbeckett.ac.uk

© Springer International Publishing Switzerland 2016

M. Dastbaz et al. (eds.), *Building Sustainable Futures*,
DOI 10.1007/978-3-319-19348-9_10

10.1 Urban Rules and Principles

Ever since Vitruvius published his Ten Books on Architecture Rowland, 2001 city builders have attempted to lay down guidelines for others to learn from and follow. Christopher Wren's unrealised plan for the rebuilding of The City of London after the Great Fire of 1666 (Summerson 2003) set out his prescriptions for the relative heights of the buildings that were to line new boulevards, or what he called the city's 'Principle Streets', and the less grand, smaller thoroughfares. The eighteenth century saw a rise in popularity of 'The Pattern Book' (Darley 2007). These were modest publications containing plans and building design details for landowners who could not afford to employ one of the growing, relatively scarce, members of the architectural profession yet who still wanted their new buildings and developments to reflect the latest fashions and tastes.

In America, city layout and design was somewhat more straightforward and ordered. This was because new towns and settlements adopted regimented grid forms by taking their cue from the incremental, geometrical mapping of the country from east to west first established by Mason and Dixon (Thadani 2010). In contrast, Cammillo Sitte's 'City Planning According to Artistic Principles' (Sitte 1889), perhaps the first proper treatise on civic or urban design, was founded on his travels to European towns and inspired by the ancient and medieval settlements he loved and admired.

Raymond Unwin's influential book 'Town Planning in Practice', published in 1909 and subtitled 'An Introduction to the Art of Designing Cities and Suburbs' (Unwin 1909), helped shape whole swathes of post–First and Second World War English residential environments, from 'Homes for Heroes' through to the New Towns of the 1950s (Hall 2014). These carefully planned places were themselves the successors of the efforts by cities such as Nottingham and Leeds to control the rapid growth of squalid Victorian development through the introduction of a series of local 'by-laws' setting out space standards for the distances between the fronts and backs of dwellings and just how many properties, and the families who lived in them, should share outside toilet blocks (Muthesius 1984). It is also the case that many dictators seem incapable of restraining their desire to meddle in matters of urban design and plan the places over which they reign according to their own warped principles and values. These interventions range from Napoleon III employing Haussmann to create the wide military-friendly boulevards of Paris, to Hitler with his plans for the redesign of Berlin in the form of the megalithic 'Germania' (Hall 2014), and The Shar of Iran's pre-Revolution commissioning of a grand master plan for Tehran (Llewlewyn-Davies 1976).

If the precedence for the writing and setting out of urban design guidelines is therefore long, if not distinguished, then it is hardly surprising that recent generations have got in on the act. Jacob's seminal work of 1961, 'The Death and life of Great American Cities' (Jacobs 1961), has only four illustrations, showing the need for small-scale city blocks, yet contains a series of principles that have become the accepted norms and benchmarks for urban designers the world over. These include the necessity to mix up the uses within a given neighbourhood, as opposed to their separation through land-use zoning, the primacy of the street, the value of people-based natural surveillance and the importance of working with and consulting local residents in the design-based decision-making process (Jacobs 1961).

Gordon Cullen's 'Townscape', also published in 1961, did not contain strict 'rules' for the design of future settlements but managed to influence a whole generation of architects and urban designers. It did so through its emphasis on the visual and picturesque components of placemaking and his core argument that we should learn to embrace the organic, piecemeal, qualities of traditional built environments (Cullen 1961).

During the early 1970s, and in the absence of any formal design advice from the then Department of the Environment, Essex County Council's 'A Guide to The Design of Residential Areas' (Essex County Council 1973) became something of a national design guide for the whole of the UK. It went on to inform, shape and inspire not only the housing schemes for the locality for which it was originally written, but also influenced much of the private sector residential development throughout the country and still has a significant impact on housing layouts today (Carmona 2001). The document included the innovative, sustainable notion that new homes ought to be constructed using locally sourced materials and the key suggestion that housing in urban locations should adopt the higher densities appropriate to their context. This, in turn, would arguably ensure the distinction between town and country could be maintained and preserved, even if this did not actually lead to the demise of the 'anywhere' type of sprawling suburban estates its authors reviled (Essex County Council 1973).

Christopher Alexander joined in this trend towards prescriptive rules with his extensive set of guidelines for everything from the design of churches and temples to attractive urban spaces, as outlined in his 'A Pattern Language' (Alexander 1977). His comprehensive and ambitious collection of 253 'patterns' included guidance on the ideal size for a neighbourhood, at 7000 people, a height limit on buildings of four storeys and the need to form clusters of housing which contain a balanced mix of people of all ages and a range of facilities in identifiable local centres.

Bentley et al's 1985 text 'Responsive Environments' built upon the work of Jacobs and Cullen by putting forward nine principles for the design of towns and cities, including the idea that places should be legible and easy for both residents and visitors to understand and navigate their way around (Bentley et al. 1985). They suggested that buildings and spaces should be robustly adaptable and urban areas 'permeable', so that, within reason, the pedestrian is able to walk wherever they wish. One of the critiques of the Responsive Environment book is that it addresses a scale of development that is relatively modest and low in density and is therefore less applicable in contexts where the value of land requires much more concentrated forms of development (Bentley et al. 1985).

Perhaps not surprisingly, given his comments on the supposed failure of modernist architects to design new buildings which members of the public, and heirs to the throne, could relate to, Prince Charles has also set out his own 'Principles' for urban design. These originally surfaced in his 1989 book 'A Vision of Britain'

(The Prince of Wales 1989) as a series of what he called in an earlier BBC documentary 'Ten Commandments' by which we could evaluate the quality of existing urban environments and the new development proposals that were supposed to improve them. These princely urban edicts included building in harmony with the surrounding context, using local materials and promoting the kind of external decorative elements the Victorians and Edwardians integrated into their building facades and which modernists decry as being false and dishonest. The Prince went on to establish his own 'Foundation for the Built Environment', which embraces an English version of what the Americans call 'New Urbanism' (Haas 2008), and does so through conferences, training events, local master planning and design charettes. These guiding rules for shaping new developments have quite literally been put into place in the Prince's own new urban extension of Poundbury on the edge of Dorchester in county of Dorset (Thadani 2010).

In the UK, three other key texts contained their own variations of these urban design principles, namely The Urban Task Force Report (1999), By Design (DETR 2000) and The Urban Design Compendium (English Partnerships 2007), in the attempt to pass on best practice to the likes of local authority development control officers and the politicians who sit on planning committees. More recently, Andreas Duanny has established the idea of the urban–rural sectional transect, within which the scale and density of development increases relative to its proximity to central areas (Duanny 2008). This in turn was an evolution of his use of Design Codes, strict design principles championed by The Congress for New Urbanism and first developed at the exclusive holiday community of Seaside in Florida (Haas 2008).

In 2009 Lehnerer published his 'Grand Urban Rules' as a summary of research undertaken at the ETH University in Zurich. He sets out 115 'significant' ingredients for the contemporary metropolis, offering up tools for 'a valuable (urban) design attitude' which supposedly 'departs from an approach that wants to control everything, and moving towards a non-fatalistic form of control between freedom and coercion' (Lehnerer 2009). This eclectic and somewhat quirky montage includes guidance on skyline view management and the design of tall buildings, 'neighbourhood compatibility' and even advice on how to ensure back gardens are aesthetically consistent with those of adjacent properties.

The innovative 'Frieburg Charter' (Daseking 2015) is championed in the UK by pressure group The Academy of Urbanism and contains some nine objectives for sustainable urbanism. This includes ' the wise use of resources, minimising additional land take up, and the encouragement of moderate degrees of urban density', as well as the insight that sustainable built environments will only come about by 'creating long-term partnerships between the community, and the public and private sectors'. These edicts are lessons that have been derived from their implementation as the guiding principles for the creation of the city's neighbourhood of Vauban—a place widely acknowledged to be a model of environmentally sensitive design that all serious urbanists need to both visit and learn from. The Charter goes on to outline 12 guiding principles grouped into the 3 categories of spatial, content and process. These include creating a 'city of short distances, with accessibility to all infrastructure networks available on foot or by bicycle', focusing on 'design quality, especially for public spaces', and the need for a 'long-term vision, incorporating awareness of the past and looking way into the future' (Daseking 2015).

Others have variously produced lexicons of key urban design ideals and principles, such as Cowan's 2004 'Dictionary of Urbanism' (Cowan 2005), Mikoleit and Purckhauer's 'Urban Code' of 'maxims, observations, and bite-size truths … that teach us how to read the city' (Mikoleit and Purckhauer 2011), and Sucher's 'City Comforts: How to Build an Urban Village' (Sucher 2003). Jan Gehl (2003) Francis Tibbalds (1988), Hugh Barton et al. (2003), Mike Jencks and Dempsey (2005) and David Rudlin and Nicholas Falk (1999) have all similarly written up their experiences, theories and principles for making sustainable streets, neighbourhoods, towns and cities.

In short, few urban designers worthy of the title, or professional organisations and agencies concerned with promoting urbanism, seem able to resist telling the world how to plan and design their version or vision of sustainable places. Many also argue that, almost by default, and certainly without the need to labour the point, good urbanism is automatically sustainable as it promotes and fosters built environments which are the result of democratic processes of engagement, are human in scale, walkable, visually vibrant and thereby inherently people-friendly. Whether these 'rules', especially if applied in a doctrinaire manner, a charge levelled at The New Urbanists for instance (Sorkin 2011), are vehicles for unlocking creativity or straitjackets which serve to constrain and limit innovation is a debate that will continue as long as designers persist in writing them down.

If it were possible to summarise the considerable consensus shared by these rule makers and bearers of tablets from on high, then it would be the notion that sustainable urban design is about the collective act of making places which sensitively respond to the local context. Such settlements would contain a mix of uses, encourage the freedom to roam, be the product of involving local communities in the design and development decision-making process, be legible and consist of robust, flexible buildings and spaces that could be adapted for new uses by future generations. They would be walkable, human-scale built environments created incrementally, whilst also being distinctive, fun and beautiful. It would be the job of the talented urbanist to bring all these elements together in order to create buildings, spaces, streets and places that not only maximise their potential, and that of the people who inhabit them, but do so in a way that adds to and enhances overall levels of social, economic and environmental sustainability.

The fact that very few places, either existing or yet to be built, can live up to these demanding criteria and idealistic principles should not necessarily deter us from aspiring to them or aiming for the goals they represent. A few recent settlements have supposedly achieved these laudable qualities and become exemplars of best practice in sustainable urbanism, such as Vauban in Frieburg, Hammerby in Stockholm and the Western Harbour of Malmo. Yet these places or the guiding frameworks upon which they were built are rarely tested and appraised in any systematic manner (Sorkin 2011). The case study which follows assesses the core best practice principles outlined above against an existing town and comes to conclusions about both the guidelines themselves and the sustainability of the place in question.

10.2 Incredible Edible Todmorden: Changing the World One Corner at a Time

Todmorden is a Market Town in West Yorkshire, England. The settlement grew rapidly in the nineteenth century, during which time it saw the development of local employment based around the production and processing of both wool and cotton. The decline of the textile industry in the twentieth century hit Todmorden hard, and the town has struggled to find a contemporary role for itself. In this period, its population fell from a peak of 20,000 to the current 14,500. The settlement could once have been described as an economy in search of a built environment. Today, it is a built environment in search of an economy and a place that is trying to be something more than a commuter town for people who work in the nearby cities of Leeds, Bradford and Manchester.

Todmorden is a fairly typical example of a northern Victorian mill town, with its neo-classical Town Hall, indoor and outdoor market, Methodist Chapel and railway viaduct. It nestles at the confluence of three valleys, in whose flood plain it was constructed, and at the western edge of the picturesque Upper Calder Valley, the moors, or 'Tops', of which are a popular draw for walkers and ramblers. Many of the significant landmark structures in Todmorden, including its impressive Town Hall, were funded in the nineteenth century by the Fielden family, the wealthy local mill owners (Birch 2006). It currently has levels of unemployment and deprivation slightly higher than the local norm (CMBC 2010) and a disproportionately greater number of elderly people when compared to other towns in the vicinity. The largest single employer provides some 80 jobs, but the majority of local employment is derived from small businesses, most of whom maintain between 2 and 10 staff.

In 1948, the well-known planner Thomas Sharp wrote a 'Plan for Todmorden' (Sharp 1948), in which he proposed the demolition of substantial parts of the town centre and its reconfiguration using modern roads, roundabouts and flat-roofed new development. The plan also suggested a large new residential area be built on the edge of town. This would be constructed at a density of 24 dwellings per acre, or at twice what it claimed to be the local average, with 'full provision' being made for 'a community centre, local shopping groups, places of worship, primary and nursery schools and playing fields' (Sharp 1984). This bold urban extension did not come about due to the prevailing economic climate of the time and the fact that it was to be insensitively sited on top of a remote and bleak hill.

Much later, in 2003, John Thompson and Partners were asked by the then Regional Development Agency (RDA), Yorkshire Forward, to work with the communities of the Upper Calder Valley, including Todmorden, on the production of a comprehensive 'Vision' for the regeneration of the area and its five major settlements (Yorkshire Forward 2003). This document contained a sophisticated and accurate analysis of the character of the Valley, exemplified by the textile-based analogy of the 'Warp' and 'Weft' of the place, as illustrated by the twisting and varied entanglement of its rivers, canal, roads and railway. Thompson's document also allocated five pages of ideas for improving Todmorden, including the development of key sites, mostly those of former clothing mills and factories, the creation of a new town square adjacent to the markets and a new mixed-use waterside development consisting of 'workshops, offices, live/work units and a variety of housing types and tenures' (Yorkshire Forward 2003).

This 'Vision' document was part of the countywide 'Rural Renaissance Programme' funded by the RDA, which also supported the writing of a 'Tourism Action Plan' (UCVR 2005) for the Upper Valley. This follow-on document identified a number of reasons why the providers of the local tourist experience, such as B&B and cafe owners and those who ran independent shops, believed people, mostly in the form of day trippers, came to visit the general area and the towns of Todmorden and Hebden Bridge in particular. These attractions were deemed to include 'a dramatic landscape with a unique settlement pattern; an important and powerful heritage; a place of inspiration and innovation, and a centre for healthy lifestyles' (UCVR 2003). Suggestions for improving the built environment of Todmorden focused on the settlement's 'buildings, streets and spaces' and identified ways in which to 'capitalise on the creation of a better image for the town'. Such enhancements would be founded on an appraisal of the 'overall levels of walkability in the town centre', the 'pro-active release of canal side mills and properties to facilitate mixed-use development' and 'the possibility of Todmorden becoming an additional centre for alternative, sustainable technologies' (UCVR 2005).

Laudable as these post-Sharp ideas and aspirations might be, few have come to fruition. Apart from the creation of a new canal side wharf, the reroofing of the outdoor market and some limited conservation work on the Grade One-Listed Town Hall, the town has seen few of its larger sites or properties being developed for new uses, experienced the long-standing dereliction of centrally located buildings, such as a cottage hospital and cinema from the 1930s, and little in the way of affordable new residential development that meets local needs.

In recent times, however, Todmorden has become known for a new way of thinking about how we live and relate to our surroundings based on the Incredible Edible Todmorden (IET) project (Warhurst and Dobson 2014).The IET initiative was started by a small group of people, mostly women, who wanted to put into practice the well-established principle of 'thinking globally, acting locally'. Their approach was significantly different. Rather than producing a strategy document or asking others, such as existing formal bodies and agencies, for permission or support, they realised the power of small, direct actions. This initially took the form of what is referred to as 'guerrilla planting', but which they prefer to call the less aggressive activity of 'propaganda gardening'. This entailed local volunteers digging up the type of 'leftover', underutilised, poorly managed and boring empty spaces to be found in all towns and cities and the transformation of much of the town's public realm into herb gardens, vegetable patches and orchards for use by the whole community (Warhurst and Dobson 2014). It included creating raised growing beds in the corner of the railway station car park, in front of the local Police Station, Job Centre, Community College and Health Centre buildings and establishing herb gardens in many spaces that were previously subject to the aesthetically dead hand of so-called 'municipal planting'.

Central to the project are a number of key principles. These include the notion that, in the face of entrenched local government bureaucracy, 'it is easier to ask for forgiveness than permission' (Dobson 2015). The aim was that this would be a truly inclusive exercise, as exemplified by the phrase, 'if you eat you are in', and that the vegetables and herbs planted around the town by local volunteers are free for everyone to harvest and pick. The founders of IET, which include a former local council leader and chairperson of national environmental quangos and a retired community development worker, set up the scheme to encompass what they refer to as three 'spinning plates' of activity—a community plate, a learning plate and a business plate—interrelated elements that must continually be kept in motion in order to achieve the desired objectives.

The result of this approach has meant that Todmorden's residents have embraced the aims and objectives of IET and begun to take more pride in where they live, coming to view their place with different eyes (Warhurst and Dobson 2014). Local schools have engaged with IET and now grow food in their own grounds. They have incorporated both growing and cooking more extensively into their day-to-day educational curricula, and many local food producers and retailers now promote and sell their products as being part of the IET 'family'.

A recent research project undertaken as part of a European Union-funded evaluation of the impact and effectiveness of the IET project was able to establish that:

> 96 % of local residents like the growing of food in public spaces
> 67 % collect food and herbs from these public spaces
> 97 % say they buy more local food today as compared to 5 years Ago
> and 57 % have begun to grow their own food following the example of IET (Trivelli 2013)

In addition, the district authority, Calderdale Council, now work closely with IET and have introduced a valleywide licensing agreement whereby local people can grow food on publically owned land. An Incredible Edible Farm has been set up on the edge of town and the local High School now plays host to a new Aquaponics Garden building and project which is one of the largest such facilities in the UK (Dobson 2015).

A big element of the success of the IET project, apart from the fact that it is an idea whose time has come, is the use of social media, websites and a coordinated strategy of actively explaining and sharing the work of the group to others and encouraging them to copy this example in their own places and communities. This objective is also achieved through tours of the town and growing sites and the preparedness of key activists to go out and talk at national and international conferences and events. This desire to disseminate their experience, philosophy and enthusiasm has meant that, in 2014, more than 1,000 people visited Todmorden to be guided around the town and engage with IET members, their work and ideas. There has been a huge growth in Incredible Edible towns and communities across the UK, with their number reaching more than 90, together with an estimated 300 being set up across the world (Warhurst and Dobson 2014).

The major UK agency for championing new ways of thinking about how to create sustainable towns and cities, The Commission for Architecture and the Built

Environment, have used Todmorden and IET as one of their 'Best Practice Case Studies'. They conclude:

Incredible Edible Todmorden succeeds because it connects people in the town through the shared growing, picking and eating of food:

- *it questions the way people think about public green space and empowers them to take responsibility for it*
- *it is based upon local action, it is about changing behaviour through strong local leadership*
- *it represents a bottom up approach to tackling wider issues of climate change which is truly inclusive—it involves everyone in the community from every background*
- *it breaks down barriers between people by focusing on something that we all love and need—food—because it is inclusive*
- *it is truly joined up thinking—it makes connections in the town so local growing is embedded into education, health and local business.*
  (CABE 2011*)*

Todmorden is a place transforming itself on the back of a simple yet very powerful idea that has quite literally been put into place by local activists. These are people who are not afraid to do things differently and to use the previously neglected resources available in their town in innovative and imaginative ways in order to challenge and change how we perceive, use and interact with urban space, whilst also providing food that is free for all.

This is why Todmorden is widely recognised as a contemporary model for other places to be inspired by, to learn from and, with the happy consent and support of its Incredible Edible community, to copy and follow—be they similar small towns, local neighbourhoods or even entire cities (Dobson 2015).

It is instructive to evaluate this particular place against the best practice criteria in sustainable urbanism set out and shared by the practitioners and commentators whose work has been examined above.

10.3 Evaluating Todmorden Today

The first best practice principle for sustainable urbanism is summarised as, *'the collective act of making places which sensitively respond to the local context'.*

There is little doubt that Todmorden has a strong sense of place. This is derived from both its natural setting and the character of its built environment. The former is the reason why walkers are drawn to its surrounding hills and valleys, and the latter comes from the local stone used by the Victorians in the building of their town at a time when they clearly wanted it to be a tangible expression of civic pride, economic prosperity and community coherence.

A positive effect of the impact of a relatively depressed local economy is that there has not been much new, modern development of the kind that has been described as being capable of being built 'anywhere' due to its cheapness and lack of design quality. A negative effect is the number of large brownfield sites on the edge of the town centre which lay dormant and the lack of investment in key buildings and spaces which could enhance the town's central area and overall image. Despite a range of plans for the town, few of the initiatives suggested have happened. The Town Council is relatively powerless, having an annual budget of less than £ 200,000, when compared to Calderdale MBC, the main local authority, who have overall responsibility for planning matters. Neither agencies seem to have the finances, imagination or entrepreneurial skill to embrace the challenges of kick-starting local development in straightened economic times.

Whilst there are a number of organisations in the town who have tried to make small-scale environmental, place-based improvements, such as 'Todmorden in Bloom' and the 'Todmorden Pride' partnership, this general level of collective local, official inaction is one of the reasons for the creation of IET. Incredible Edible could be said to be a response to the local context through its creative reuse of neglected, leftover spaces and potential sites for growing. It is certainly an example of volunteer-based community action, albeit one which, in theory, could have taken hold in any town, hence its easy transferability to other places, neighbourhoods and communities.

The fact that Todmorden's image and reputation has been transformed on the back of the work of IET is a mark of the clever and sensitive manner in which this project has responded to prevailing conditions and challenges and it has been used as a way of trying to address more difficult, overarching problems such as social cohesion and place-based identity (Dobson 2015).

The second best practice principle is focused on '*the provision of mixed uses*' within a given place and locality.

As a compact, relatively dense town nestling in the narrow valley bottoms of a former flood plain, the development and growth of Todmorden in the nineteenth century saw terraces of workers' houses being built in close proximity to, indeed often in the shadow of, the mills and factories in which their occupants worked. Shops, pubs and places of workshop were similarly nearby and remain so today. With the decline of local industry and the closure of most if not all of the mills, came the necessity to seek employment out of town, further down the valley, up the road or along the railway line.

In this way, in comparison with earlier times, Todmorden is much less of a mixed-use place than it once was. As clearly identified in the Upper Calder Valley Renaissance Vision (Yorkshire Forward 2003), the opportunity provided by the sites vacated by former industries once more presents the chance to create development that consists of mixed uses, even though the economic factors have not been in place to foster this, or hardly any, type of development.

So far there are no 'gated communities' in Todmorden. This might be due to the dearth of recent new residential development and relative lack of prosperity in the town. It is also noticeable that if the economically buoyant want to exempt themselves from being close to the less well-off then in this locality they do so by living above the town and not within it. This is evidenced by the expensive high spec vehicles which fill the busy town centre railway station car park and belong to those who commute to Bradford, Leeds and Manchester and who tend to live in converted farmhouses, cottages and barns in remote and secluded villages on the Moors or 'Tops'. These expensive conversions are physically and financially out of the reach of the vast majority of Todmordians. This economic apartheid, which after all is Afrikaans for 'separate development', means that those who live above a certain contour line invariably look down on the town.

The third principle is that of *'permeability'* or the opportunity, within the bounds of reason and safety, to encourage the freedom to roam, so that people can easily access their place and its constituent parts.

One of the recent successes of IET has been that of persuading the local Calderdale Council to allow them to create a temporary community garden on a brownfield site the authority own in the town centre. Half of the site has been given over to a herb, vegetable and fruit garden, called 'Pollination Street', and a grassed over green space with picnic benches. This space is also used as the venue for the annual IET Harvest Festival, which sees more than 500 people gather to eat locally produced food and to see and hear about the activities of the Incredible Edible volunteers. The other half of the site has been boarded up in the usual way, but even this has been adopted by children from the local Asian community who use it as an informal cricket ground upon which to practise their skills. Whilst there is evidence that the local community are separated by topography and house prices within the area, they have managed to come together by colonising the available space in the town, even if they are not necessarily invited or meant to do so.

The fourth urban design guideline is that a good, sustainable built environment should be *'the product of involving local communities in design and development decision making'*.

Unlike other Yorkshire Forward Renaissance Programme master plans (Yorkshire Forward 2008), the UCVR vision was the product of a sophisticated consultation process, as opposed to the typical exercise undertaken in other regional settlements in which a 'Town Team' made up of elected members and business people became the client body for future plans.

The UCVR visioning process involved gathering the views and aspirations of local people via a series of consultation days, one of which was held in Todmorden Town Hall. However, it is obvious that many of the urban design-based proposals in the end document are those of the talented consultants themselves expressing their own ideas through skilfully executed and engaging drawings (Yorkshire Forward 2003), rather than design interventions put forward by local residents. The various action plans outlined in the 'Vision' were supposed to be implemented by subject-specific themed groups, only one of which, the 'Sustainable Transport Group', based in Hebden Bridge and not Todmorden, survives to be active today. The Tourism Action Group did meet and produce the aforementioned Tourism Action Plan, but apart from commissioning a visitor survey and persuading CMBC to reinstate the role of a local tourism officer, their objectives have unfolded more organically than through the agency of an adopted plan or strategic implementation process.

Over the last eight years, a range of attempts have been made to redesign the spaces, sites and buildings on or adjacent to Todmorden's centrally located Bramsche Square. Town and district councillors have been represented on at least two regeneration boards or panels whose aim has been to work towards identifying ways of unlocking the town centre's development potential. Due mostly to the poor state of the local economy, these plans have also failed to come to pass. The opportunity to develop a Todmorden Neighbourhood Plan is being explored under the provisions of the Localism Act (DCLG 2012). This would involve yet further community participation in setting local development priorities and the chance for the town council to wrest control of planning matters back into their own hands. As Neighbourhood Plans can embrace specific themes, it would be surprising if Todmorden's did not embed the local food growing economy within its core objectives and policies.

The IET ethos of just getting on with it, without asking for permission, is partly founded on long-standing frustrations around raised expectations generated by a raft of existing bodies and agencies who, over the years, have promised local residents the hope of regeneration but have rarely been able to deliver. With its self-selecting steering group of local volunteers possessing a range of pertinent skills and creative enthusiasm, IET could be argued to have developed an implementation strategy for effective local change which does not depend upon, or is impeded by, the usual type of formal consultation. In fact, they have done so in a town whose populace could be said to be suffering from 'consultation fatigue'. This experience serves to demonstrate that having set guidelines for shaping future development is all well and good but that, where appropriate, these principles might need to be bypassed or broken.

The fifth guidance is the one which states *'places should be legible'* and thereby easy for both residents and visitors to understand and get to know. As a compact small town nestling in the bottom of three narrow valleys, which represent the only ways in and out, Todmorden generally meets this criterion. It is aided by the prominent design and location of its Town Hall, which is carefully sited at the confluence of these three arteries, some prominent landmark buildings at strategic locations on the edge of the central area and the way in which development has also traditionally clung to the sides of these valleys.

However, it is not necessarily clear where the town centre begins and ends and this became a contentious issue in a recent planning application for a fourth supermarket on a large gateway site along the road from Halifax. The developers argued that the scheme should be deemed as being 'out of town' and therefore subject to less stringent planning rules and obligations. They did so even though they also maintained that it was near enough to the town centre for their customers to walk down the busy road to continue their shopping by visiting local independent shops and the markets. Their application was a thrown out on the basis that a new store some 20 % larger than the existing medium-sized supermarket, which clearly is on the edge of town, would have a detrimental effect upon the vitality and viability of the existing provision.

As with many places, the town also has an ad hoc collection of welcome signs at key entry points, erected at different times and which send out a range of messages.

More recently, IET have commissioned a large sign for the east-bound Platform Two of the railway station, which says' Welcome to our Incredible Town'. This is a superb piece of design by a talented local graphic artist who has managed to capture the character of the settlement and its people in an exemplary way.

The sixth principle aimed at achieving sustainable urbanism is that places should *'consist of robust, flexible buildings and spaces that could be adapted for new uses by future generations'*. With the decline of its founding industries, Todmorden also saw the demolition of most of its textile mills, some of which were very large structures, and the removal of a substantial proportion of its Victorian housing, especially those properties in a poor state of repair. Some robust, well-built structures, such as a town centre Methodist Chapel opposite, and indeed larger than the Town Hall, were cleared away in the 1960s.

There are a few instances of buildings being converted to new uses, such as a Victorian Art School on the edge of Centre Vale Park becoming a community centre, one or two mills being turned into flats and former shops becoming cafes. Like other places, many of the cleared sites, especially those within the town centre, have become surface car parks, facilities which local retailers say are essential to the success of their businesses and about which local residents complain at having to pay supposedly high parking charges collected by the Borough Council, revenue which is not automatically fed back into the local community.

The most successful example of the restoration and creative reuse of a local building is that of the Unitarian Church. Perched on a hillside overlooking the town centre, this magnificent example of gothic ecclesiastical architecture has been deconsecrated and restored by The Historic Chapels Trust. Although it is still used for inter-faith weddings and funerals, it now also acts as a conference centre, art gallery, music venue and community gathering place, not least for the volunteers who maintain the growing beds of IET who store their tools and equipment there.

It is doubtful whether the likes of the well-built Methodist Chapel on Bridge Street, or the Dale Street Co-op building with its spacious first-floor ballroom, would be demolished today. This in the context of a town whose popular nineteenth-century Indoor Market building has spent many years waiting for its leaking roof to be repaired and a place that, thanks to IET, has also now found a new and exciting way of pro-actively reusing the type of leftover spaces, if not the local buildings, to be found in all towns and built environments.

The seventh principle is that *'good places should consist of walkable, human-scale environments built incrementally'*. Conceived, like many northern towns, in an era when the main forms of transport were the steam locomotive and the horse, Todmorden is a place that is both walkable and built at a human scale. The vast majority of the workers in its former mills would have walked to their nearby place of employment. They lived, as they do today, in well-constructed rows of stone terraces, laid out in a gridlike fashion. The original inhabitants of these solid properties would not have needed to travel far for provisions or access to amenities, including the fresh air available on the tops of the moors. And it is the case that, until relatively recently, a considerable number of local residents rarely, if ever, ventured out of the town and its valleys.

The notion that a good place is one that can be walked across within the space of 10 min has become something of a mantra for those urbanists aiming to help foster sustainable settlements (Thadani 2010). This is a measure of sustainability that Todmorden achieves almost effortlessly, despite the need to improve the quality and maintenance of local footpaths such as the town's canal towpath. Observation reveals that many children still walk to one of the town's seven schools, few adults are far from a local pub or shop, and more people walk to the railway station than drive there. Alexander would be pleased to see that, church spires apart, there are no buildings taller than four storeys and there is relatively little new development that is unattractive or does not now use locally sourced stone in its design and construction.

As for the ideal of incremental, rather than comprehensive, development, the town fulfils this goal by default, in as much as the lack of investment, certainly over the last four decades, has meant that there are no examples of large, over powering or domineering schemes in the town, be they residential or another use. It lacks an example of the ubiquitous business park full of the poorly designed crinkly tin sheds, a large-scale public housing scheme, or any cheap off the peg office buildings. This has meant that much of the original character of Todmorden survives into the present to be appreciated by residents and visitors alike, and does so to the point where the absence of these forms of contemporary development, and the preservation of a sense of place that goes with this attribute, is cited by those who are now moving into the town from elsewhere as one of the reasons for their locational choice.

Finally, high-quality, sustainable places should be *'distinctive, fun and beautiful'*. The other principles and urban aspirations in this list of shared themes and issues are perhaps easier to achieve and more obviously apparent than these three illusive qualities.

Distinctiveness, in the form of built environments that are unique and memorable, can be hard to find in the age of the 'Clone Town' high street (New Economics Foundation 2004) and rapid globalisation. Todmorden is certainly in possession of a distinctive natural setting. Although many of its buildings, including the Town Hall, Unitarian Church and shop fronts, are examples of Victorian architecture composed of prefabricated elements available from contemporary building catalogues and journals, the physical fabric of the town retains much of its original character and charm. When cleaned up and restored to their former glory these buildings provide an attractive collective backdrop and powerful traditional townscape.

The recent success of the town's markets, which now hold two busy days selling second-hand goods, helps to make the settlement fun for both locals and visitors. Recent community art festivals, including puppet and lantern-based parades, have also made a significant contribution to putting the place on the map and making it more enjoyable. This is aided by the fact that a number of recent television travel programmes, themed around canals and railways, have featured Todmorden as a good place to visit and pass through. Whilst the teenagers from the local High School might beg to differ, the town has recently become a more fun place, with a growing number of cafes, bars, new shops and even art galleries. It can also be said to be an interesting, lively and inspiring place for the 1,000 people who chose to engage in IET-inspired 'vegetable tourism' and -guided tours every year.

Todmorden may not have the organic beauty of a hilltop town in Tuscany, or the planned order of a model settlement like nearby Saltaire. It does have a stunning natural setting, best experienced by a walk in the town's Park, which has unrivalled views of the local hills and rocky outcrops, and it is still in possession of a powerful form of visual integrity that springs from a unified built environment shaped within a relatively short period in time. Although the recent demolition of some key buildings and lack of new development on important sites have arguably led to the dilution of this aesthetic cohesiveness, certain parts of the place have the kind of townscape-derived beauty that Cullen and others might recognise as a high-quality built environment with many of the attributes of a sustainable place.

10.4 Conclusions

This evaluation of the small town of Todmorden against established and shared best practice guidelines for sustainable urbanism has served two purposes. It has shed light on the usefulness and legitimacy of these criteria whilst also enabling them to be used to appraise the constituent components of a built environment that struggles to be economically and socially viable—albeit a community which has also managed to develop a particular model of environmental sustainability that other places have enthusiastically embraced and copied.

Even though they are relatively few in number, these specific urban design guidelines can be deployed as a helpful way of assessing places. They could be used, for instance, as a quick method for a town or neighbourhood's inhabitants to undertake an initial overview of their settlement's core qualities and as a starting point for a wider community-based discussion as to how to their place might deal with its past, present and future. If a town like Todmorden is subject to a range of forces, both internal and external, which serve to shape its buildings, spaces and the people who use them, then the central issue is the extent to which a local community such as this is at the mercy of change or in control of it.

Through the creation of the innovative IET project, whose evolution has been both planned and spontaneous, this town has changed the way its residents perceive their home and how they work together to bring out its potential. It is doubtful if prior to becoming 'Incredible', a deliberately tongue in cheek yet knowingly aspirational and confident exercise in the branding of place, Todmorden would have met many of the principles explored here. Whether the establishment of this significant initiative is a happy accident, a co-incidental and creative meeting of local minds in a specific place at a judicious time, remains to be seen.

It is also worth noting that, whilst there is an apparent wealth of urban design guidelines available which aim to spell out the qualities our places should aim for, adopt and share, there is a distinct lack of guidance which sets out the *processes* by which they can realistically become more sustainable by design. Most urban design theorists and practitioners agree that good urbanism is, and must be, inherently sustainable, but relatively little research exists on how to get there from here.

If, as might be hoped by the idealistic urbanists who have set out and created these criteria, the people residing in a town such as Todmorden can learn to better understand its assets, qualities, opportunities and distinctiveness by using these principles to reflect upon their home and its future then much might be achieved. Above all, urban designers aspire to foster a sense of place in those settlements which do not currently possess this illusive quality and to preserve and enhance it where it already exists. This exercise in benchmarking has shown these best practice principles to be well grounded and useful. As such, they may yet prove to be powerful tools in the creation of better, more sustainable built environments.

References

Alexander C (1977) A pattern language. Oxford University Press, New York
Barton H, Grant M, Guise R (2003) Shaping neighbourhoods for health, sustainability and vitality. Spon, London
Bentley I et al (1985) Responsive environments. The Architectural Press, London
Birch R (2006) Todmorden, people, places and events. The Woodlands Press, Todmorden
Calderdale Metropolitan Borough Council (2010) Todmorden ward profile. CMBC, Halifax
Carmona M (2001) Housing design quality. Spon, London
Carmona M (2004) Public places: urban spaces, the dimensions of urban design. The Architectural Press, London
Commission for Architecture and the Built Environment (2011) Case study: incredible Edible Todmorden
Cowan R (2005) The dictionary of urbanism, Streetwise, Tisbury
Cullen G (1961) Townscape, London. The Architectural Press, London
Darley G (2007) Villages of vision, five leaves, London
Daseking W (2015) Frieburg: principles of sustainable urbanism. J Urban Regen Renew 8(2)145
DETR (2000) By design, department of communities & local government, London
Department of Communities and Local Government (2012) The National Planning Policy Framework. DCLG, London
Dobson J (2015) Achieving equity: access to good local food for all. J Urban Regen Renew 8(2)122
Duanny A (2008) The urban transect, Duanny Plater-Zyberg and Company, www.dpz.com
English Partnerships (2007) The urban design compendium, English partnerships and the housing corporation. Homes and Communities Agency, London
Essex County Council (1973) A design guide for residential areas. Essex County Council, Chelmsford
Gehl J (2003) New city spaces. The Danish Architectural Press, Copenhagen
Haas T (2008) New urbanism and beyond. Rizzoli, New York
Hall P (2014) Cities of tomorrow. Wiley-Blackwell, London
Jacobs J (1961) The death and life of Great American cities. Random House, New York
Jencks M, Dempsey N (2005) Future forms and design for sustainable cities. The Architectural Press, London
Lehnerer A (2009) Grand urban rules. 010 Publishers, Rotterdam
Llewellwyn-Davies (1976) The Tehran masterplan. Llewellwyn-Davies, London
Mikoleit A, Purckhauer M (2011) Urban code. MIT Press, Cambridge Mass
Muthesius S (1984) The english terraced house. Yale University Press, Yale
New Economics Foundation (2004) Clone town Britain, NEF, London

Rudlin D, Falk N (1999) Sustainable urban neighbourhood. The Architectural Press, London
Sharp T (1984) A plan for Todmorden. Todmorden Town Council, Todmorden
Sitte C (1889) City planning according to artistic principles, in Cammillo City: the birth of modern city planning, Collins C (2006) Dover Architecture, Dover
Sorkin M (2011) All over the map: writings on buildings and cities. Verso, London
Sucher D (2003) City comforts. City Comforts Inc., Seatle
Summerson J (2003) Georgian London. Yale University Press, Yale
Thadani DA (2010) The language of towns and cities: a visual dictionary. Rizzoli, New York
The Prince of Wales (1989) A vision of Britain. Doubleday, London
The Urban Task Force (1999) Towards an urban renaissance. Spon, London
Tibbalds F (1988) Making people friendly towns. The Architectural Press, London
Trivelli L (2013) Final report on incredible sdible Todmorden. University of Pisa, Pisa
Unwin R (1909) Town planning in practice. Fisher Unwin, London
Upper Calder Valley Renaissance Co. (2005) A tourism action plan for the upper calder valley. UCVR, Hebden Bridge
Vitruvius (BC25) in Rowland, I (2001) Vitruvius: ten books on architecture. South Western University, Texas
Warhurst P, Dobson J (2014) Incredible! Plant veg and grow a revolution. Urban Pollinators, Kibworth Beauchamp
Yorkshire Forward (2003) Upper calder valley renaissance: the valley vision. Yorkshire Forward, Leeds
Yorkshire Forward (2008) Renaissance market towns manual. Yorkshire Forward, Leeds

Chapter 11
Landscape Urbanism and the Building of Sustainable Futures

Alan Simson and Silvija Krajter Ostoić

Change is an inherent aspect of human civilisation, and it is rarely comfortable. Cities across the globe are subject to constant change, and no urban area is likely to be immune from the forces that bring this about. Indeed, as the twenty-first century progresses, this change is going to have to be far more radical than it currently is, if we want our towns and cities to be sustainable, to create or maintain a healthy setting for people, and be places that attract and retain innovative investment and development.

In the UK, most of us are now urban people, even if we still believe that we retain some vestige of the countryside within us. It has been suggested that over 90 % of us in the UK now have an 'urban lifestyle'. In the UK, 1851 was the year when urban inhabitants officially outnumbered rural ones, but we only have to go back maybe two generations on mainland Europe to find an ancestor who was either involved in farming or relied upon the land to provide them and their family with a living. Much has changed within the short period of one century; the Industrial Revolution may have been the catalyst for urbanism, but the twentieth century was the period that really made the difference (Simson 2008).

According to the United Nations (http://www.un.org/esa), only 14 % of the world's human inhabitants live in cities around 1900. As late as 1960, two-thirds of humanity still lived in rural areas, but, by 2007, for the first time in history, the world hosted more urban dwellers than rural. By 2025, the UN estimate that only one-third of the world's population will live in rural areas, and it has been suggested that some 60 % of the urban areas that will exist in 2030—only 15 years from now—have yet to be built (Secretariat of the Convention on Biological Diversity 2012). Such is the pace of urbanism worldwide, and this will have a corresponding impact on resources, biodiversity, sustainability and human health and well-being.

A. Simson (✉) · S. Krajter Ostoić
Leeds Sustainability Institute, Leeds Beckett University, Broadcasting Place, LS2 9EN Leeds, UK
e-mail: a.simson@leedsbeckett.ac.uk

© Springer International Publishing Switzerland 2016

M. Dastbaz et al. (eds.), *Building Sustainable Futures*,
DOI 10.1007/978-3-319-19348-9_11

It is likely that the pace of urbanisation will accelerate as towns and cities respond to change—changes in population levels, both increases and decreases, ethnic composition, climate change, changes in industry and employment and changes in people's expectations and demands. This will in turn affect population densities, lifestyles and the design, sustainability and governance of our towns and cities. Such changes are already occurring in Europe, where there have been noticeable realignments over recent years in the physical growth and sprawl of some of our urban areas, although over 60 % of European residents are said to inhabit urban areas (European Commission Communication 2013b).

Whilst urban sprawl is often thought of as a classic American problem, which indeed it is, the European Environment Agency 'Urban Sprawl in Europe' report (European Environment Agency 2006) suggests that European urbanism is going in exactly the same direction as that in the USA, albeit for different reasons. In the USA, sprawl has generally occurred as a result of industrial decline, with people vacating the inner city for a better quality of life in the ever-increasing suburbs. Detroit, Michigan, is the classic example, having 60 % less population in 2010 than it had in 1950. The collapse of the motor industry in Detroit towards the end of the twentieth century caused the population of that city to fall over 25 % between 2000 and 2010 alone (US Census Bureau 2011). In Europe, town and city centres have tended not to have suffered in this way, although changes in people's retail habits to online shopping have affected many of the traditional shopping streets, as has the rise in peri-urban shopping centres and retail outlets.

European sprawl is something of a paradox however. In most countries—other than in the UK—national populations are either stable or declining, which would suggest that, at best, cities should be either static in area or actually shrinking. They are not however—indeed, many cities are appreciably expanding, and the prime reason for this is a desire for people to seek a better, quieter, healthier and greener quality of life, which is hard to find in many city centres—even some of the historical European city centres (European Environment Agency 2006). It can be found in the greener suburbs however, and the fact that the registration of private motor vehicles in Western Europe exceeds the registration of births by a factor of 4:1 suggests that even with the high quality of most European public transport systems, people are opting for up-to-date personal transportation, rather than for children.

In addition, in some parts of Europe, the larger cities are where there are good job opportunities and where the higher education institutions are to be found, and thus, they attract significant numbers of people to engage with these activities, with a corresponding pressure for housing, transportation and other developments. In these circumstances, issues of 'green' can sometimes be viewed by city authorities and other interest groups as being counterproductive to providing such new developments economically.

Whether the American or the European example is selected, the results are much the same. Urban sprawl has been eating away at the rural and 'natural' areas of the countryside, partially through direct building and construction, as people opt for the 'greener' lifestyle of sub- or ex-urbia. The extension of communication infrastructures to serve such development, such as new roads, road widening, new or the reinstatement of railways and cell phone masts—the classic 'grey infrastructure'—has also taken its toll of the countryside and compromised the sustainability of towns and cities as a result.

The American response to this predicament was to invent the concept of 'green infrastructure', defined in 2006 by Benedict and McMahon *as*

> a strategically planned and managed network of wilderness, parks, greenways, conservation easements and working lands with conservation value that supports native species, maintains natural ecological processes, sustains air and water resources, and contributes to the health and quality of life for America's communities and people (Benedict and McMahon 2006).

This concept was not entirely new, as it could be argued that the concept, if not the phrase, could be traced back to the nineteenth-century Quaker industrial philanthropists in the UK, who firmly believed in 'green infrastructure' as an integral part of the success of their model settlements, the idea subsequently being propounded by Ebenezer Howard and the Garden City Movement in the early part of the twentieth century. The actual term 'green infrastructure' seems to have first been deployed in 1994 in the USA, when a report was presented to the Governor of Florida on land conservation strategies, which suggested that 'natural' systems were equally important aspects of infrastructure, if not more important.

As is often the case however, even the existence of a fairly comprehensive definition does not inhibit a broad interpretation of a concept. Many government agencies, at the Federal, State and city level in the USA deployed the term 'green infrastructure' to cover a multitude of policies and ideas essentially to do with conservation and surface water management including, in 2007, its use by the US Environmental Protection Agency to describe management practices that were used to achieve a low-impact development strategy of built engineered structures for storm water management.

In Europe, the publication in 1990 of the EU's *Green Paper on the Urban Environment* (European Union 1990) attempted to confront the issues of urban sprawl by extolling the virtues of urban densification as a way of creating more mixed land uses and related multifunctional livable open space environments. Subsequent policies were promoted by the Green Paper throughout the Member States to achieve this aim, but it became apparent that, far from halting the spread of urban areas, they were increasing with renew vigour. Thus, the EU had an interest at about the same time as the USA in refining and redefining the concept of green infrastructure in Europe and for similar reasons, the concept being promoted as a comprehensive green infrastructure network across the Continent to confront, contain and deal with the issues of urban sprawl and its effect upon 'nature' and biodiversity.

Green Infrastructure is multifunctional, in that it can perform several functions in the same spatial area at the same time, in contrast to grey infrastructure that generally performs only one function, such as drainage or transportation (European Environment Agency 2011). Green Infrastructure is not in opposition to 'grey infrastructure' however. Grey infrastructure enables the transportation of raw materials and assists the production of goods and services and the distribution of finished goods to markets, be they local or further afield. Grey Infrastructure also facilitates the provision of basic social services and benefits to the local population and thus is fundamental to successful urbanism. The key to successful urbanism in the future would seem to hinge on an ability to meld the green infrastructure with the grey, even if this mix and its application to both policy and delivery vary from country to country and place to place.

The EU is well aware that the future of Europe depends on having successful, sustainable cities and city regions (Potočnik 2013). Further, the report *'Building a Green Infrastructure in Europe'* (European Commission 2013a) very much promoted the idea of a European Urban Green Infrastructure, including urban green space, street trees and other urban forestry planting and sustainable urban drainage (SUDS). Research carried out recently, however, for COST Action FP 1204 in 13 EU counties has shown that there is a wide variety of national responses to the concept of both green infrastructure and urban green infrastructure, the role it plays in national policy making and its inclusion in national urban futures (COST Action FP 1204 2013).

In some ways, we have been here before, albeit with the concept of urban forestry. It is generally agreed that the concept of urban forestry was originally conceived in Toronto by Jorgenson in 1965, quickly adopted by the urban green movement in the USA and introduced into Europe in the 1980s. It proved to be a difficult concept to sell to decision-makers and urban designers however, due mainly to its lack of a tight definition, but the European Union were successful when, in 1997, they commissioned COST Action E12 Urban Forests and Trees (1997–2002) to investigate the potential of developing a European approach to urban forestry, as opposed to trying to emulate and apply the American version. The Action subsequently established urban forestry in the eyes of the European Union as a specific scientific domain—in other words, it came of age in Europe—and has been making significant progress ever since (Konijnendijk et al. 2005). Might it be an opportune time therefore to consider whether a European version of the concept of Landscape Urbanism could be devised as the vehicle to meld green/urban infrastructure, grey infrastructure, blue infrastructure, urban forestry and urban design into a mechanism that enabled a different approach to be put into train to both retro-fit existing urban developments and deliver new, sustainable urban forms across the European Continent?

Castells (1993) was right when he said that *'the major cities throughout Europe constitute the nervous system of the economic and political body of the continent. The more nation-states wane, the more cities emerge as the driving force in the making of a new European society'.* That said, such cities have to be sustainable, efficient and cost-effective in the impecunious times in which we live, and it may be that a redefined Landscape Urbanism approach might well be the vehicle to deliver this.

Although the concept of 'Landscape Urbanism' first surfaced in the USA in the late 1980s amongst university academics, its first official public outing was in April 1997 at a conference at the Graham Foundation in Chicago. This conference set out to consider the issues that were arising in many towns and cities in the USA at that time which were suffering from a prolonged period of industrial decline, the subsequent creation of a multitude of brownfield sites and the exodus of populations from town and city centres in favour of a better quality of life in the ever-expending sub-urban areas. This was followed by a touring exhibition that promoted the idea that Landscape Urbanism was a theory of urban planning that suggested that the best way to organise these failing post-industrial towns and cities was through the design of the city's landscape, rather than through the design of its buildings (Thompson 2012).

More specifically, the exponents of the concept of Landscape Urbanism believed that the cities of the future offered little in the way of permanence in terms of design, infrastructure or ecology were likely to be constantly changing and that sustainable urban futures could only be assured by structuring a city through the union of both the urban and the rural landscapes into a 'green infrastructure'. Such an approach was deemed to be *'a response to the failure of traditional urban design and planning to operate effectively in the contemporary city'* (Corner 2006). This was epitomised in Frank Lloyd Wright's plan for Broadacre City (1958), within which everyone could afford to have a house on an acre (0.4 ha) plot of land and get around by private car. According to Register (2002), this set the seeds in the USA of an intellectual and aesthetic justification for urban sprawl. Fishman (1990) was especially critical of the gridiron layout, deeming it to be *'boundless by its very nature, capable of unlimited extension in all directions ... it is destroying the freedom of movement and access to nature that were its original attraction'.*

Landscape Urbanism was particularly critical of the ideas behind the concept of 'New Urbanism', deeming them to be seeking permanence through the revival of traditional, outmoded urban forms and being anti-car. Further, they considered that, all too often, 'landscaping' in towns and cities tended to promote a form of the 'romantic' or 'picturesque', rather than up-to-date modern, contemporary urbanism, whilst pursuing an undeliverable version of nostalgic 'naturalness', based on the mistaken assumption that we can somehow bring back past ecosystems by removing invasive species and replanting native ones. This over simplistic view of the world, according to the promoters of Landscape Urbanism, ignored the fact that environmental stability is an illusion (Botkin 1990). Landscape Urbanists firmly believed that they were in quest of a new urban aesthetic that had the potential to be the solvent that would break down the barriers between the professional disciplines—Tatom (2006) stating that the intellectual promise of landscape urbanism is *'to integrate the conceptual fields of landscape architecture, civil engineering and architecture for the design of the public realm'.*

Moderate critics of Landscape Urbanism have considered it to be only a réchauf-féd version of Ebenezer Howard's early twentieth-century 'Town-Country' idea which, at the time, was a new hybrid landscape type that brought together the advantages of both town and country life, allegedly without any of the dis-benefits, and was the subsequent foundation of the Garden City Movement in the UK. Fiercer critics deem it to be a concept only dreamt up to impress an academic elite, to be full of impenetrable jargon, dubious philosophy and unhelpful imagery, and thus fail to communicate adequately to the very people who might make it a viable approach to sustainable urbanism—politicians and decision-makers, fellow professionals or the general public—and therefore is of limited use in the urgent quest to plan and deliver sustainable urban futures. Even Charles Waldheim, one of the key promoters of Landscape Urbanism, willingly concedes *that 'the urban form promised by landscape urbanism has not yet arrived'* (Waldheim 2010).

It cannot be denied that, whilst there are many aspects of the concept of Landscape Urbanism that deserve detailed scrutiny and support in the quest for a more enlightened approach to developing sustainable futures, a wholesale application of the American concept of Landscape Urbanism to address the current European issues with, and aspirations for, developing sustainable urban futures would seem not to have many significant advantages and would have an uncomfortable fit with the policies being promoted and pursued by the EU. The American landscape and the relatively recent development of the towns and cities in the history and culture of that country are in stark contrast to the history, culture and development over many centuries of the European landscape, both urban and rural.

The European Union is very firmly of the view that the concept of 'Natural Capital' has to be an integral part of the drive towards smart, sustainable and inclusive growth, an EU priority for 2020 (European Union 2010)—something that would not be acknowledged in the current Landscape Urbanism ethos. Thus, whilst there are commonalities on both continents with innovative projects successfully confronting the issues of sustainable development and sustainable futures, and some of the ideas behind Landscape Urbanism undoubtedly have a significant part to play in this, perhaps a European version of Landscape Urbanism needs to evolve, as it did with urban forestry over a decade ago?

The idea that the future economic, environmental and social success of Europe is inexorably bound up with nature-based solutions and viable sustainable towns and cities has been high on the European Union's agenda for many years. References to such issues can be found in many of the debates and discussions that have taken place within the European Commission for several decades, but it was with the Habitats Directive and the establishment of Natura 2000—an EU-wide network of nature protection areas promoted as a form of green infrastructure—that such matters became enshrined in law as Directives (European Commission 1992).

Under paragraph three of Article 249 EC, a Directive is deemed to be binding upon each Member State to which it is addressed, whilst leaving the National Authorities the choice of form and methods for implementing the Directive in question. The fact that the European Union takes the issues surrounding 'nature' and green infrastructure very seriously can be established by the fact that, between the issuing of the Directive in 1992 and 2006, a total of 59 cases for infringement of the Directive, in one way or another, had been referred to the ruling of the European Court of Justice.

In spite of the above and the increased opportunities that this initiative has provided for the creation and preservation of areas of 'green'—some 20 % of European Union territory has been protected as part of the Natura 2000 initiative—the European Union concluded that significant progress had not been made in integrating biodiversity into other sectoral policies relating to the conservation and restoration of biodiversity and ecosystem services in the wider European Union countryside, or with reinforcing the compatibility of regional and territorial development with biodiversity in the European Union. Perhaps the policy was too focused on the countryside, and the concept of green infrastructure needed to be broadened to appeal to urban areas?

A revised concept of green infrastructure was originally set out in the report *'Towards a Green Infrastructure for Europe'* (European Union 2007) and backed up a year later with a special report entitled *'Building a Green Infrastructure for Europe'* (European Environment Bureau 2008), which expanded the concept of green infrastructure to encompass issues that also affected urban areas, including climate change, economic viability, the health and well-being of the population and sustainable development. Several definitions of 'Green Infrastructure' have emerged from the European Union, but it is now generally deemed to be defined as:

> a strategically planned network of high quality natural and semi-natural areas with other environmental features, which is designed and managed to deliver a wide range of ecosystem services and protect biodiversity in both rural and urban settings (European Commission 2013a).

Even though it has been estimated that in excess of 1000 km^2 of land is swallowed up annually in Europe by housing and grey infrastructure (European Commission 2013), the European Union is well aware that the concept of green infrastructure/urban green infrastructure does not constitute a 'one size fit all' approach applicable to all cities or city regions. It acknowledges that green infrastructure can make a significant contribution in the areas of regional development, climate change, sustainability and human health, [and] happiness and well-being, and that what is needed now is to ensure that it becomes a standard part of spatial planning and territorial development, and that it is fully integrated into these policies. It therefore advocates green infrastructure to be applied through the national, regional and local applications of existing European Union initiatives, such as the Integrated Sustainable Urban Development Initiative in the 2014–2020 Cohesive Plan, the Cohesion Fund and the Connecting Europe Facility, to name but a few.

For example, innovative initiatives have developed in Europe where a clash between green infrastructure and grey infrastructure has taken place, particularly where the movements of wildlife have come into contact with human-made barriers such as highways and railways. The first wildlife crossings were created in France in the 1950s, but these become much more sophisticated in the late 1960s/early 1970s and have now been constructed extensively in the Netherlands, Germany, Switzerland and France to reduce the conflict between wildlife using the green infrastructure and roads, highways and other aspects of 'grey' infrastructure. The Netherlands can boast of over 600 such structures, the longest being an ecoduct called the Natuurbrug Zanderij Crailoo, which is over 800 m long and 50 m wide and carries green infrastructure over a railway line, a business park, a river, a roadway and a sports complex (Danby 2004). Similar ideas are now emerging in Canada and the USA.

It has been acknowledged, however, that the development of green infrastructure/urban green infrastructure is at a crossroads in the European Union (European Commission 2013). In spite of the fact that more and more green and urban green infrastructure projects have been carried out over the past 20 years or so, and a wealth of experience has been accumulated that demonstrates that a flexible approach is both sound and cost-effective, the quality of human health and well-being within our towns and cities is generally decreasing (European Environment Agency 2013). It may be, however, that the potential of green/urban green infrastructure to deliver increasingly cost-effective solutions and improvements to our urban problems can be further increased by developing appropriate technology and processes, particularly in relation to the design and functioning of our towns and cities. Not only do our *'Cities need green in sizes S, M, L and XL—otherwise the human ecosystem is incomplete'*, as Peñalosa has so poetically suggested (Peñalosa 2014), but also we need 'intelligent' resource-efficient buildings and grey infrastructure that can deliver the environmental, sustainable, social and health benefits that are so often lacking in the current approaches to urban design in Europe (European Commission 2012). Might this be the opportunity for Landscape Urbanism to raise its profile in Europe?

In 2010, *Topos,* an influential international journal published in Munich to review matters to do with landscape and urban design, devoted a themed issue to the concept of Landscape Urbanism (Topos No 71 2010). The editor, Robert Schäfer, prefaced the issue by saying that the aim was to *'become more familiar with the ambiguous concept of landscape urbanism'*. It only seemed to contribute to the bewilderment of fellow professionals however as to how Landscape Urbanism as practised could contribute creatively to European sustainable futures. But does it need to? It could be argued that significant progress is already being made across Europe as towns and cities heed the advice from the European Union from their Resource Efficiency Roadmap (European Union Communication (2011), which suggested that our cities need to share their ideas and experiences, as the failure to protect our natural capital and to give a proper value to ecosystem services will need to be more radically addressed as part of the drive towards smart, sustainable and inclusive growth, which is a priority for the European Union (Europe Union 2010).

A number of significant networks are already in existence, sharing both 'good practice' and, more importantly, 'next practice'. For example, there are networks of 'green' cities across Europe that are working to establish such practice and share knowledge in both local land use planning and habitat protection (see http://www.capital-biodiversity.eu/44.html). One significant network, the Local Action for Biodiversity (LAB) programme (http://www.iclei.org/details/article/local-action-for-biodiversitybiodivercities-lab.html), whilst having a worldwide remit, is championed by European cities such as Amsterdam, Barcelona, Bonn and Zagreb. Using the Local Action for Biodiversity platform, these cities work to share knowledge regarding urban biodiversity management and policy implementation, as well as local ecosystem services' assessment reports targeted towards specific local and national needs.

The Council of European Municipalities and Regions (CEMR) has created a reference framework for sustainable cities integrating the task of biophysically greening cities into a broader process of sustainable economic development, which is included in an online toolkit. The main objective of the sustainable cities framework is to increase communication and knowledge sharing within and between European cities that are dealing with similar environmental policy challenges (see http://www.ccre.org/en/activites/view/25). Thus, there is an increasing amount of evidence that cities are sharing expertise and policy tools to biophysically green their urban infrastructure across a broad spectrum, embracing everything from local initiatives to trans-European Union-orchestrated networks.

As has been discussed earlier in this chapter, the pace of urban change across Europe has been accelerating over recent decades, with a range of new ideas and concepts being promoted and a variable quality of results being achieved. What should be borne in mind, however, is that as new thinking takes place and new scenarios are established, they do not wholly replace the previous ideas but simply add another layer to our depth of knowledge. This applies as much to the concept of urban forestry as it does to anything else.

It has been argued by Simson (2008) that urban forestry began to emerge at the end of the twentieth century as a philosophical approach to having an overview of the planning, design, establishment and prospective management of the whole mosaic of a town's, a city's or even a region's urban and peri-urban green space, and all the implication that that has to engaging with the associated social, economic, and environmental issues and strategies that are to be found there. Urban forestry is a 'big idea', and now that it is established and recognised in Europe, can it adapt and respond to the myriad of changes that are occurring across the urban stage of the continent, and if so, are there any good examples to consider? Indeed there are.

One area in Europe that has proved that such an approach is both feasible and effective is to be found in the Emscher region, the northern part of the Ruhr area in the state of Northrhine-Westfalia in Germany. The area was once the heartland of mining and heavy industry in Germany, but once that declined after the end of World War II, a landscape of industrial wasteland, spoil tips and waste tips remained, often situated directly next to residential areas. Successful land recycling depends on a corresponding demand for real estate and land, but the 'lunar landscape' of this part of the state of Northrhine-Westphalia did not encourage economic investment. Thus, from 1989 until 1999, the Internationale Bauausstellung Emscher Park, a vast regenerative enterprise conducted by the State Government, achieved great success by demonstrating that the standard approach to regeneration is not always the most appropriate—it is sometimes necessary to evolve a new kind of approach. Such an approach in the Emscher region involved linking up new urban development with a new urban forest to revitalise the ecological, economic and urban fortunes of the area through the creation of collaborative partnerships with the 17 local authorities that governed the area, private industry, professional associations, environmental groups and local communities. In this mission, it was very successful, and today, the Emscher Landscape Park is one of the greenest urban areas in Europe. It covers an area of 450 km^2 and stretches for 80 km from the Rhine near Duisburg in the west to Bönen in the east. Since the end of the International Building exhibition in 1999, more than 500 additional projects have been jointly planned and implemented by 20 local authorities, two local authority districts, three district governments, the State of Northrhine-Westphalia, the Regionalverband Ruhr, the Emschergenossen-schaft and the Lippeverband (the latter two enterprises both being water companies). This support and activity has confirmed the success of the extremely innovative approach taken to urban regeneration in this region and has added significantly to the profile of the Emscher Landscape Park as a trailblazer in promoting the concept of Landscape Urbanism in Europe (Simson 2005).

Innovative developments have continued apace in this region. At the end of 2007, the 35 cities and three counties that comprise the Ruhr area adopted the 'Concept Ruhr', an initiative to further advance the sustainable urban and regional development across the whole area. This focuses upon five 'ruhrbasics' as guidelines for sustainable development over the next 10 years, one of which is 'ruhrcities'—the quality of the city centres and the city districts. The original regenerative developments were a catalyst for the subsequent large financial investments currently taking place and being planned by the international private sector, so that the future Concept Ruhr developments will not require any public funding for their implementation, including the sports, leisure and tourist facilities that are being planned, as well as the green and sustainable industrial developments that are flooding into the area (Seltmann 2007).

It can be argued therefore that such a far-reaching, comprehensive, integrated and successful project as the Emscher Landscape Park, with its viable mix of urban green infrastructure, urban forestry and sustainable habitation, infrastructure and industrial development, combined with a more radical approach to all levels of governance, could be deemed to be a powerful example of the potential of a Landscape Urbanism approach to sustainable development and urban futures in Europe.

More recent examples involving green governance initiatives and the implementation of urban forestry to improve the quality of city life have been taking place in Montpellier, France. Montpellier is a city with about 260,000 inhabitants, and the population grew by some 22.5 % between 1990 and 2012, challenging the existing green space. The initiatives shown and progress made by this city with its adopted green vision and revised forms of governance, whereby the city administration works closely with citizens by establishing long-term partnerships, is a convincing example of Landscape Urbanism in practice.

Since the 1970s, the city of Montpellier has been working constantly on improving the quality of life of its inhabitants. In 1994, the city adopted a charter for the improvement of its urban and peri-urban environment, including aspects addressing the protection and improvement of existing urban landscape. Since the 1990s, the city has been working on promoting public participation in the governance of the city in the form of surveys, interviews, events, interactive websites and participation in projects and a wide range of communication tools.

The city has 741 ha of public green space that covers 13 % of the total area. In addition, there are over 80,000 street trees, and although the planning and management of urban and peri-urban green spaces are responsibility of the municipality, data and information on the green spaces and urban and peri-urban forests are publicly available and constantly updated.

The city has adopted '*The management reference to improve urban planning*' (AURA), which is essentially a handbook and a strategic tool based upon some

30 indicators and organised into 9 areas, that act as guidelines for the sustainable development of the city. These indicators relate to a number of issues associated with green space, including the maintenance of the ratio of the green space area and the total area of the city, the proximity of green space, the provision of connectivity between green spaces within the district, the number of plant species per ha, and the ratio of the total canopy cover generated by trees to the open area in any given sector of the city. The city's strategy is to open private green space for public use, and thus, from the 1970s onwards, the Montpellier authorities have been buying private properties for this purpose (Besse et al. 2014).

Other innovative initiatives have been taking place across Europe that could also be deemed to be assisting the development of a European Landscape Urbanist approach to sustainable futures. One of the best examples perhaps is the annual European Green Capital Award, which was set up as a result of an initiative taken on 15 May 2006 in Tallinn, Estonia, by 15 European cities—Tallinn, Helsinki, Riga, Vilnius, Berlin, Warsaw, Madrid, Ljubljana, Prague, Vienna, Kiel, Kotka, Dartford, Tartu and Glasgow—together with the Association of Estonian Cities. This award was officially launched by the European Commission in 2008 and aims to credit and acknowledge cities with more than 100,000 inhabitants that are actively involved in the protection of nature and biodiversity through the planning, design and application of sustainable urban development. Cities that apply for the award are assessed on 12 indicators of their sustainability:

- Local contribution to global climate change
- Transport
- Green urban areas
- Noise
- Waste production and management
- Nature and biodiversity
- Air
- Water consumption
- Waste water treatment
- Eco-innovation and sustainable employment
- Environmental management of the local authority
- Energy performance

The European Union recognises the importance of promoting and exchanging good practice amongst these cities, thus influencing the generation of 'next practice' and rewarding their efforts to provide increasingly sustainable living areas whilst reducing their impact on the global environment. Most importantly, these cities are seen as role models for other cities to emulate and for raising the awareness levels concerning urban sustainability issues. Whilst other intercity networks do exist—for example the Local Government for Sustainability (ICLEI), the Covenant of Mayors, URBACT and EUROCITIES—where cities can discuss and exchange good practice, this award acts as an incentive for cities to 'get their act together'. It is prestigious, as rewarded cities get more publicity and more tourists, and people like to live and work there, thus helping to improve the local economy. Stockholm was granted the first European Green Capital Award in 2010, followed annually by Hamburg, Vitoria-Gasteiz, Nantes, Copenhagen and this year (2015) the city of Bristol. In 2016, the mantle will be handed over to Ljubljana. Much can be learned by considering the activities and applications of the winning cities (see http://ec.europa.eu/environment/europeangreencapital/).

In 2012, for example Vitoria-Gasteiz, the capital city of the Basque Autonomous Community in the province of Álava, Northern Spain, won the Green Capital Award. It has more than 30% of public green areas, and all citizens live within 300 m of public zones and green spaces. These are interconnected into a green network that extends over the entire city and connects with the Green Belt. Between 1974 and 2010, the area of green space doubled, and the city now boasts 20.2 m^2 of green area per inhabitant—46 m^2 if the area of the Green Belt is taken into account and there are in excess of 130,000 street trees. The concept of the Green Belt started in the 1990s and currently comprises 727 ha. The project is still ongoing, and also embraces restored previously degraded areas.

The proposed target size of the Green Belt area is almost 1000 ha, and this is significant for both its environmental (Natura 2000 and wetland area protected by Ramsar Convention) and its social services, which encompass a variety of recreational opportunities, including more than 47 km of pedestrian and bicycle routes. In 2010, it was visited by 651,000 people, whilst 32,700 people took part in the educational and environmental awareness activities. The urban forest covers about one-third of city area, representing 465 m^2 of forest per inhabitant. It is mostly publicly owned, and more than 90% is natural forest. Several forest areas are included within the Natura 2000 Network. This Green Belt project was selected as best practice in the 2000 and 2004 Dubai International Award for Best Practices in Improving the Living Environment.

In 2003, three new by-laws were approved, one of which relates to the management and protection of the urban forest and trees. Digital inventory was created to facilitate and improve the management and maintenance of the green urban zones. The city pays great attention to the connectivity between the natural systems of the city and tries to minimise habitat fragmentation through, for instance, the consolidation of the Green Belt.

The city implements campaigns and environmental awareness programmes, such as 'Adopt a tree and grow with it', which has been in place since 1996, as well as an 'Ecological Horticulture Programme', which is a workshop linked with community gardens and also raises awareness of endangered species, such as bats. The city has developed educational material for teachers and children aimed at significantly raising the levels of environmental awareness.

Vitoria-Gasteiz is part of a number of international networks of cities, such as the Covenant of Mayors, which it joined in 2009. It also signed the Green City Declaration in 2005, during the San Francisco Urban Environmental Accords of the United Nations Environmental Programme (UNEP). The Green City Declaration drew together some 21 actions related to seven key areas—energy, waste reduction, urban design, urban nature, transportation, environmental health and water—with the explicit goal of implementing as many actions as possible until a review of progress held in 2012 (City Green Star Programme). More information on the sustainable urbanism stance taken by Vitoria-Gasteiz and its contribution to the concept of European Landscape Urbanism can be obtained from http://www.vitoria-gasteiz.org/we001/was/we001Action.do?idioma=en&accion=anilloVerde&accionWe001=ficha.

Nearer home, the UK City of Bristol has won the European Green Capital Award in 2015. The city, with a population of about 500,000 people, is one of the greenest cities in the UK and the country's most energy- and carbon-efficient city. About 20% of the population walk to work each day, and over 90% of the population live within 300 m of an area of green infrastructure.

A green infrastructure creates a network of multifunctional, interconnected green spaces throughout the city, and the value of planning, maintaining and enhancing connectivity for recreation, active travel and adaptation to climate change and biodiversity is fully recognised. The green infrastructure covers approximately 1600 ha—31% of the city area—and comprises eight nature reserves and more than 400 parks. Eleven of these parks have won United Kingdom Green Flag Award status, and the city won the gold award in the Entente Florale competition in 2012.

A statutory land use planning policy relating to the Strategic Green Infrastructure Network shapes land use decisions and developments in Bristol. Urban sprawl has been limited by cooperation with neighbouring municipal areas. This 'West of England Partnership' ensures that strategic housing, transport and green infrastructure are coordinated with adjoining authorities and that growth is focused, as far as possible, on existing centres and brownfield land. Over the past 10 years, 98% of business development and 94% of housing development have taken place on brownfield sites, creating high-density, mixed-use neighbourhoods.

Bristol has been chosen as the first city in the world to launch a new international initiative called One Tree Per Child. This was launched in February 2015 by Olivia Newton-John, a well-known Australian environmentalist and entertainer, and will involve 36,000 children from the 130 primary schools in the city planting at least one tree each. There is no cost attributable to the schools—the costs are covered by the city authorities. The Mayor of Bristol has said that *'Bristol's role as European Green Capital for 2015 makes us the ideal city to adopt this vision. Planting trees is a great way for school children to connect to the environment and their local community. As a child's tree grows, their commitment to the environment and their local community grows as well'.*

(George Ferguson, Mayor, pers comm.)

Governance and community engagement are key issues for this city. The city actively engages with communities in land use planning and the planning, design and management of green areas and has devolved significant decision-making powers to Neighbourhood Partnerships. In line with UK Government legislation (The Localism Act 2011), Bristol is in the vanguard of Neighbourhood Planning in the UK, with several Neighbourhood Planning Forums in place and over 25 local residents planning groups. Additional information that supported Bristol's application for Green Capital Status and the promotion of the concept of landscape urbanism in the UK can be obtained from two websites—http://www.bristol.gov.uk/sites/default/files/documents/planning_and_building_regulations/planning_policy/local_development_framework/Adopted%20Bristol%20Core%20Strategy.pdf and http://www.bristol.gov.uk/page/bristol-parks-and-green-space-strategy.

The European Green Capital Award for 2016 has been awarded to Ljubljana, the capital of the Republic of Slovenia and home to about 280,000 inhabitants. The city has gone through very significant changes over the last decade or so on its way towards promoting and achieving sustainability, and the reward for making such significant progress with this is gaining the title of European Green City Capital.

Major improvements have been made in local transport and bringing the city centre back to being dominated by pedestrians. Ljubljana has made equally important progress related to green space and today can boast of 560 m^2 of green area per inhabitant or 542 m^2 of public green space. In the compact part of the city, there is 106 m^2 of green area for each inhabitant, 66 m^2 of which is in the public realm. Almost all residential areas lie within a 300 m radius from public green space. More important than the amount of green space per capita is the even distribution of green space across the city. Green wedges act as links between the city centre and the hinterland and are considered to play a significant role in the climate regulation of the city. In the last few years, five new parks covering a total of 40 ha have been added, mostly as a result of the renovation of degraded areas, such as the Stožice Sport Park which replaced an old gravel pit.

The backbone of the Ljubljana Green System is the Path of Memories and Comradeship that was started in 1946. It is continuous, tree-lined avenue with 7000 trees and is very popular for sports and recreation activities. The Path lies along the line of the former barbed wire barrier which was erected during the World War II, was completed in 1985, and is continuously upgraded. Today, it is 34 km in length and comprises 166 ha of green space.

The internationally awarded project 'Renovation of the Banks and of the River Ljubljanica' is a great example of urban rehabilitation in Ljubljana, and the project received a European Prize for Urban Public Space in 2012. The River Ljubljanica runs like an artery through the heart of the city, along which various opportunities are provided for citizens. These include walking and cycling areas, and cultural and socialising opportunities, such as reading points or so-called the 'Library under the Trees'. Another project 'Revitalising the Embankments of the River Sava' has added 20 ha of public space and will be enriched with new sports and recreational infrastructure.

Despite the increase in urbanisation, Ljubljana has almost 50 % of its entire area covered with natural forests. The international project EMoNFUr developed guidelines for managing such urban forests that are recognised as being more under pressure than forests outside urban areas. A monitoring methodology was proposed to better understand the processes taking place in urban forest ecosystems, which is very necessary for their preservation. When we think about cities, biodiversity is not usually at the top of the agenda, but, in reality, urban areas can harbour valuable sites important for biodiversity conservation. In Ljubljana, there are several Natura 2000 sites, and in total, more than 20 % of the area has nature protection status. In addition, 1400 ha of forest has recently been declared as special-purpose forest.

Urban planning policy regarding green and open space areas aims at protecting the existing vegetation cover that is currently almost 80 % of the city and keeping the minimum of 50 % of green space on parcels planned for construction.

Ljubljana collaborates with non-governmental organisations and volunteers in small projects aimed at improving the quality of life in local communities by the active involvement of citizens. One of these projects is the 'Labyrinth of Art', which comprises trees planted by citizens where each person has an opportunity to become responsible for their tree. Another example is providing garden allotments for citizens on land owned by the city or temporary gardens on unused agricultural or construction land. The City co-finances many civil society projects related to the removal of invasive species, the implementation of measures for the conservation of endangered species and the provision of interpretative trails. Citizens, especially children, have been involved in raising awareness activities related to endangered species in Ljubljana. Additional information on the green infrastructure, governance and success of the Ljubljana story can be obtained from http://www.ljubljana.si/en/green-capital

An investigation into both the winners of, and the cities shortlisted for, the European Green Capital Award scheme, where ideas of green infrastructure and urban green infrastructure have to be incorporated into both their policies and their application for the award, has yielded valuable insight into both urban green infrastructure policy and application. These key European cities place a high value on the urban setting for people, places and investment as the nucleus of European human culture, and several common themes emerge from an interrogation of the above studies in their pursuit of sustainable development and, arguably, European Landscape Urbanism.

They all exhibit, for example, an integral approach to planning for sustainability and having criteria and indicators in place for monitoring and evaluating the progress of these indicators, as well as trying to minimise the loss of biodiversity and green space due to urbanisation and development. They have positive policies in place to revitalise old industrial areas and wasteland in order to provide new services for their citizens, usually for recreational purposes.

Another feature of these Green Capital Award Cities is their emphasis on the revitalisation of riverbanks wherever possible and the protection of wetlands. This is primarily for biodiversity, nature conservation purposes and flood alleviation, but it also assists with improving the connectivity between their green space and, in some cases, their green and blue infrastructure. This pursuit of connectivity also assists with policies to meld green infrastructure with their urban forestry in order to help with delivering a sustainable approach to urban development.

A key feature of these Green Capital Award Cities is the desire to include their citizens, whenever possible, in the planning, design, implementation and management of green space projects and a belief in devising and applying revised forms of governance to assist in delivering such policies. Governance can, of course, mean many things to different people. In its broadest sense, governance is '*any effort to coordinate human action towards goals*' (Rayner et al. 2012). Governance involves certain actors, whilst at the same time other may be excluded, depending on the rules of the game, and it takes certain resources such as money, time, human skills and expertise, to achieve the desired goals.

Governance is a broad and 'elusive' concept, according to some (Arnouts et al. 2012) and it needs to be deconstructed to be analysable. There is consensus in the literature on what the key attributes (concepts) are and the processes that characterise good governance. Generally, governance is considered to be good if all the relevant stakeholders are involved in the process, the decision-making is transparent, and there is accountability of actors, especially the decision-makers, the rule of the law and predictability. It also implies the effective and efficient management of resources, and a fair and equitable allocation of resources and benefits (Food and Agriculture Organisation 2011). Governance often involves coordination and cooperation amongst state actors, the private sector, the public sector, the voluntary sector and civil society.

Discussions relating to the governance of sustainable development, green infrastructure and urban forestry are undeniably complex issues. So are the societal challenges that sustainable development, green infrastructure and urban forestry are addressing, such as improving the quality of life of urban residents, stopping biodiversity loss, climate change adaptation, the provision of various ecosystem services, fair and equal distribution of benefits to constantly growing urban population, and many more. Each and every one of these challenges cannot be simply solved by one actor. Society, in the broadest sense, has to be involved as a whole in the discussions as to what goals are going to be set, what activities are expected to enable these goals to be achieved, who would be responsible for these activities, and what inputs would be forthcoming from the various sectors and disciplines.

Also, global issues such as climate change cannot be limited to the administrative borders of a city, a region or a state. These affect society on various levels, and hence, decisions for their mitigation have to be made on various levels. Therefore, such governance occurring on these various levels is called multilevel governance (Hooghe and Marks 2001).

The European Union is a typical example of multilevel governance. This means that some decisions are made on a European Union or supranational level (e.g. decisions on common currency), whilst other are left to be decided by Member States (on national or subnational levels). For example, responsibility for decisions on forest policy rests with each Member State, and the same applies to the governance of green infrastructure. The European Commission's communication *Green Infrastructure (GI)- Enhancing Europe's Natural Capital* (European Commission 2013b) provides a definition on what green infrastructure is, why it is important and how it contributes to other European Union policies, as well as what should be done at a European Union level to encourage the implementation of green infrastructure.

The European Union's guidelines and recommendations of what is considered to be good practice related to the implementation of green infrastructure are, however, non-legally binding (if the Natura 2000 network is put aside), or put simply, there is no redress if these recommendations are not implemented. But since green infrastructure is contributing to goals of other policies such as regional development, climate change, biodiversity and nature conservation, the European Union can support and promote its implementation through funding mechanisms. Other way of steering good practice is by providing incentives in terms of awards meant for those who act as role models, such as European Green Capital Award, as discussed previously, and the recently established European Green Leaf Award—see http://ec.europa.eu/environment/europeangreencapital/europeangreenleaf/index.html for details of this award.

Despite the importance of urban forest and green infrastructure governance in providing many benefits to urban populations and thus major contributions to sustainable development, research on urban forest governance is scarce (Krajter Ostoić and Konijnendijk van den Bosch 2015; Lawrence et al. 2013). James et al. (2009) in their review paper identified governance as an emerging theme and several key research questions for the field. A search of SCOPUS on 26 February 2015, one of the major bibliographic databases, on 26 February 2015 showed that when looking for papers containing both terms 'urban forest' and 'governance' in their title, abstract and key words, the search results in only 16 hits, whilst a search for 'green infrastructure' and 'governance' resulted in 19 papers found.

Lawrence et al. (2013) a provide comparative framework for describing and comparing urban forest governance, based on the policy arrangement approach that comprises four elements—discourses, actors, rules and resources (Arts et al. 2006). The paper employs five various case studies from Belgium, Italy, Sweden and UK to test the framework which proved helpful for the description and comparison between different modes of urban forest governance and across different contexts. Analytical frameworks such as this cannot, however, provide information about the quality of the governance. Even though the five case studies presented in the paper are not representative at the European level, the authors identified a trend towards more 'governance with government' in comparison with the previous trend of 'governance by government', also known as top-down governance. This means that urban forest governance has become more inclusive in terms of public participation in decision-making, at least it is based on these examples.

Similar conclusions have been made recently by a research group within the European Union's FP7 project Green Surge (ENV.2013.6.2-5-603567; 2013–2017), who produced a report on the governance of urban green spaces in 20 European Union cities (Buizer et al. 2015). The aim of the group is to identify and conceptualise innovative participatory governance arrangements regarding the management of urban green infrastructure, based on the policy arrangement approach. They came to the conclusion that, besides traditional types of governance arrangements where government has the major impact on decision-making (governance by government), there are also examples of governance with government and interestingly but sporadic, informal, spontaneous activities without governmental influence (governance without government). However, green space projects involving non-governmental actors are often initiated and led by governmental actors.

It is not easy to make firm, all-embracing conclusions on how urban forest or green infrastructure governance works across Europe as a whole because it can differ widely between and within the individual countries (Britt and Johnson 2008). What we do know is that, in spite of the fact that Europe is in the vanguard of developing new models of governance in the pursuit of sustainable development—one of the key differences between Europe and the USA in exploring the concept of 'landscape urbanism'—there is a need for more research on urban green infrastructure/urban forest governance, what modes of governance exist and what achieves the best results on the ground for successful sustainable development (Krajter Ostoić and Konijnendijk van den Bosch 2015; Bentsen et al. 2010).

Moving forward from the identified knowledge gaps in terms of the state of the art in studying the melding of the concepts of urban forestry and urban green infrastructure, COST action *GreenInUrbs* has been initiated by the European Union, partially to address these gaps. The COST Action started in 2013 and gathers together researchers and practitioners from over 35 countries. Amongst the many goals of the Action are '*to provide indicators and/or thresholds to be included by policy makers in local, national or international regulations about green infrastructure and urban forests*' and '*to develop guidelines for GI planners and managers on how to implement green infrastructure approaches with an emphasis on linking the environmental and social services of urban forests*'. For additional information, see http://www.cost.eu/COST_Actions/fps/Actions/FP1204.

One of the working groups of this COST Action is dedicated to specifically studying the governance issues associated with urban forests and (urban) green infrastructure and its potential contribution to the delivery of sustainable development. Information from all Member States is being collected and analysed to enable guidelines for policy- and decision-makers to be produced by 2017 relating to the role of urban forests when planning, designing, establishing and managing green infrastructure as a metaphysic for sustainable urban development—and Landscape Urbanism?

Whilst predicting the future is not an exact science, in spite of the popularity of reading our horoscopes in our daily newspapers, one thing can be relied upon and that is that one of the biggest problems that we face internationally, nationally and locally is the matter of urban change. As has been discussed earlier in this chapter, towns and cities across the continent of Europe will be subject to constant change over the coming generations, and it is likely that no area will be immune from the forces that bring this change about. Indeed, as the twenty-first century progresses, it is very likely that the pace of change will quicken, and this will bring about a number of readjustments to urban areas (Brotchie et al. 1995), including the approach to creating and maintaining sustainable development.

Since the publication of the Green Paper on the Urban Environment in 1990 (European Union 1990), urban densification has been advocated as the way to create more mixed land uses and related multifunctional and livable open space environments, whilst protecting existing areas of open space within urban areas and saving peri-urban green space and 'natural' areas from urban encroachment—in short, sustainable developments. All too often, this has not been successful however as in spite of the fact that contemporary urban design policies at both international and national levels advocate versions of the Compact City as a strategy to address urban sprawl and achieve a more sustainable approach to urban development, European cities are expanding at an increasing rate.

Neuman (2005) has called in to question the whole concept of the compact city, suggesting that the evolution of the planning profession at the end of the nineteenth century was largely due to the need to improve the health and well-being of urban inhabitants by relieving the overcrowding of the industrial towns and cities by letting in more light, air and 'green', the results of which have been largely successful, until now. Faced with the current issues of increasing urban sprawl, the political response now seems to want to reverse these ideas and 'recompact' the city. He also questions whether cities can really be sustainable at all, as they have always been dependent upon their hinterlands and distant counterparts for food and trade.

There is also a growing amount of research that suggests that contemporary urban design is not creating the 'livable cities' that it claimed it would. Adverse health issues are significantly on the increase for urban inhabitants. Air pollution from motor vehicles for example is on the increase across Europe and is causing many premature deaths, and in the UK the use by children of the outdoors has shrunk by 90 % over the past 30 years, partly influencing an increase in childhood obesity. Thus, as has always been the case, those people who can move out of the city have been doing so, with the result that the urban footprint has been increasing significantly, which cannot be deemed to be 'sustainable'.

The advent of the concept of green infrastructure, initially as a counter to the loss of 'nature' and 'natural landscapes' and, more recently in conjunction with the concept of urban forestry as a counter to the more negative aspects of urban development, has been of some success. There is also a growing canon of research that proves that the inclusion of trees in and around the city has great benefits. We now know that trees can improve our health and well-being, improve learning, increase property values, provide focal points to improve social cohesion, improve air quality, offset carbon emissions, promote biodiversity, limit the risk of flooding, cool our towns and cities, promote inward investment and job creation and even make us drive more safely (Konijnendijk et al. 2005).

Both urban design and green infrastructure are at a crossroads, however, in Europe, as the quality of urban life continues to decline, cities continue to expand as a result, sustainability is at best static, whilst green infrastructure, urban green infrastructure and urban forestry are still perceived, in some circles anyway, to be a 'green cosmetic', rather than the first principal that they really are. Might the time be opportune therefore for a new concept to emerge that champions sustainable development by bringing together built form, grey infrastructure, blue infrastructure and the broad concept of green infrastructure into an all-embracing, transdisciplinary approach to urbanism, and might that concept be Landscape Urbanism, or a European version of the idea?

There are those who say that we really do not need another '-ism' for sustainable development to be the norm. We already know enough to enable us to deploy the concept, and there are clearly some very good examples across Europe where this is the case—in the European Green Capital Cities for example. That said, there does arguably need to be further refinements of our approach to sustainable development, not least out definition of 'nature'.

As discussed previously, the original ideas behind green infrastructure were to limit the detrimental effect of urban development and grey infrastructure upon 'nature', an idea that was later modified to assist with bringing 'nature' into the urbs for the benefit of the urban population. What this suggested, however, was that human being were not really part of nature, but at best interested onlookers and that somehow we could emulate aspects of the natural world in and around our towns and cities. Clearly, this is not the case, nor has it been for some while.

Just as urbanism is changing, so is the 'natural' world. We need to accept and embrace the fact that, in the twenty-first century, there is an increasing variety of hybrid ecosystems developing across our landscapes and that they will influence sustainable development. They are the result of the interacting forces of urbanisation, globalisation, and climate change and are made up of organisms that have been brought together by the removal or modification of barriers that have kept them apart for maybe millions of years. Such hybrid eco-systems are not only to be found in and around our towns and cities, but also in many of the landscapes that have been disturbed by intensive agriculture, industrial processes and mining, and it is unrealistic to assume that we can turn the ecological clock back to bring 'nature' back into our cities any more than we can turn back the economic clock that created these disturbed landscapes in the first place. Barker has described these eco-systems as comprising 'recombinant ecology'—in other words 'learn to love the alien' (Barker 2000).

Thus, just as there is a need for us to accept a hybrid series of contemporary eco-systems to assist us with developing a sustainable approach to our landscapes, it could be argued that we also need to accept an all-embracing, transdisciplinary hybrid approach to sustainable urban development—or better still, see them as different sides of the same coin.

The word 'landscape' is interesting, for as Jackson has stated:

> Landscape is a compound word. As far back as we can trace the word, land meant a defined space, one with boundaries … Scape once meant a composition of similar objects … landscape is synthetic space, a man-made system of spaces superimposed on the face of the land. (Jackson 1984)

Thus, there is a need emerging for a responsive concept for planning, designing, constructing and managing sustainable development that is transdisciplinary, cost-effective, and easy both to understand and communicate to all levels of the community. For a city to be successful, it has to have a critical infrastructure in place. This comprises sufficient energy, food, water, viable transport, telecommunications, public and emergency services, health and viable finance/economic activity. We now know, however, that this is not enough—what is also needed is a viable urban forest, a green infrastructure, a blue infrastructure, a hybrid ecological infrastructure, a mosaic of multifunctional green space, liveable communities and transboundary connections.

To guarantee a viable sustainable approach to European urban development, these infrastructures should perhaps be amalgamated into a transdisciplinary 'integrated infrastructure', whereby development cannot be deemed to be sustainable if an integrated infrastructure is not deployed? Such a concept could well be described as the European approach to Landscape Urbanism, where the landscape structure is not seen as superior to or a prerequisite of successful urban development but an equal partner in the quest for sustainable development and viable revised governance, so that our towns and cities regain their rightful position as centre for human creativity, innovation and quality of living. A fanciful theory perhaps—but as Lewin stated '*There is nothing so practical as a good theory*' (Lewin 1951).

References

Arnouts R, van der Zouwen M, Arts B (2012) Analysing governance modes and shifts—governance arrangements in Dutch nature policy. For Policy Econ 16:43–50

Arts B, Leroy P, van Tatenhove J (2006) Political modernisation and policy arrangements: a framework for understanding environmental policy change. Public Organ Rev 6:93–106

Barker G (eds) (2000) Recombination in urban areas: implications for nature conservation. English Nature, Peterborough

Benedict MA, McMahon ET (2006) Green infrastructure: linking landscapes and communities. Island Press, Washington D.C.

Bentsen P, Lindholst AC, Konijnendijk CC (2010) Reviewing eight years of urban forestry & urban greening: taking stock, looking ahead. Urban For Urban Green 9(4):273–280

Besse F, Congliaro M, Fages B, Gauthier M, Mille G, Salbitano F, Sanesi G (2014) Montpellier—green city. Unasylva 242(65):23–28

Botkin D (1990) Discordant harmonies: a new ecology for the 21st century. Oxford Press, New York

Britt C, Johnston M, (2008). Trees in towns II: a new survey of urban trees in England and their condition and management. Research for amenity trees. Department for Communities and Local Government, London

Brotchie J, Batty M, Blakeley E, Hall P, Newton P (1995) Cities in competition. Longman, Melbourne

Buizer M, Elands B, Mattijssen T, van der Jagt A, Ambrose B, Gerőházi É, Santos A, Møller MS (2015) The governance of urban green spaces in selected EU-cities. Policies, practices, actors, topics. http://greensurge.eu/working-packages/wp6/files/Buizer_et_al_2015_D6.1_GREEN_SURGE_The_governance_of_urban_green_spaces_in_selected_EU_cities.pdf. Accessed Jan 2015

Castells M (1993) European cities: the information society and the global economy. J Econ Soc Geogr 84(4):247–257

Corner J (2006) Terra fluxus. In: Waldheim C (ed) The landscape urbanism reader. Princeton Architectural press, New York, pp 21–33

COST Action FP1204 (2013) Green Infrastructure approach: linking environmental with social aspects in studying and managing urban forests (GreenInUrbs), http://www.cost.eu/COST_Actions/fps/Actions/FP1204 WG3 National case studies: (www.greeninurbs.com). Accessed 20 Jan 2015

Danby D (2004) A green latticework. http://www.worldchanging.com/archives/001093.html. Accesed 13 Jan 2015

European Commission (1992) Habitats Directive, Council Directive 92/43/EEC

European Commission (2012) Connecting smart and sustainable growth through smart specialisation. ISBN 978-92-79-27345-2

European Commission (2013a) Building a green infrastructure in Europe ISBN 978-92-79-33428-3

European Commission (2013b) Green infrastructure (GI)—enhancing Europe's natural capital, a communication from the commission to the European Parliament, the Council, the European economic and social committee and the committee of regions. COM 249

European Environment Agency (2006) Urban sprawl in Europe, EEA Report No 10/2006. ISSN 1725-9177

European Environment Agency (2011) Green Infrastructure and territorial cohesion: the concept of green infrastructure and its integration into policies using monitoring systems. Technical report No 18/2011. ISBN 978-92-9213-242-2

European Environment Agency (EEA) (2013) Joint EEA-JRC report No 5/2013 environment and human health. ISSN 1725-9177

European Environmental Bureau (2008) Building green infrastructure for Europe. A special report 2008/017

European Union (1990) Green paper on the urban environment. COM (90) 218 final. ISSN 0254-1475

European Union (2007) Towards a Green Infrastructure for Europe—developing new concepts for integration of Natura 2000 network into a broader countryside. EC Study ENV.B.2/SER/2007/0076

European Union Communication (2010) EU's priority Europe 2020: a strategy for smart, sustainable and inclusive growth. COM 2020 final

European Union Communication (2011) Resource efficiency roadmap COM 0571 final

Food and Agriculture Organisation of the United Nations (FAO) (2011). Framework for assessing and monitoring forest governance, Rome. http://www.fao.org. Accessed 13 Jan 2015

Fishman R (1990) America's new town: megalopolis unbound. Wilson Quarterly, Washington, DC, pp 25–48

Hooghe L, Marks G (2001) Types of multi-level governance. EIoP 5(11):1–24. http://eiop.or.at/eiop/pdf/2001-011.pdf. Accessed 26 Feb 2015

Jackson JB (1984) Discovering the vernacular landscape. Yale University Press, New Haven

James P, Tzoulas K, Adams MD, Barber A, Box J, Breuste J, Elmquist T, Frith M, Gordon C, Greening KL, Handley J, Haworth S, Kazmierczak AE, Johnston M, Korpela K, Moretti M, Niemelä J, Pauleit S, Roe MH, Sadler JP, Ward Thompson C (2009) Towards an integrated understanding of green space in the European built environment. Urban For Urban Green 8:65–75

Konijnendijk CC, Nilsson K, Randrup TB, Schipperijn J (eds) (2005) Urban forests and trees—a reference book. Springer, Berlin

Krajter Ostoić S, Konijnendijk van den Bosch CC (2015) Exploring global scientific discourses on urban forestry. Urban For Urban Green 14:129–138

Lawrence A, De Vreese R, Johnston M, Konijnendijk van den Bosch CC, Sanesi G (2013) Urban forest governance: towards a framework for comparing approaches. Urban For Urban Green 12(4):464–473

Lewin K (1951) Field theory in social science: selected theoretical papers. Harper and Row, New York, p 169. (In: Cartwright D (ed))

Neuman M (2005) The compact city fallacy. J Plan Educ Res 25:11–26

Peñalosa G (2014) Happy city—transforming our lives through urban design (Mongomery C (ed)). Farrar Straus Giroux, New York

Potočnik J (2013) Cities will define our future. Speech given to the 7th European conference on sustainable cities and towns, Geneva, 17 April 2013. http://www.europa.eu/rapid/press-release_SPEECH.13-326_en.htm. Accessed 15 Jan 2015

Rayner J, Humphreys D, Perron Welch F, Prabhu R, Verkooijen P (2012). Introduction. In: Rayner J, Buck A, Katila P (eds) Embracing complexity: meeting the challenges of international forest governance. A global assessment report prepared by the global forest expert panel on the International forest regime. IUFRO World Series, vol 28. International Union of Forest Research Organisations (IUFRO), Vienna

Register R (2002) Eco cities—building cities in balance with nature. Berkley hill Books, Berkley. ISBN 1 893 163 37 7

Secretariate of the Convention on Biological Diversity (SCBD) (2012) Biannual report on the administration of the convention on biological diversity. http://www.cbd.int. Accessed 14 Jan 2015

Seltmann G (2007) Renaissance of an industrial region: 'Internationale Bauausstellung Emscher Park'—achievements and future model for others. http://www.ris.osaka-.a.c.jp/jp/events/point/P.Selltmann.pdf. Accessed 13 Jan 2015

Simson AJ (2005) Urban Forestry in Europe: Innovative solutions and future potential. In: Konijnendijk C, Nilsson K, Randrup T, Schipperijn J (eds) Urban forests and trees—a reference book. Springer, Berlin

Simson AJ (2008) The place of trees in the city of the future. Arboric J Int J Urban For 31:97–108

Tatom J (2006) Urban highways and the reluctant public realm, p 181. In: Waldheim C (ed) The Landscape Urbanism Reader. Princeton Architectural Press, New York

Thompson IH (2012) Ten tenets and six questions for landscape urbanism, landscape research, vol 37, No1. Routledge, London

US Census Bureau (2011) Population distribution and change—Detroit, Michigan 2000-2010. In: Mackun P, Wilson S (eds). http://www.census.gov. Accessed 13 Jan 2015

Waldheim C (2010) On landscape, ecology and other modifiers to urbanism. Topos 71:20–24

Chapter 12
Sustainable Architecture Assemblages

Maria Theodorou and David Turnbull

12.1 Part 1: Thesis

12.1.1 Architecture's Lively Matter: A Thinking Device

Author: Dr Maria Theodorou, senior lecturer at The Leeds School of Architecture, director of the School of Architecture for All (SARCHA).

We can start our *exegesis* of architecture by the way of a commonplace, by employing an old fashion and almost 'boring' understanding of the discipline as a type of manipulation of ideas and matter. In such an understanding, architects seem caught in a platonic conceptual device between the immaterial realm of ideas and the material construction of the real; it is the mediation of a projection (project, drawing) that translates and transfers their inner vision onto the external word to materialize in the form of a building or any other architectural structure.

This understanding appears reductive and outdated, especially in the current technological condition of sophisticated design software, processes of scripting and state-of-the-art construction technologies and 3D printing. Architects visceral relation with the machines creates a thinking assemblage of human brain and computational programmes in a nonlinear process; human bodies of architects and technicians, software, hardware and materials continually reconfigure ideas and matter and reshape the conceptualization and experience of our world. As a matter of fact, our planet is having a redesign. To paraphrase Latour (2008): Will Prometheus ever be cautious enough to redesign the planet in a sustainable way?

M. Theodorou (✉)
Leeds School of Architecture, Leeds Beckett University, Room B502 Broadcasting Place, Woodhouse Lane, LEEDS, LS2 9EN, UK
e-mail: m.theodorou@leedsbeckett.ac.uk; david.turnbull@atopia-research.org

D. Turnbull
The Irwin S. Chanin School of Architecture, ATOPIA Research Inc., Cooper Union, New York, USA
e-mail: david.turnbull@atopia-research.org

© Springer International Publishing Switzerland 2016

M. Dastbaz et al. (eds.), *Building Sustainable Futures*,
DOI 10.1007/978-3-319-19348-9_12

Latour's (2005a) actor network theory (ANT) and theories of new materialism, especially Jane Bennett's Vibrant matter (2010), provide a valuable conceptual framework in thinking the redesigning of the world currently unfolding; their authors put under scrutiny the category of 'matter' understood as inert entity; they, thus, challenge the standard architectural approach to 'matter' as an entity that has only instrumental value for the design and construction of various types of structures and infrastructures. Sustainable design guidelines and materials specification—although they address performance—still largely remain within the Cartesian object—subject and anthropocentric conceptual framework.

Assemblage theory from Deleuze to De Landa (2006) to Latour (2005a) thinks through the organization of complex nonhierarchical systems. If we were to understand buildings as assemblages of human, technical, technological, organic and inorganic matter that are in a constant condition of vibrancy and emergence, another approach to architectural design unfolds. It was in fact Manuel De Landa's new materialism that opened up the field in that direction; De Landa framed the understanding of performativity in architecture and provided the conceptualization of emergence, computational generative design and its potential to integrate sustainable principles. Current architecture theory and practice has nevertheless been dominated by the discussion of materialism understood as a different technical—technological way of manipulating matter and data. Sustainability policy in industry focuses on a consensual model of sustainable architecture practice; it ignores the complexity of assemblages implicated in the process. This approach has already been contested; the current debate on architecture sustainability has acknowledged the fact that its terrain is made up of heterogeneous and conflicting views (Guy 2013).

Assemblage and new materialism theories emerged in other disciplines (Glynos 2012) and signalled an 'Affective Turn in Social Science', as Chantal Mouffe phrase it in the recent Westminster CSD symposium (27 February 2015). Architecture is nevertheless always quick to exploit the rich terrain of interdisciplinarity to import concepts in its toolkit of invention. Our cities can be conceptualized accordingly and understood as complex assemblages in a condition of constant emergence. Design for a sustainable future cannot but take onboard this complexity.

Theorists have also coined the term 'anthropocene' (Turpin 2013) to indicate our era as an identifiable geological strata on which human activity has imprinted its presence by constantly reconfiguring the natural and built environment. If sustainability takes as a point of departure the human's relation to nature and the human's technical technological exploitation of natural resources, new theories of the environment (Harrison 2014) rethink and challenge this understanding: humans, nature and technologies cannot sit comfortably within given and distinct boundaries.

New materialism, assemblages, the anthropocene reframe in a fundamental way the humans' relation to both nature and technology. This cannot but impact our views on the very constitution of the concept of sustainability (Goodbun and Jaschke 2012). The critique to new materialism, namely, that it produces a flat ontology in which humans are just one component in the assemblage and thus bear no responsibility whatsoever for their actions is counterbalanced by Latour's (2005b) matters of concern; the latter provides a tool for thought and action. But let us take a breath and go back to our opening outdated paragraph on architecture.

Is this fascinating redesigning/rethinking of the world in which architecture partakes, which makes the platonic conception of the architect as manipulator of matter and ideas boring? Is this the reason that the long-dropped aesthetic judgment, based on platonic beauty, has been substituted by the category of the interesting? For interesting is what catches the attention of the bored individual.

Boredom is a condition of acute anxiety, manifested mainly in teenage years when adolescents are caught in the limbo between limited past resources (memories) and scarce future projections (visions). This teenage frustrating limbo has a name: it is called the present. Boredom is this condition of being caught in the present, being contemporary with one's own time. Boredom is a condition of modernity (Svendsen 2005) but as Latour argues, 'we have never been modern' (1993) and thus we return to it.

> Nothing much is left of the scenography of the modernist theory of action: no male hubris, no mastery, no appeal to the outside, no dream of expatriation in an outside space which would not require any life support of any sort, no nature, no grand gesture of radical departure—and yet still the necessity of redoing everything once again in a strange combination of conservation and innovation that is unprecedented in the short history of modernism. (Latour 2008)

This boredom of the platonic model is the anxiety that we need to bear as we cannot go back to the past reassuring dualism, back in a world that we were familiar with. Ordered categories of ideas and matter, each belonging to their realms, which humans—architects included—could traverse and manipulate. We cannot also run towards the future since there has been a rapid expansion of fields; predictions and visions are difficult and almost impossible. Since we dwell in the present, we have become, albeit not teenagers, but contemporaries.

The contemporary, according to Agamben (2009), is the one 'who firmly holds his gaze on his own time so as to perceive not its lights, but rather its darkness' As contemporaries, we can no longer be blinded by the shining materials and convoluted forms of impressive architectural achievements, or be concerned with the painstaking writing of universal sustainable architecture specifications and marvel at green buildings; our 'off-cells' peripheral vision is adjusted so that we already discern the shadows that creep and crawl and are inseparable from the shining lights in our cities. We have named them shanty towns or slums in the global south, or unaffordable housing in London.

Everyone is keen to escape the grim embrace of the present, to work out a vision of the future. The proliferation of initiatives, conferences and publication on future cities is telling and frantic. The current publication *on Building Sustainable Futures: Design and the Built Environment* is no exception; it partakes in this anxiety of the present. And this chapter reaffirms that the concept of sustainability has been already challenged, contested, and it is in the process of re-thinking.

The intertwining of future with sustainability was already there in plain view in the title of the Brundtland Report (1987), and the standard historical reference to the definition of sustainability, first introduced by the (United Nations General Assembly 1987). Agamben (2009) reminds us 'to be contemporaries not only of our century and the "now," but also of its figures in the texts and documents of the past'. It is worth following his advice and rehearse the report's title, 'Our Common Future: Towards Sustainable Development', and its well-known definition:

> Sustainable development is development that meets the needs of the present without compromising the ability of future generations to meet their own needs. It contains within it two key concepts: the concept of 'needs', in particular the essential needs of the world's poor, to which overriding priority should be given; and the idea of limitations imposed by the state of technology and social organization on the environment's ability to meet present and future needs.

Sustainability was born as a Siamese twin of development. The UN text launches a cautionary remark, addressing the anxiety of its own time, the 1980s present. The concept of sustainability appears to cast a shadow on the notion of development, as an undisturbed process of modernization, but it immediately suppress it by the evocation of the future. The Brundtland Report signalled the moment of a reversed trend. As Saskia Sassen's (2011, p. 24) statistics chart has revealed retrospectively, the equal distribution of resources was about to end and the redistribution of wealth meant a bumpy road ahead for sections of population that have enjoyed the implementation of welfare policies after the WW2 and were to be subjected to brutal expulsion.

In the 1992 Earth Charter Preamble, sustainability alone carries the future of a just and peaceful global society about to enter the twenty-first century.[1] Rising concerns about development and the revival of de-growth movement were sidestepped by the decoupling of sustainability from development; the former could deliver the future vision of cultural mix-bliss societies on a peace-ridden mother-earth; but already the anti-globalization movement was brewing. UN, however, never decoupled sustainable development in any of its subsequent documents. The pairing of sustainability with development has been interpreted as an oxymoron (Till 2012) but in fact, sustainable development is the name of a different logic, which makes perfect sense within the current pervasive mentality of control.

If, according to Michel Foucault, nineteen century early twentieth disciplinary state took care of its citizens and governed their bodies, current states do not care to impose order and discipline. As Gilles Deleuze vividly described in its 1990 'Postscript', the disciplinary state has been substituted by a state that manages and controls. Agamben (2014) argues that the state is not interested in preventing crisis, but it is instead prepared to manage any crisis; the aim is not '…to maintain order but … to *manage disorder'*. Sustainable development is not an oxymoron but the lexicon of the management of a crisis, which has already happened. How can one keep living standards if, as we are told and experienced, we already live in failing

[1] "We stand at a critical moment in Earth's history, a time when humanity must choose its future. As the world becomes increasingly interdependent and fragile, the future at once holds great peril and great promise. To move forward we must recognize that in the midst of a magnificent diversity of cultures and life forms we are one human family and one Earth community with a common destiny. We must join together to bring forth a sustainable global society founded on respect for nature, universal human rights, economic justice, and a culture of peace…".

systems, in a permanent condition of environmental, financial and humanitarian crisis? The fitting concept seems to be 'survivability' rather than sustainability.

The sustainability concept cannot be dissociated from the political context which invented it and still supports it. In Gilles Deleuze's control societies (1990), 'This is no longer a capitalism for production but for the product, which is to say, for being sold or marketed'. Sustainable development is promoted as a product to be marketed, not as a problem (in the sense of a project) to be solved. Gilles Clément (2013, p. 261) in his fierce critique of the 'green business' writes:

> At no point has the problem of human suffering in the 'planetary garden' been confronted, but many give the impression of doing so. First of all, they communicate. The primary task of partisans of sustainable development is to establish a communication strategy. Words and images. Images above all. Make the aestheticized claim of a planet in crisis with photographs that are marvelous and tragic, taken from a bird's eye views, or from afar—make books, speeches, proclaim good intentions—to change the climate, we will find the means. The technology of the twenty-first century is repositioned under the sign of sustainable development

Architects are not always attracted to theories and philosophers, but they are educated to become practitioners and understand sustainability as a technical issue, i.e., as a problem at hand to be solved by designing sustainable structures that can save resources. Architects contribute in building up a new green economy (Clément 2013) and related industry. The coupling of sustainability to development makes perfect sense: New sustainable building materials are produced, new sustainable design principles employed. A new architecture field of study has opened up. Sustainability was inserted in architecture education, shaping the curriculum.

The concept of sustainability seems rather settled within the technical discourse, even though there seems to be a controversy; as already mentioned, the consensual model of sustainability rules/guidelines, to be applied in practice-led building design, has been challenged by a call for a more inventive, flexible, fluid approach which emerges within the sustainability agenda in both practice (Guy 2013) and education (Austin 2013).

Sustainability is a much contested term, if we take away its camouflaging as a technical problem that asks for a specific solution. A problem—in the Greek sense of the word *problema*, has always a double meaning; it is something to put in front of one's eyes as a project (projection) and at the same time, it operates as a shield behind which one takes refuge (protection; Derrida 1993).[2] When sustainability is apprehended, within architecture, as a technical problem to be solved by prescribing appropriate design guidelines, the sustainability project resides within the safe haven of a problem intended as 'protection'.

[2] '…problema can signify projection or protection that which one poses or throws in front of oneself, either as the projection of a project, of a task to accomplish, or as the protection created by a substitute, a prosthesis that we put forth in order to represent, replace, shelter, or dissimulate ourselves, or so as to hide something unavowable –like a shield (problema also means shield, clothing as barrier or guard-barrier) behind which one guards oneself in secret or in shelter in case of danger' (Derrida 1993, pp. 11–12).

Nevertheless, criticism of sustainable architecture is manifested in an unlikely place, which appears to share the premise of sustainability, namely, the condition of depleting resources. The recent discussion on scarcity (Till 2012) is used to critique sustainable development and explore alternative ways of practice within the current austerity and scarcity conditions. When Jeremy Till (2012) argues that 'Sustainability thinking assumes that scarcities are inevitable and can be quantified, and that the way to handle these scarcities is through programs of reduction and control'; he, in fact, confirms the intertwining of the discourse of sustainability with the Deleuzian current state of the pervasive control logic.

It is worth noting that when building performance and post-occupancy evaluations point to a failure of sustainable design, the tendency is to blame the unpredictability of the human factor (Herring and Roy 2007), i.e. the buildings occupants whose behaviour defies control. Latour's theory of 'distributive agency' in which the source of action, the 'actant', can be either human or inhuman and the understanding of Bennett's matter as a vibrant condition, can provide a reframing of the sustainable architecture practice and education and bring forth the political aspects of alternatives.

'In its broadest sense, the strategy for sustainable development aims to promote harmony among human beings and between humanity and nature' reads the conclusion of the Brundtland Report (1987). And this is a reminder that not only there is no going back to the platonic conception of architecture and its dualism of matter and ideas, but we cannot rely either on a settled relation between humanity and nature.

Our past conceptual resources are depleted already, and we are condemned to the anxiety of the intense reflection of the present. This is not 'a form of inertia or passivity, but rather implies an activity and a singular ability'. (Agamben 2009); this is the task that the authors of this chapter have undertaken in presenting and discussing instances of their practices and experiences in the present tense.

12.2 Part 2: Praxis

12.2.1 Waterbanks: A Case Study

Author: Professor David Turnbull, The Irwin S. Chanin School of Architecture at The Cooper Union for the Advancement of Science & Art, Design Director of PITCHAfrica and the Waterbanks Initiative.

Waterbanks is an Initiative of PITCHAfrica, a not-for-profit Design and Research group that I co-founded with Jane Harrison, and where I am Design Director. We work across the African continent, with some of the most disadvantaged communities in the World.

The premise is simple. Given the impact of Climate Change on weather patterns, rainfall and other forms of precipitation might it be both necessary and desirable that building types are developed that can harvest and store as much water as possible, providing enough capacity that the increasingly erratic time divisions between the 'rains' can be ameliorated. Could this provide an incentive for the widespread provision of locally based, dispersed and decentralized infrastructures for water supply? And, finally could this model, which demands a level of community based stewardship that piped, linear, centralized systems do not require, provide the circumstances for invention in relation to the political processes that govern water-use generally, and ultimately 'survival'?

Many of the challenges that arise in communities stricken by poverty and disease stem from a lack of clean water and a lack of access to a feasible water supply. Development initiatives rightfully focus on the urgency of water provision but the extreme complexity of the social issues that surround water provision are all too often beyond the capabilities and resources of the organizations providing the technological expertise. Additionally, the international development community has focused on ground water resources since a UN edict in the 1980s. Groundwater and surface water supplies are often seriously contaminated and heavily fluoridated or contain concentrations of toxins such as arsenic. Borehole wells drilled into underground aquifers frequently break, and without engineers or tools, they remain broken. Aquifers sink and are not replenished. According to the UNEP, 60 % of the borehole wells drilled in 2004 failed within a year. More than 320 million people living in Africa do not have access to clean water. Many of these are children, particularly young girls, who cannot go to school because they have to walk miles to collect water from rivers or streams, dirty water, instead. Dirty water kills around 4500 children a day.

Many people do not know that there is 13 times the amount of water needed, falling as rain, across the African continent, and that most people who do not have access to clean water live in areas with more than 300 mm of rainfall annually. In many areas of the Sub-Saharan region suffering from extreme poverty and challenging conditions in relation to water access, nutrition and sanitation, have up to a metre of rain in the year, some much more. The distribution pattern of rainfall throughout the year is changing as a consequence of climate change, but erratic rainfall patterns—the arrival of the 'rains' earlier or later than expected, deluge followed by drought—makes rainwater harvesting and water storage more important than ever.

We are committed to demonstrating the overlooked potential of large-scale community-based rainwater-harvesting initiatives, based on a local, dispersed, decentralized and nonlinear approach to infrastructural construction; significantly, the potential of rainwater harvesting as a catalyst, instigating social transformation within a community by making carefully designed, low-cost buildings, with significant water storage volumes and integrated water filtration systems. We call these structures 'Waterbanks'. We have been focussing on schools, as in our experience 'Waterbank Schools' can become a significant resource in the community, addressing multiple issues simultaneously. In areas where Land Rights and land ownership battles are common, schools are built on land that is less contested. Schools have principles that embrace concepts of shared ownership, governance and accountability that transcend tribal differences. Within the precinct of a school, the pursuit of gender equality is more achievable. Schools also have the capacity to attract sympathetic, like-minded people in community-based organizations, foundations and non-governmental organizations, who can provide the funding and social support that is necessary in confronting some of the most intractable social issues that underpin poverty and impact many people, girls and women in particular, profoundly.

We insist that Waterbank School Buildings are never 'just' schools, and we have found that it is possible to build twice the volume, with integrated water storage and filtration for the same cost as a typical four classroom school. So that while working with completely generic space standards and normative budgets that are familiar to every organization working in the International Development field, every school can have a cistern and clean water supply, can be a tool for teaching children about environmental issues, an attractor and catalyst for the community and effect environmental and social change. The discipline of building at the lowest costs possible while providing more space, light, ventilation, protection and a consistent water supply at school involves us in a process of design and development with local partners, NGOs, community leaders, teachers, students and their families that builds social engagement and an enduring commitment to people and place. In the past 10 years, we have designed and built a collection of high-yield rainwater-harvesting building types, invented and patented modular ceramic water filtration systems, and a structure that integrates rainwater harvesting and water storage into a sports stadium—bringing Africa's passion for Football together with its greatest needs, water, nutrition and sanitation. The work depends upon a return to some of the fundamentals of architecture and involves spatial, structural, material and technological invention at multiple scales, from the nano/micro-scale behaviour of water percolation through ceramic water filtration membranes to the scale of buildings and accumulations of building types, to climatic, geographic, social and ecological patterning. A key idea behind all the work relates to a term, 'reverse innovation', used in the sense of looking back historically in order to look forward more effectively, but also in support of a reciprocity between advanced research and poor technique that makes the designs easily replicable, increasing impact.

How Did We Start?

In 2004, we were working on a speculative project in South Sudan, linking access to water to the catalytic power of football. This project for a football field that could harvest rainwater was exhibited in New York at The Van Alen Institute in 2006. It addressed water access, food security, health, education, gender equality and community development, simultaneously. We called it PITCHAfrica. Once the word was out, we established important relationships with Grassroots Soccer and Play Soccer International who were having real success using football to effect social change. The following year we were in Copenhagen, for the Metropolis Biennale, organized by the Copenhagen International Theatre, KIT, and the Danish Architecture Centre, and collaborating with the Homeless World Cup, an organization that is using the power of street football to end homelessness. While we were at the Tournament (Fig. 12.1), Jane and I did calculations for the rainwater harvesting yields for the small street football stadium that we were sitting in. The results were extraordinary, in excess of 1 million l per year in a semi-arid region, and the space

Fig. 12.1 The finale of PITCHAfrica's friendly game at the Homeless World Cup in Copenhagen 2007

under the stands was more than adequate for classrooms, clinics or other community facilities. We knew then that a small stadium designed to catch the rain could be incredibly transformative.

We organized a friendly game with players from Kenya and Uganda, made a video and drawings (Fig. 12.2). We gave every African team documents explaining how a PITCHAfrica stadium worked. In early 2008, we met the Annenberg Foundation Board and a few months later received a significant grant providing backing for project development and testing. We built prototypes, and we conducted extensive research into water filtration options focussing on ceramic filters which can be made locally rather than expensive imports (Figs. 12.3, 12.4, 12.5, and 12.6). We filed patents. We made projects. In July 2010, the Annenberg Foundation supported work culminated in the construction a full-size demonstration model of a PITCHAfrica stadium in Los Angeles, launched during the World Cup by South African actress Charlize Theron. This was televised in the USA and created considerable international attention (Fig. 12.7). On the back of this event, we set up PITCHAfrica as a social enterprise in its own right. The rainwater-harvesting stadium was the flagship, and a wide variety of rainwater-harvesting structures, water filtration devices and irrigation tools were designed and ready to go. We wanted to develop PITCHAfrica rainwater-harvesting demonstration centres in Southern Africa, West Africa and East Africa, but getting projects implemented was tough-going. We had become increasingly convinced that finding the right scale for an initial community-based project was crucial. We knew that the cost of the stadium building, although

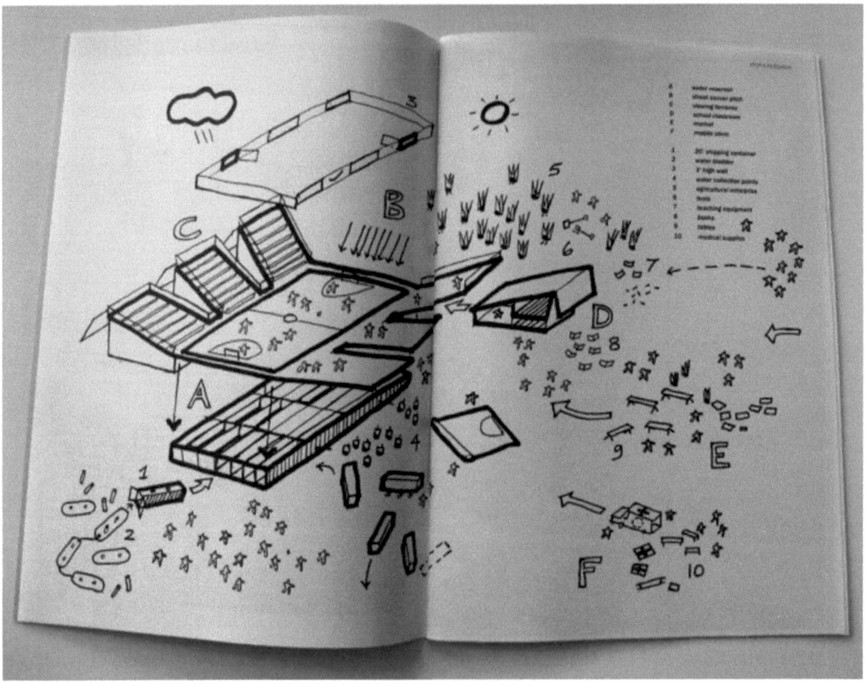

Fig. 12.2 Assemblage explained—a sketch prepared during the Homeless World Cup 2007

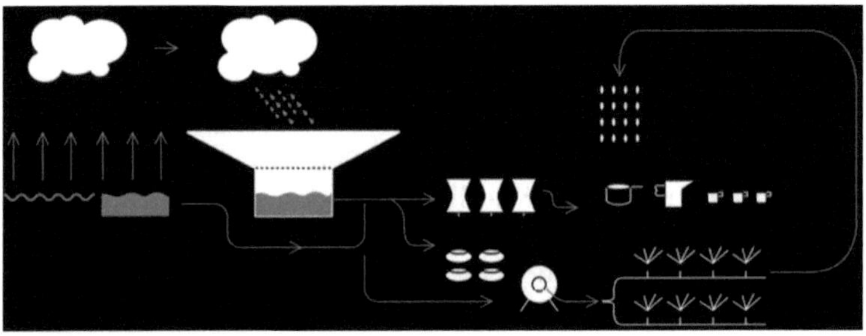

Fig. 12.3 Diagram: rainwater harvesting, storage, filtration and use

very low in relation to comparable buildings in the USA or Europe had the potential to destabilize community development by putting too much investment in one place at the expense of neighbouring areas, biasing migration patterns and creating problematic population growth, especially true in pastoralist communities. Starting small had always seemed more appropriate and a structure that could be copied easily and cost not more than 38,000 €, or US $ 50,000 a more manageable target.

We struggled to work out viable partnerships in Southern Africa, trying to put together projects in Kwazulu Natal, ZA, in 2011, and further north in Ndirande, Malawi, in 2012, making grant applications and multiple project proposals. But we

Fig. 12.4 In the workshop, fabricating the Waterbank prototype with Lucien Peebles. Activist, Rebecca Gomperts, founder of Women on Waves and Women on Web and PITCHAfrica advisory board member, filming

Fig. 12.5 Testing a corner assembly—with Lucien Peebles and Jane Harrison from PITCH Africa

failed. I was invited to be a visiting professor of design and innovation at The African University of Science and Technology, a Nelson Mandela institution in 2012, and we produced a strategic plan for a Waterbank Farm on the Campus in collaboration with AUST faculty including Prof. Wole Soboyejo and Prof. Albert Ayeni. In East Africa, we had more success. Since late 2007, we had been in contact with a number of important Kenyans including Mohamed Ahmed and KHSSA, Kenyan Homeless Street Soccer Association, and Bob Munro at MYSA, the Mathare Youth Soccer Association, who were doing great work. I had met Wangari Maathai at the Cooper Union in New York and initiated a dialogue.

Fig. 12.6 Prof. Wole Soboyejo inspecting the modular water filtration unit prototype, used for testing flow rates and yields.

Fig. 12.7 PITCHAfrica launch event in the Port of Los Angeles 2010 concluding the initial research period. This event coincided with the World Cup in South Africa

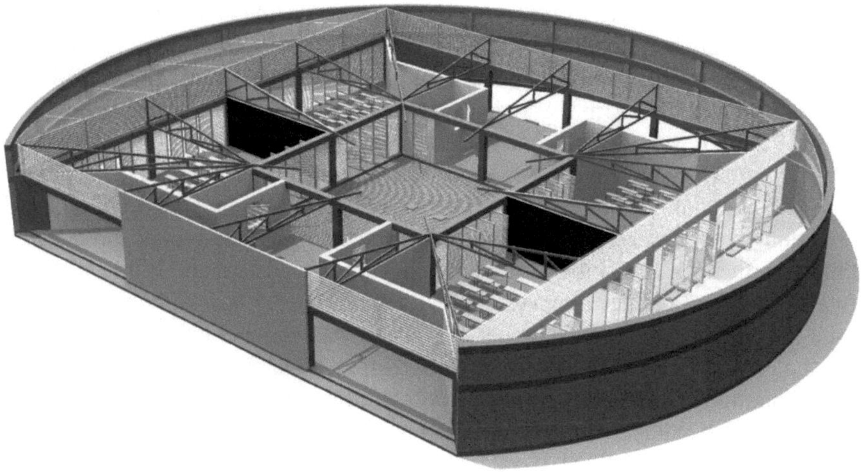

Fig. 12.8 Waterbank School—digital model

Then, after an introduction from colleagues at the Mpala Conservancy in Laikipia, in early 2011 we received a call from Dr Liz Rihoy, director of the Zeitz Foundation, based on a neighbouring Conservancy, Segera, and I went to Kenya for the first time. Kenya is an ideal place to implement community-based rainwater-harvesting initiatives. The geography and climate are perfect, and the presence of the United Nations Environment Program Headquarters in Nairobi is an asset. Dr Rihoy felt there was a great fit between our work and the Laikipia Unity Cup, a programme the Zeitz Foundation had developed. The LUC and Laikipia Unity Program use a Football Competition as a setting for environmental education. She was keen to see how we might bring PITCHAfrica to Laikipia. For us, a collaboration with the Zeitz Foundation was an excellent opportunity to realize the designs and technologies we had developed and demonstrate their transformative potential. Later that year, we sent Daniel Gastfriend, as a PITCHAfrica intern, to work with the Zeitz Foundation and I made my first visit of many in Segera. Daniel spent time with the school community at Uaso Nyiro, near Nanyuki in Kenya's semi-arid central highlands, and also worked on a Zeitz Foundation project that has since become the SATUBO, SAmburu, TUrkana and BOrana, traditional crafts enterprise. Then, in early 2012 the Zeitz Foundation obtained funding from Guernsey Overseas Aid for a school building at the Uaso Nyiro Primary School and asked us to propose a design that demonstrated PITCH principles. This was an opportunity to adapt one of our designs (Fig. 12.8).

In Kenya, there is seven times the amount of rain fall than is needed by the population. Rain is an astoundingly misunderstood and underused resource and one that asks that we change our unsustainable relationship with water to a balanced one. A Waterbank is designed to transform our relationship with water, if 'business as usual' advocates extracting water from underground, non-renewable aquifers, and building schools that deflect rather than catch and store rainwater easily. Waterbanks reverse this, bringing the water to the centre of the building, manifestly, demonstrating how this resource can empower whole communities and lead to systemic change and that everyone can have clean water. And so followed the construction of the first Waterbank School which was built at the Uaso Nyiro Primary School where Daniel had been working, in 2012. The building is a square, 24 × 24 m set within a circular perimeter wall. The plan is very simple, by necessity.

The Waterbank School is designed to harvest the maximum amount of rain with minimum materials and effort in a central courtyard with underground cistern. Rain falling on the roof drains with 95 % efficiency into the cistern and is drawn off and filtered by ceramic water filters when used, removing 99.9 % of pathogens. Water is used for drinking, daily meal preparation, hand-washing and irrigation. Surrounding the central cistern are protected, well-ventilated classrooms, teaching gardens and a community workshop, for the parents and for community activities, enabling the school to become a catalyst for transformation. Even in this semi-arid region, the school delivers 350,000 l of water annually. For example, the school initiates region-wide rainwater harvesting, filtration and conservation agriculture efforts. The roof of the cistern is all-school gathering place and environmental theatre where essential knowledge about practices and technologies is shared across ethnic and language divides. The school educates 360 children and provides water for 680 children, from seven tribes, pastoralist communities living on less than $ 1.40 per day. It serves as a community education centre for 4000 and as a demonstration school for a region of 400,000 people. On the day Nelson Mandela died and people across the continent were mourning, we got message from Njenga Kahiro, Community Liaison Officer at the Zeitz Foundation to say the tank was full to overflowing for the first time. It was a poetic moment, as if both tears and rain had created a supply of water and renewed life for a community. The school was named Greenest School on Earth by the US Green Building Council, and 'Waterbanks' was listed as one of the top 100 sustainable innovations by SUSTAINIA, Denmark, in 2013 (Figs. 12.9, 12.10, 12.11, and 12.12).

Also, in 2013 we participated in the Buckminster Fuller Challenge and won the inaugural 'Interface Support Award' for 'Waterbanks'. As part of the very rigorous challenge process, we had to write short sentences that described how the Waterbank School addresses Buckminster Fuller's seven criteria for innovation. Being forced to make succinct claims in short form is always useful, and this is what we said:

1. *Visionary*: All new schools in poor communities where it rains can produce clean water and fresh food for children and catalyze community transformation.
2. *Comprehensive*: The solution is simple, affordable and cross-cutting, directly addressing rural poverty, ill-heath, lack of education and environmental degradation. Systemic change is driven by a geometrical, material and social reconfiguration of standard school design.
3. *Anticipatory*: Anticipating global aquifer depletion, with 1 billion people worldwide without clean water and 60 million children out of school, our relationship with water must change. The Waterbank School directly addresses this need.
4. *Economically responsible*: The school is an artificial watershed. Each element works hard, structurally and socially, using the fewest resources, to deliver the

Fig. 12.9 Waterbank School, 'harambee' organized by the Zeitz Foundation, engaging the school community, digging the foundations

Fig. 12.10 Waterbank School, early morning roll-call

Fig. 12.11 Rainwater harvesting and water filtration workshop with the school community, taste testing

Fig. 12.12 Waterbank School, in use: a prize-giving in the courtyard, the classroom, and work in the garden

Fig. 12.13 The Waterbank Campus at the Endana Secondary School under construction

greatest impact on a child's health and a community's transformation. All buildings will need to be Waterbanks.

5. *Feasible*: It is built simply using local skills, resources and technologies, supporting self-sufficiency and avoiding dependencies.
6. *Verifiable*: The school is built and a post-occupancy evaluation underway. Attendance has risen by 25% to 95%. Instances of waterborne disease have dropped to zero.
7. *Replicable*: With simple guidelines anyone can build a Waterbank School.

The principles seem to be gaining some traction, and more Waterbank School structures were completed in the region in 2014: The Endana Secondary School in Laikipia in Kenya's Central Highlands is the location of the first Waterbank campus (Figs. 12.13 and 12.14), a collection of Waterbank building types have been built in 2013–2014 and are connected together by conservation agriculture plots, creating a model school that harvests and stores more than 2.5 million l of water annually. The Waterbank building types demonstrated here include PITCHKenya, a 1500-seat rain-harvesting sport stadium with integrated classroom block and environmental centre, a Waterbank Girls Dormitory for 100 girls, a Waterbank Canteen and Dining facility and various Waterbank Latrine blocks (Figs. 12.15, 12.16, 12.17, and 12.18). Each Waterbank building type is designed to integrate water harvesting, storage and potable water filtration. The form of each building is designed to maximize the volume and retention efficiency of the water harvested, storing the supply underground, at the heart of each building, for use by the building occupants where a dedicated water supply can help support specific needs. The Waterbank Girls Dormitory provides a clear example of how this works. The building is designed to accommodate 100 girls in three dormitories that face onto a protected courtyard garden. The building can harvest up to 360,000 l annually creating a dedicated supply

Fig. 12.14 The Waterbank Campus, aerial view showing the three principle structures: the stadium, canteen and girl's dormitory and secondary buildings, the boys dormitory, staff housing and latrines

Fig. 12.15 PITCHKenya under construction. Every surface is arranged to maximize rainwater harvesting yields

Fig. 12.16 The Waterbank Girls Dormitory under construction

Fig. 12.17 The Waterbank Canteen with removable fabric walls used in the rainy seasons

for the young women to address comprehensive needs including drinking water, water for sanitation, showering and the washing of clothes and water for irrigating the garden that will be developed to provide a sanctuary with nutrition-rich plants and natural remedies. It is a safe place, away from the boys and men in the school community, in a walled compound, with a secret garden, and a small house for the matron, a surrogate mother, who looks after the girls (Figs. 12.19 and 12.20).

Fig. 12.18 Waterbank Campus Latrines: four blocks in a compact arrangement including open air courts for drying laundry

Fig. 12.19 Waterbank Girls Dormitory: a protected compound around a triangular garden courtyard

In January 2014, I walked around the Endana site with the governor of Laikipia County (Fig. 12.21). He was incredibly enthusiastic about the work and really understood the potential of the Waterbank and PITCH building types. There was a mood of excitement about the project and the idea that water-harvesting centres

Fig. 12.20 The Matron's House in the dormitory courtyard on top of the underground water storage tank

Fig. 12.21 PITCHAfrica Design Director, David Turnbull at the Waterbank Campus construction site with the Governor of Laikipia, discussing cloud formations

could also be known as centres for community and peace building and an example that could be followed in other schools. Now, in early 2015, The Waterbank Campus is nearing completion and the conservation agriculture beds are being prepared. The ceramic water filtration systems have not yet been installed in the respective

Fig. 12.22 Waterbank Campus: a weekend visit from an enterprising herdsman's goats reveals a water management issue

Waterbank buildings but when completed will provide potable water throughout the campus (Figs. 12.22, 12.23, 12.24).

Each Waterbank structure has been developed to address a specific set of issues that are central to improving livelihoods and health in rural regions suffering from a lack of access to clean water. These issues range from water and food security, to gender rights, environmental education and sport for development support. The campus has been developed to welcome community participation and support the spread of knowledge about sustainable lifestyles throughout the region.

What Are We Doing Now?

We have stated work on a Waterbank District or Urban Quarter in the Niger Delta where the annual rainfall is close to 1.5 m per year, but where the groundwater is heavily contaminated. This district is intended as a demonstration of the Waterbank approach in an urban setting, and the design process involves the development of residential and mixed-use building types as well as schools, clinics and institutional structures. Historical antecedents can be found in Vitruvius' descriptions of the construction of cisterns in antiquity, Roman 'Impluvium' courts, French 'Lavoirs' or English 'Dewponds'. The filtration systems are made simply using clay, wood-dust and colloidal silver, based on open-source documents prepared by Ron Rivera, improved and tested in laboratories at Princeton University and the African University of Science and Technology in Abuja. The first Waterbank school was an island, and isolated singular building, the Waterbank campus is an archipelago of types, the Waterbank district is a dense urban accumulation or cluster. Our research continues by building and continuing to explore the ways and means for scaling the project.

Fig. 12.23 Waterbank Campus: a menstrual hygiene workshop with BeGirl reusable sanitary pad designer, Diana B. Sierra in the PITCHKenya environmental training room

Fig. 12.24 PITCHKenya with the stands full on opening day: the mexican wave

This work is only possible with a lot of goodwill, and with funding or in kind support from the Annenberg Foundation, Zeitz Foundation, The Buckminster Fuller Institute, Interface Inc., AUTODESK and The Clinton Global Initiative. We are working towards the launch of WATERBANKS OS in 2015, an open-source operating manual on the design, construction and use of Waterbanks, starting with a Waterbank school. A central part of this effort is to communicate how the stored water can be used and regulated most effectively so that throughout the dry seasons, there is a consistent water supply and that the water does not run out. Feedback from the school buildings that have been completed at Uaso Nyiro and Endana will play a key role in the Waterbanks story and the success of the open-source variants. The challenges that lie beyond the self-evident complexity of building anything in remote parts of the world without roads, with limited resources and skills, are all concerned with governance, the way that groups of people organize themselves to manage the water resource, to draw-off only the amount of water that is needed and understand the dynamic interrelationship of water captured with water stored and water filtered and used, in uncertain times with less predictable weather; in this sense, Waterbanks pose a question.

12.3 Part 3: Stasis

12.3.1 Architecture Knowledge Assemblage the Sustainable Present of Architecture Practice and Education

A discussion with David Turnbull and Maria Theodorou on ATOPIA[3] and SARCHA[4]

The discussion focuses on the experiences that both authors share in practicing architecture by taking a stance towards the profession and seek alternative ways of architecture production and of the production of architecture knowledge. Can their experimentations be understood as 'sustainable' architecture practice, in other words, can 'sustainable' architecture be produced only within a different mindset that generates another type of practice and education?

12.3.2 The Architecture Knowledge Assemblage

Maria: We have met long ago in 1995 at the AA in London but it was only 10 years later in 2005 at Princeton that we had the chance to discuss our common interest in developing and experimenting with alternative ways of practicing architecture. I remember you presented ATOPIA and first ideas on the PITCHAfrica project in your backyard and I was trying to sketch out an independent structure (SARCHA), a hybrid between school and practice, to carry out research on topics

[3] ATOPIA founding members are David Turnbull and Jane Harrison www.pitch-africa.org.

[4] School of Architecture for All (SARCHA) non-profit was initiated by Maria Theodorou and Larry Cool in 2006 www.sarcha.gr.

that were somehow considered unintelligible. It seems there were a number of things that had changed within the architecture profession between 1995 and 2005 that created a different mindset for both of us.

David: Yes, I think that there had been many changes, personal and professional, not least the transition for us, from the UK to the USA. In 1996 I was invited to teach at Yale as the Eero Saarinen Professor, a big step, and the beginning of a trans-Atlantic life that continued until 2002 with Jane's appointment at Princeton University. This provided a period of stability in which we have built our practice, and developed the ATOPIA project as an umbrella for many design and research activities that challenge conventional expectations about what it means to be an Architect. SARCHA's conception as a 'School of Architecture for All' also questions the authority and the education of the architect in addressing complex city conditions. In common with many of my colleagues my teaching started to reflect the impact of the 90s, the 'digital decade', critical structural changes in the global economy, manifest in urban structure and building design, and an increasingly urgent ecological imperative.

Why ATOPIA…? ATOPIA is an idea that can be discovered equally in the fields of Architecture, Aesthetics, Economic Theory and Science Fiction. In accord with Vittorio Gregotti's essay 'On Atopia' in his collection, Inside Architecure (1996), we saw spatial diffusion and the failure of the local as a category as symptomatic of a generalized globalization of means and methods. We witnessed the exacerbated thinness and dematerialization of architectural effects, and the collapse of any sense of social purpose. We agreed with Gregotti that an increasingly pervasive 'oriented atopia' should be challenged.

This was the perfect alibi for our work. Atopia is an idea about space and time; borderless, enabled by advanced communications media and digital tools, and omniscient; the perfectly flawed but ubiquitous setting for invention, technological innovation and design, an explicit 'anti-utopia'. The publication of German sociologist, Helmut Willke's 'ATOPIA' (2001) reinforced the obligation that operating on atopia, in 'atopian constellations', refusing the cult of the individual and the romanticism of individualism, had become for us. The more recent publication of 'The Atopia Chronicles' (2012) by Science Fiction writer Matthew Mather dramatizes fundamental choices that impact the survival of the human race; choices that are not options for us. This made the definition of two paths and two 'businesses' a necessity for us, both working with the paradoxes, ambiguities and contradictions of atopia: ATOPIA Innovation (for profit), consulting with Institutions, Development Companies & Foundations; ATOPIA Research (not for profit) working on global challenges with NGOs, Community Based Organizations (CBO's) and Corporate Social Responsibility Programs (CSR). PITCHAfrica and Waterbanks originate from the not for profit path.

In the same time-frame you were developing SARCHA, what sparked the idea?

Maria: SARCHA was conceived as an architecture school open to all with the aim of conducting independent research on pressing issues on architecture and the city. This was a much needed experiment in 2005 since the teaching of architecture still reflected a 'post-critical' condition. The fascination with new technologies of design and construction overshadowed any political, social and financial issues that were already present in the cities but largely ignored by architects.

In 1997, I curated the Greek Ministry of Culture pavilion at the UNESCO International Conference on "Education and Public Awareness for Sustainability". However, the notion of sustainability as a vision 'of a peaceful and just global society in the 21st century' (Earth Charter 1992) was already crumbling. In 2003, during my visit as a Council of Europe cultural policies expert in Tbilisi, the capital of Georgia in Caucasus, I was confronted with the vision of the future to come: Iveria a luxury hotel in the heart of the city was turned into a vertical IDP's (Internationally Displaced Persons) camp. Instead of a report on cultural policies, I drafted a short article on Iveria published in the Council of Europe journal (Theodorou 2003), which later became a longer piece for the Anarchitektur publications (Theodorou 2007). It seemed that the cultural, smooth, just and peaceful sustainable future had a sinister turn. The present was manifested as a moment of the 'political' defined by political theorists such as Chantal Mouffe (1993) as the moment of rupture of an existing condition which has not yet substituted by a new one. Routlegde book series on the thinking of political that run from 1994 to 2005) featured a number of publication that informed my thinking. Yannis Stavrakakis, author of the *Lacan and the Political* book (1999), became an interlocutor, contributing to the thinking and naming of SARCHA; he sits on the advisory board ever since.

SARCHA's focus on the political aspects of architecture could not be fully grasped by architecture educators in 2005. It seemed the schools in academia were operating on a standard repertoire of topics related to the agenda of the 90s; however, this trend was soon to change. In 2005, only a handful of marginal groups and individuals were meeting in Europe to discuss Architectural Resistance and similar topics of concern; today everyone in architecture schools appears aspired to architectural activism and radical pedagogies.

If in the notion of sustainability, the 1992 Earth Charter envisioned a 'peaceful global society', the emergence of the 'political' signaled another disturbance: a number of thinkers had already pointed out a problem with the concept of society and there were developing further Deleuze's notion of assemblage to conceptualize emergent complex systems of organization and address issues of governance and self-governance. Hardt and Negri's *Multitude* (2004) was followed by Latour's *ANT: Actor-Network-Theory* (2005). I remember Bruno Latour's book on *Making Things Public: Atmospheres of Democracy* was out in 2005, and it was part of our discussion; Manuel De Landa's *A New philosophy of Society: Assemblage Theory and Social Complexity* followed a year later (2006). Do you think that what we started in 2005 emerged within this theoretical framework?

David: We have a number of seminal figures whose thinking has inspired us, who we talk with, or who we feel an affinity for. Clearly, Bruno Latour is one of these figures, 'We have never been Modern' (1993), his reflections on 'Design' with particular reference to Peter Sloterdijk, and the ZKM exhibition catalogue: 'Making Things Public' (Latour B, Weill P 2005) have been particularly important, as they have been for many in my generation. I also love the work of

Michel Serres. Love being the keyword, because my relationship with it is more emotional than intellectual. My library is the library that you might expect, a collection of volumes brought together obsessively since the early 1970s, reflecting conversations that were taking place within and outside Architecture, in London at the AA and internationally, but with particular emphasis and completeness in the collecting when ideas resonated; Serres, Latour and Deleuze, De Landa, and Felix Guattari's Ecosophy (Guattari 2000). It is clear to me that without this theoretical framework, and insights found more recently, the ATOPIA project would not exist….and SARCHA.

12.3.3 The Sustainable Present of Architecture Practice

Maria: yes, indeed. SARCHA was envisaged as an assemblage; it operates as an open structure and everyone can join with no subscription fee. To remain independent, SARCHA plugs its network of associates, their skills, abilities, knowledge, technical, technological equipment and physical locations, into available institutional or independent structures, operational mechanism and devices to maximize the potential of humans, things, and processes involved. It currently has 277 associates worldwide, who can freely choose topics to set up projects on issues of concern. Migration and xenophobia, city and economy, the sharing of resources and self- governing are among the research projects SARCHA has undertaken and implemented with limited funding (SARCHA 2011). SARCHA's architecture practice + education structure understands 'sustainability' as an insightful anatomy of the present conditions (Theodorou 2013).
Next time we crossed paths, was in December 2008 in Athens Greece. This was an important year signaling the start of the economic crisis. In USA the real estate bubble had burst and Athens experienced the first disruption of an hypnotic state in which democracy and the city functioned as spectacles of undisturbed consumerism. By that time, we had both set up our alternative structures (ATOPIA and SARCHA) to implement our ideas and it seems they were timely. The financial crisis, that was to become a permanent condition paired with the environmental and the humanitarian crisis, manifested in migration flows and expulsion of populations, has created an opening for rethinking architecture as a professional practice. How does ATOPIA deal with such issues?
David: OK, I think that rethinking architecture, specifically 'how and why' became obligatory. The experience of working on a project in Sri Lanka after the Boxing Day Tsunami of 2004 dramatized the need for new models, and practices in which the 'client' disappeared as a figure, to be replaced by 'issues' or in Latour's terms, 'matters of concern' (2005). We were working on a plan for a small University on the Southeastern coast of Sri Lanka, seriously damaged by the Tsunami, and then again by the rains that followed. The remains of the campus were south of the extensive land-mine fields from the Civil War, but subject to the activities of insurgents from the South. The Vice Chancellor was

a Tamil. Girls studying at the University now found it too dangerous to walk or cycle from home. The terrestrial communications networks needed to be rebuilt or augmented. We were hoping for a viable plan for what we called 'the smallest university with the biggest informational footprint', a university with almost no architecture in a conventional sense, for which a combination of electrical engineering and curriculum design would provide the foundations. A relational architecture of places and processes, tools and techniques, (informational) networks and (ecological) loops was required, inflected by the impact of War and Environmental Catastrophe. The lesson could not have been more stark. The question was then how to carry this insight forward to become more generally applicable in complex situations that may not be described in relation to an overt crisis, but where there is an inherent instability that must be understood, and there are multiple 'actors'. This is true for our research on the increasingly depleted Ogallala Aquifer in the USA, as much as it is for the PITCHAfrica and Waterbanks initiatives.

Maria: Political theory and Assemblage theory shaped the constitution of the School of Architecture for All—SARCHA and its first research project on the category of the Unbuilt. The "Negotiating the Unbuilt: Space and Democracy" symposium (partially funded by the ESRC UK) coincided with the Athens December 2008 political events and framed the context in which ATOPIA presented its PITCH Africa projects (Theodorou 2011). What was the relation of your project with the theme of the symposium?

David: 2008 was a crucial year for us. We had just received funding from the Annenberg Foundation for PITCHAfrica. We had established some principles and were working on preliminary design proposals, drawing diagrams and thinking through what a 'place of assembly' might be, as it pertains to complex constituencies, unpredictable social conditions, difficult technical considerations, poverty, gender equality, nutrition, education, health and access to water, in a context marked by tribal conflict, with contested territories, persistent land rights battles, the legacy of colonial governmental practices, protocols and corruption. At that stage, our project was literally unbuilt, and explicitly addressing structures that must be paradoxically 'unbuilt' and yet constructed—assemblages.

Maria: I remember one recurrent issue that we discussed in a number of occasions; it touched upon the relation of your project to philanthropy and charity work conducted in Africa by a number of organisations. Philanthropy has been criticized, and rightly so, for keeping a condition as it is, providing only temporary relief. The 'empowerment' approach has been also taken up before by organisations in Africa in order to smooth the ground for collaboration when the help of locals was indispensable for the successful implementation of various projects (Theodorou 2011). How do your projects relate to both philanthropy and empowerment?

David: Its complicated and, understandably, this question haunts everything that we do, and comes up every time that either Jane or I present the work. PITCHAfrica is now the business name for ATOPIA Research Inc. which has tax-exempt status as a charitable organization addressing environmental issues

and humanitarian concerns, at least that is how it is expressed in the documents that were used to achieve that status. However, we are acutely conscious of the predicament that is produced by the terms and the dilemma that we must sit with, by necessity; the proximity of 'humanitarian' to 'development' and 'security' being critical. Every aspect of a project is fraught with complexity, the funding, contractual negotiations, community relations, document production, dissemination, project management, monitoring and evaluation. We had many arguments with one funder whose use of militaristic expressions like 'in country', or 'boots on the ground' rendered the Foundation/ NGO relationship particularly problematic. The question of motivation is always present in any donor's relationship to their donation and the activities that it supports, as it is at all levels in the communities that we have worked with. As we all know, giving and receiving gifts is never straightforward. We cannot escape, but we can rewrite the terms, and chose which issues to focus on. This is why we are so concerned about cost and replicability—in a sense, privileging the unexceptional and the typical as a goal. There are too many examples of philanthropic investment coupled with designing 'like you give a damn' that produce remarkable situations, even beautiful buildings, that are only possible in a unique nexus of political power, wealth and expertise. We are also very interested in the questions that the notion of 'success' produce—how is a project successful, or rather, when, and according to who? We have learnt that sticking with the questions, having respect for ambiguity, and understanding the agency of words and images in different contexts is necessary and useful. So, we might use the word 'empowerment' or 'empower' on our website, we might talk and write about 'campaigns', but we will avoid 'victim' photography, preferring a documentary approach.
Maria: Have you looked at the history of water shortages in Africa and the traditional ways of coping with this issue? Or did you just take a practical approach, namely: what is the problem? How can we solve it? In other words did you address only the effects of a problem?
David: We are very interested in the history of peoples' relationship with water all over the world and in the history of technologies that have been developed to bring water from one place to another so that it can be used. We have also collected photographs, texts, diagrams and drawings of buildings, infrastructures and technologies for domestic water supply and irrigation that have been used in the past in many countries on the African continent. Devices for collecting or redirecting surface water resources being the most obvious. The construction of dams at many scales being another. There are also amazing examples of rainwater harvesting building forms in Benin, Ghana and the Niger Delta in West Africa, and on the Tanzanian coast in the East. Roman settlements in North Africa and the Nilotic region were built on a patch-work of cisterns.
Maria: to continue with the name giving; at the initial stage of your project the 'PITCH Africa' was your catch phrase, and it seems it had a celebratory—joyful tone, solving water shortages and using sport. Was there a particular reason for this? And why change it in 2010, and use the term Waterbanks? Does this relate to the promotion of the project in the west rather than in Africa?

David: We have not thought about our use of the name 'Waterbanks' as a change. PITCHAfrica started as a project that was exhibited at the Van Alen Institute's 'The Good Life' in 2007 (curated by Zoe Ryan) bringing 'Africa's greatest passion, Football, together with its greatest need, Water' as a demonstration of the potential of hybridity. It was intentionally optimistic, indeed joyful. When we launched the next stage of the project in Los Angeles we created PITCHAfrica as a social business, not just a project, and funding for this allowed us to develop 'Waterbanks' at a variety of scales, 'Rainchutes' as demonstration products for teaching groups about rainwater-harvesting, and 'Filterbanks' as a way of promoting ceramic water filtration at a scale where it could be effective with large populations. Since then we have discovered that the idea of a water-bank communicates broadly and very effectively. People get it…and I think that is true because a bank full of water has value, can include reserves. Water can be drawn-off, or out. The supply can be replenished. It flows, it can be metered, measured, systems of allowances can be set. The language of banking is a common language. Equally a bank can be owned, and then things get interesting.

Maria: Water has become a commodity to be sold by hedge funds; does the name 'Waterbanks' reflect this change and point to the current exploitation of basic goods?

Saskia Sassen in her recent talk (in LSE symposium in India on governing urban futures 2014) entitled *Who owns the city*, explained how city and land are privatized; large chunks of cities and land are being bought by big corporations and this will affects the water and food supply. From money shortages to water/food shortages. Does the name Waterbanks highlight this process?

David: We have actually discussed our approach with Saskia Sassen and Richard Sennett in the context of a workshop on *Democracy and Climate Change* at NYU in October last year; specifically an idea that 'rain' may be a last manifestation of the commons—the crucial questions concerning the ownership of the water once it has been captured, and how it is valued and managed, and who is responsible for this.

Maria: In the future will all buildings will need to be waterbanks—in Africa or elsewhere?

David: We think so. Let's consider the water crisis in California as an example, or desertification globally, and climate change generally. Erratic weather patterns, depleted aquifers, increasingly saline coastal wetlands and waterways, diminished snow cover in mountainous regions, high energy and maintenance costs, all point to the potential of nonlinear, decentralized, dispersed approaches to water infrastructure, and local places of assembly with high yield, high capacity reservoirs or cisterns: Waterbanks.

Maria: what was the difference in working in Africa and in the West? University approach rather than community approach, what was the resistance from local people and the difficulties in reaching them?

David: Africa is a continent and we have been working in four countries, Kenya, Nigeria, Malawi and South Africa, with early stage work elsewhere but too early for me to say too much. Nevertheless it is clear that the challenges, and opportunities vary considerably depending on the context and the partnerships, and other informal relationships that are necessary in order to accomplish as much as possible. We have a very strict rule that makes University partnerships difficult, even after the common realization that University—Community partnerships can be very effective. This rule concerns the relationship between construction cost and overall project cost. Projects that involve Universities in the US and probably the UK carry considerable overheads, up to 60 % of any grant application, and travel and accommodation costs alone can be more than the construction cost of one of our buildings. For me to make a site visit in rural Kenya, costs 10 % of the cost of the Waterbank School building…clearly this has a big impact if our stated goal is replicability. We use Skype and dropbox for supervision, cell-phone photographs and video clips, email and Facebook chat. We can do this because while there is no water supply, poor or no sanitation infrastructure, and extreme poverty, there is also incredible beauty and a robust wireless communications network. Kenya is the home of Mpesa, mobile banking, and almost everyone can get to a cell-phone, even if they do not own one. East African time is the same as Central European time, so communications typically take place in the morning in the US, afternoon in East Africa. Drawing protocols are British, as a post-colonial inheritance, but beyond the fundamentals of setting out, the drawings are only a guide or will be unread. Marked up photographs, sketches and other indications of design-intent are more effective. A comprehensive drawing set dictating all on-site processes would be meaningless. The work progresses like a conversation, or call and response.

Maria: Education, school and schooling are integral to Waterbanks; what is the role of school in the local communities?

David: Schools are the central meeting place in rural communities, and the only place where land rights and tribal relations can escape the ambiguities and conflicts that exist everywhere else.

12.3.4 The Sustainable Present of Architecture Education

Maria: Sustainable design has entered schools' of architecture curriculum at both Undergraduate and postgraduate level. Do you share the view that the role of education is to question existing conceptualizations rather than turning them into subjects of learning? Can we use the word sustainability when systems have already failed? It seems the vision of a global just society as envisaged in 1992's concept of sustainability has evaporated, turning 'sustainability' into an empty signifier. For a large portion of earth's populations, the concept of sustainability, as keeping up existing living standards and conserving resources for future generation, is not applicable; the majority of people just strive to reverse the trend of constant failure. It is 'survivability' at the present tense that matters.

David: In design education, toward sustainability however that is defined, I favor a tripartite structure that draws on Felix Guattari's elision of Ecology and Philosophy into a singular and urgently needed Ecosophy. His identification of three ecologies, mental, social and environmental, for me, provides the template. The provocation, by Bruno Latour among others, that a site can be considered as a place of assembly and exchange, of humans and non-humans, energy and matter, with multiple scales, and time horizons, provides the alibi. Overlaid on this, in response to the irrefutable 'call to order' that the ecological imperative presents, it is becoming clear that new habits of thought and action are necessary, that need names, surrogates or avatars so that they can be communicated. As a short-hand I use the 'rainmaker', the 'spider' and the 'fly' as characters, or figures. The rainmaker designs the atmosphere, transforming the weather, the spider builds networks, the fly is an expert in non-linearity. For pedagogic purposes I consider all distinctions useful, and attempt to embody their ways of being, individually, in order that we might arrive at a better understanding of who we are and what we can do. Never-the-less, just as mental, social and environmental ecologies can never be separated meaningfully, the ecologically savvy architect must embody all three characters, simultaneously.

Finally, our experience with Waterbanks really underscores a realization that architectural education must address present day realities much more forcefully, must be willing to look long and hard at the world, as it is…..and that work is conversational. A conversation is more than a dialogue. Conversations use networks to build networks that, however complex, facilitate effective action. Conversations are not arguments, but may involve disagreement. Conversations are not consensus builders, but they can be inclusive: person to person, issue to issue, human to non-human, and back again, as an endless, recursive, call and response.

Maria: It seems architecture education would be less and less about the design of sustainability; the focus is slowly and steadily shifting into training architects as contemporaries. The contemporary is able to discern the darkness emanating from its own time rather than seeing its blinding lights (Agamben 2009). Focusing steadily on understanding the present conditions would enable architects to reconceptualizing their practice and education. Sustainability as a vision of the future is inevitably under scrutiny in the current intensity of the present.

References

Agamben G (2009) What is the contemporary. What is an Apparatus and other essays. Stanford University Press, Stanford

Agamben G (2014) From the state of control to a praxis of destituent power. http://roarmag.org/2014/02/agamben-destituent-power-democracy/. Accessed 23 March 2015

Austin W (2013) The anti view. Five views on sustainability in architectural education AJ. http://www.architectsjournal.co.uk/students/five-views-on-sustainability-in-architectural-education/8651398.article. Accessed 10 March 2015

Bennett J (2010) Vibrant matter: a political ecology of things. Duke University Press, Durham
Clément G (2013) The emergent alternative. In: Harrison AL (ed) Architectural theories of the environment: posthuman territory. Routledge, London, pp 258–277.
De Landa M (2006) New philosophy of society: assemblage theory and social complexity. Continuum, London
Derrida J (1993) Aporias. Stanford University Press, Stanford
Glynos J (2012) Body, discourse, and the turn to matter. In: Bahun S, Radunovic, D (eds) Language, ideology, and the human: new interventions. Ashgate, London, pp 173–192
Goodbun G, Jaschke K (2012) Architecture and relational resources: towards a new materialist practice. Archit Design 82(4):28–33
Gregotti V (1996) Inside architecture (trans. Peter Wong & Francesca Zaccheo; Foreword by Kenneth Frampton). MIT Press, Cambridge
Guattari F (2000) The three ecologies. (trans. Chris Turner). Athlone, New Brunswick
Guy S (2013) Pragmatic ecologies. In: Harrison AL (ed) Architectural theories of the environment: posthuman territory. Routledge, London, pp 138–150
Hardt M, Negri A (2004) Multitude: war and democracy in the age of empire. Penguin Press, New York
Herring H, Roy R (2007) Technological innovation, energy efficient design and the rebound effect. Technovation 27:194–203
Harrison AL (ed) (2014) Architectural theories of the environment: posthuman territory. Routledge, London
Latour B (1993) We have never been modern. (trans. Catherine Porter). Harvard University Press, Cambridge
Latour B (2005a) Reassembling the social: an introduction to actor–network theory. Oxford University Press, Oxford
Latour B (2005b) From realpolitik to dingpolitik or how to make things public. In: Latour B, Weill P (eds) Making things public: atmospheres of democracy. MIT Press, Cambridge, pp 4–31
Latour B (2008) A cautious prometheus? A few steps toward a philosophy of design (with special attention to Peter Sloterdijk). In: Hackne F, Glynne J, Minto V (eds) Proceedings of the 2008 Annual International Conference of the Design History Society—Falmouth, 3–6 Sept 2009, e-books, Universal Publishers, pp 2–10
Latour B, Weill P (eds) (2005) Making things public: atmospheres of democracy. MIT Press, Cambridge Mass
Mather M (2012) The atopia chronicles (Montreal). Phuture News Publishing, Quebec
Mouffe C (1993) The return of the political. Verso, London
Sassen S (2011) Beyond Inequality: Expulsions paper, 9th annual conference (session 5 2011-09-14). The Center on Capitalism and Society Columbia University. http://capitalism.columbia.edu/files/ccs/Session5_Wang-Sassen.pdf. Accessed 23 March 2015
Stavrakakis Y (1999) Lacan and the Political. Routledge, London
Svendsen L (2005) A philosophy of boredom. Reaktion Books, London
Theodorou M (2003) The Iveria hotel Tbilisi. Naturopa 100:24
Theodorou M (2007) Architecture and activism. In: AnArchitektur (ed) Theorizing architectural resistance. Anarchitektur, Berlin, pp 21–29
Theodorou M (2011) Architecture and economy: the ethics of empowerment. In Stergiou L (ed) AAO: ethics/aesthetics. Papasotiriou, Athens, pp 141–163
Theodorou M (2013) The city in common: SARCHA's Pilot Study for the Gerani area in the center of Athens. In: Maloutas T, Kandylis G, Kautantzoglou R, Petrou M, Solioutis N (eds) The Athens city center as a political question in Greek. National Centre for Social Research EKKE publications, Athens, pp 295–317
Till J (2012) Scarcity contra austerity. Places, London, 10 Aug 2012
Turpin E (ed) (2013) Architecture in the Anthropocene: Encounters Among Design, Deep Time, Science and Philosophy Open Humanities Press http://openhumanitiespress.org/architecture-in-the-anthropocene.html. Accessed 23 March 2015
Willke H (2001) ATOPIA: Studien zur atopischen Gesellschaft. Suhrkamp Verlag, Frankfurt

Documents of organisations (on line)

Earth Charter (1992) http://www.earthcharterinaction.org/content/pages/Read-the-Charter.html. Accessed 23 March 2015

SARCHA (2011) CCR city common resource: Athens_Gerani (2010–2011) pilot study. http://www.scribd.com/doc/97463886/CCR-Athens-booklet-EN2012-SARCHA. Accessed 23 March 2015

United Nations General Assembly (1987) Report of the World Commission on Environment and Development: Our Common Future. Transmitted to the General Assembly as an Annex to document A/42/427– Development and International Co-operation: Environment. http://www.un-documents.net/ocf-02.htm. Accessed 23 March 2015

Chapter 13
Sustainable Communities and the New Patchwork Politics of Place

Quintin Bradley and David Haigh

The pairing of community and sustainable development has dominated the international policy agenda for at least three decades with its assertion that the imperatives of capital accumulation can be balanced for the needs of social reproduction (Raco 2005). As a framework of state strategy, the concept of sustainable communities has come to define a particular mode of governance in which the responsibility for ameliorating the impact of unfettered growth is devolved to place-based voluntary and community associations (Mayer 2000). The community provides a model of sustainability in which the economics of collective consumption and the politics of community action can be engaged in the planning and stewardship of local development. The strategies of sustainable communities that result combine the market zeal of spatial liberalism with themes of redistributive justice and equality, finding in the concept of community both a model of resilience and self-reliance and conversely a dynamic of mutual aid and co-operation (Clarke and Cochrane 2013).

This chapter identifies these competing strands in government strategies for sustainable communities in England and particularly the programme of neighbourhood planning introduced from 2011 in which the sustainable community was positioned as the regulator of development and a reassuringly familiar substitute for the welfare state. We argue that through this programme responsibility for achieving environmental and social sustainability was largely abandoned by the state and relegated to the domestic networks of the community. We explore the definition of sustainability that emerged from communities and their neighbourhood plans, one in which the priorities of environmental quality and the welfare needs of social reproduction were pursued through a Hobson's choice of economic growth or self-reliance. In attempts by neighbourhood planning groups to establish innovative

Q. Bradley (✉) · D. Haigh
School of Built Environment and Engineering, Leeds Beckett University, Northern Terrace, Leeds LS2 8AG, UK
e-mail: q.bradley@leedsbeckett.ac.uk

D. Haigh
e-mail: d.p.haigh@leedsbeckett.ac.uk

© Springer International Publishing Switzerland 2016
M. Dastbaz et al. (eds.), *Building Sustainable Futures*,
DOI 10.1007/978-3-319-19348-9_13

strategies of participation and community management, we evidence the continuance of claims of redistribution and spatial equality in the concept of sustainability and in this unequal geography of community initiatives, we chart the development of a new patchwork politics of place (Dikec 2012).

The chapter begins in an analysis of the community as a representational space that can be drawn on to support conflicting strategies of social, environmental and economic sustainability. We then review the definition of the sustainable community that emerged under the Labour government in England from 2003 when economic development goals were fused with those of social welfare. We trace the uncoupling of concerns for social justice from this agenda and the increasing reliance on the community for any hope of a sustainable future. We then introduce research with communities engaged in neighbourhood planning and identify three competing dynamics that point to the future of the sustainable community.

13.1 Sustainable Communities as Fetishised Space

The sustainable community has a noble pedigree in place-based projects of visionary design and the fusion of nature and nurture in 'self-contained and balanced' urban living (Reith 1946). The antecedents of this view lie in a renunciation of capitalism, in the collectivist communities of Charles Fourier and Robert Owen, in Peter Kropotkin's (1912) celebration of mutual aid, and in William Morris' (1890) anarchist naturalism. In this radical vision of sustainable communities, economic life was to be localised and organically fused with the rhythms of social reproduction and an idealised natural world (Harvey 2000). State development strategies continue to present the locality, neighbourhood and community as a bounded and discrete place where sustainability can be achieved in apparent isolation from its global connections. No repudiation of capitalism is envisaged, of course. Instead, the sustainable community is conceived as a coherent collective amid a global market—a homogenous space of moral order in a world without scruples (Rose 1999; Clarke 2009). Much discussion has taken place over the failure of the abstract spatial construction of community to portray the relational processes of lived space and the 'hydra-like' transnational networks that cross spatial boundaries and render localities heterogeneous and diverse (Amin 2004, p. 38). Critics have pointed to the porous nature of the boundaries of place, and the relational, subjective connections of its residents (Allen and Cochrane 2010; Massey 1994). The evidence for these arguments is so compelling that in this section we want to explore what the state finds so politically expedient in the notion of the community as a coherent space of sustainability.

The theorist Henri Lefebvre argued that spatial abstracts such as community are created to provide the appearance of homogeneity for the deployment of state power and capital (Lefebvre 1991, p. 287). Abstractions of space dissolve difference and provide a unit that can be planned, divided, valued and exchanged to promote the economic growth and serve as a smooth surface for political economies (Wilson

2013). The fetishising of community gives territorial identity to spatial practices that promote this homogeneous imaginary. The community appears as state space mobilised to sort and catalogue subjects and to act as the place-holder for intended social identities. It operates as a space of representations; as Lefebvre said (1991, p. 287) 'to look upon abstract space as homogenous is to embrace a representation that takes effect for the cause'. Lefebvre's famous triad of space as perceived, conceived and lived enumerates the production of space as an abstract, and as a set of spatial practices, and as a representational space in which 'space is at once lived and represented, at once the expression and the foundation of a practice' (Lefebvre 1991, p. 288). Spatial abstractions serve as a representation that refers to, or stands in the place of, specific social practices. The abstraction has 'authority and prescriptive power' not only to represent these practices but also to construct them as behavioural norms, as 'representations derived from the established order: statuses and norms, localised hierarchies and hierarchically arranged places, and role and values bound to particular places' (Lefebvre 1991, p. 311). The attribution of political meaning to spatial practices hails populations with a specific address in a relational scale of a political space (MacLeod and Jones 2007; Painter 2009). Community is an abstraction through which lived space is produced according to a representation of its socio-spatial content or spatial practices. The task of mobilising these spatial practices to deliver the policy goals of sustainable development entails the assemblage of a representation of community as a territory of social reproduction (Dalla Costa and James 1980). Sustainable communities can be mapped onto gendered and globalised power relations, evidenced in the artificial divide between waged and non-monetary labour, and measured in the unpaid work of domestic and neighbourly care (Mitropoulos 2005). Demarcated from public life as the private and domestic sphere, the community acts as a metaphorical reserve for social practices that can be engaged to regulate the excesses of free market zeal and its accompanying cycles of market failure (Williams et al. 2003).

The spatial practices attributed to, and represented by, community have been identified as those of 'neighbouring' (Bulmer 1986), or reciprocity (Williams and Windebank 2000), and come attached to corresponding assertions of sustainability as a set of ambiguous and contradictory policy tools. Market theorists have conflated reciprocity with the civic virtues of enterprise and responsibility and social capital (Fukuyama 1995), a note of credit that can be drawn on the collective capital of the community and that enables its bearer to improve their labour value in the formal economy (Bourdieu 1986; Putnam 2000). The same practices of reciprocity recall the working-class communities studied in the 1950s and 1960s whose networks of mutual aid were portrayed as the building blocks of collective provision and socialisation (Hoggart 1957; Young and Wilmot 1962). The reciprocity celebrated in these studies was a non-monetary household economy, founded on the geographical immobility of women (McCulloch 1997), that developed in the absence of alternative means of surviving. It was described as 'an extended subterranean chain' of services enabling delayed, transferred or indirect repayment of good deeds where help given in the past was accumulated credit invested in the future (Bulmer 1986,

p. 112). Networks of community reciprocity are still of primary importance to those outside the paid labour force dependent on informal care from family, friends and neighbours (Forrest and Kearns 2001). The appeal to community romanticises a subsistence economy of good deeds as the building blocks of capitalist enterprise or the resourceful collectivism of mutual aid and presents these as normative practices of responsible citizenship and sustainable substitutes for statutory services (Staeheli 2003).

The sustainable community has proved an essential strategic tool in a wider reordering of state institutions and state power into new assemblages of distributed governance (Swyngedouw 2004). Community organisations and voluntary associations have been recruited as the minor scale in the private and quasi-private networks now delivering state services. Local initiatives in community self-management, asset transfer and social enterprise present the community as an island of sustainability, while around it, public assets and public contracts are packaged out in profitable tranches to transnational companies and financial institutions (Fyfe 2005). The engagement of community organisations in this outsourcing of political governance appears to have recruited once radical visions of sustainability to an agenda of market liberalism. Community campaigners, however, have an uncanny ability to move almost seamlessly from co-option to a contentious opposition to state strategies (Newman 2012). The twinning of sustainability with social reproduction provides a contemporary reflection on the urban social movements and neighbourhood groups of the 1960s and 1970s whose sustainable community, as Castells (1978) argued, was the site of a class struggle over the distribution of the surplus and the organisation of the collective means of consumption. Sustainability entailed a repudiation of market values and promised an economy where the requirements of social reproduction were prioritised (Mollenkopf 1981). These community initiatives were inspired by an ethic of domestic and neighbourly care extended into the public sphere and took the social relations of reciprocity and neighbouring as a model of co-operation from which to construct more sustainable societies (Abel and Nelson 1990). This is what Raymond Williams called 'the positive practice of neighbourhood' that aims to foster the social relations of community as a model for the collective organisation of society: 'the basic collective idea' that 'the provision of the means of life will, alike in production and distribution, be collective and mutual' (Williams 1967, p. 326). It was a model that inspired left-wing labour councils in the 1970s and 1980s to experiment with the local management of redistributive services in devolving elements of planning and delivery to the community. They piloted the decentralisation of service budgets and the establishment of elected neighbourhood committees and area panels that worked with local managers to determine community priorities (Cockburn 1977; Burns et al. 1994). The most purposeful of these local government devolution programmes in the 1980s Walsall mobilised a representation of community reciprocity and mutual aid as the 'gift relationship' (Titmuss 1970) to reassert the foundational values of the welfare state and redesign services to achieve an ambitious equalities agenda. The community as a territory of social reproduction was made here into a representation of socialised collective consumption, an ideal model of redistribution and justice. Jeremy Seabrook (1984, p. 22) describing the Walsall initiative wrote:

> Those values [...] the mutuality and the sharing, the sense of a collective predicament, the imaginative understanding of other people's suffering—were precisely those which shaped the idea of the welfare state

The sustainable community emerged from these initiatives as a political representation of social welfare founded on the attribution of normative meaning to the practices of gendered care and the economies of social reproduction. This identification of sustainability with the goals of social welfare became central to state policy when the still 'New' Labour government published its Sustainable Communities Plan in 2003 with the goal of reducing spatial injustice. The next section discusses the representations of community that guided this plan and explores how the spatial practices of social reproduction were fluidly adapted to the behavioural norms of spatial liberalism.

13.2 Sustainable Communities and Sustainable Development

The sustainable communities plan (ODPM 2003, p. 3) was a spatial development framework that promised to achieve equity in the geography of economic growth and 'put the needs of communities first'. A definition of sustainability acquired through quality of life indicators of economic activity, transport connectivity, health, crime and educational achievements provided the performance regime that enabled central government to devolve operational responsibility to the regions, municipalities and neighbourhoods. The rationale for a focus on communities was firmly linked to the engagement of citizens in helping government deliver its aims (Griggs and Roberts 2012). Institutions of community governance were established in the New Deal for Communities programme and neighbourhood management projects, and local residents were engaged in community boards to ensure service efficiency and the delivery of national targets (Painter et al. 2011).

The initial aim of the sustainable communities plan was to restore a degree of territorial equality in the face of a decade or more of deindustrialisation and public service erosion. As the plan developed, however, it became clear that deprived communities were being cast as the obstacle to achieving 'lasting, rather than temporary solutions' to spatial inequalities (ODPM 2003, p. 5; Wallace 2010). Programmes intended to raise the enterprising capacities of residents were prioritised for funding in the belief that territorial inequality and social injustice could be rectified by market mechanisms rather than through the redistribution of public services (Lupton 2013). Participation was equated with paid employment and empowerment with engagement in the private market (Durose and Rees 2012). In transformative projects such as the House Market Renewal programme, local residents were presented as impediments to economic growth, and mass demolition of low-cost housing was seen as the strategy to resolve the problems of poverty (Allen 2008). Labour's 2008 White Paper, *Communities in Control* appeared addressed only to affluent communities with its package of consumer rights to take ownership of public assets. The devolution of power promised in community control required no strategies of renewal or redistribution. Instead, it entailed a transfer of responsibility from the state to the community, cast as an economic actor in charge of its own destiny and responsible for its own well-being (Hall and Massey 2010).

Under the coalition government, which took office in 2010, the sustainable development of communities was effectively applied as a weapon against the collective provision of a redistributive state (Featherstone et al. 2012). The abolition of all national regeneration programmes was accompanied by severe reductions in funding for local government, while cuts in welfare spending were dignified with a narrative around the need for communities to take greater responsibility for their own sustainability (Hastings et al 2013; Lowndes and McCaughie 2013). An overhaul of planning policy imposed a mandatory 'presumption in favour of sustainable development' on communities to signify their commission as homogenous spaces for market growth (National Planning Policy Framework 2012). The introduction of neighbourhood planning under the Localism Act 2011 aimed to lock-in the consent of communities to private market development (DCLG 2011a) and was backed by a framework of economic incentives to orientate neighbourhoods towards the property market (Haughton and Allmendinger 2013). Neighbourhood planning allocated responsibility to communities for achieving sustainable development within growth targets set by strategic authorities and subject to the plans of private developers. Their neighbourhood planning powers could not limit the amount of growth but could influence its location and design by establishing the local policies that development would be judged against. Subject to a light touch examination, and ratified by popular referendum, a neighbourhood plan could become a statutory development document, nested within and conforming to the strategic plan of the local authority and national planning policy.

To enlist place-based groups in the regulation of sustainable development, neighbourhood planning provided a significant new governance framework for community engagement in spatial planning. The Neighbourhood Planning Regulations (2011b) established criteria for the designation of an urban neighbourhood as a strategic planning area, and a clear process by which place-based community groups might establish neighbourhood forums and be recognised as representative bodies. Neighbourhood planning, then, established the community as a legal entity and despite its many restrictions, appeared to offer some new political opportunities to develop a sustainable strategy of place. In areas across the country, therefore, communities saw in neighbourhood planning the potential to harness the practices of spatial liberalism to the representation of social reproduction (Clarke and Cochrane 2013).

The first neighbourhoods to produce their own plans were able to mitigate the impact of large new housing developments by parcelling it up into acceptable smaller sites (Thame 2012) and change planning policy to enable more affordable housing to be built in rural areas (Upper Eden 2012). By the beginning of 2015, over

1300 neighbourhood plans were under production and most were in market towns and rural parishes, reflecting the inequalities of the market in the distribution of the requisite economic, cultural and social capital needed to write a plan. Within this total were over 150 urban areas and among these many disadvantaged inner city neighbourhoods that had previously benefited from neighbourhood management or House Market Renewal programmes and had a history of participation and community action to draw on in developing their neighbourhood forums. The interest in neighbourhood plans in these deprived areas was often evidence of the proactive involvement of local authorities who saw a potential resource in the wake of the withdrawal of regeneration funding and in the face of budget cuts that had reduced services to the statutory minimum (Lowndes and McCaughie 2013). Neighbourhood planning became almost the only available framework through which the politics of collective consumption could be pursued.

The sections that follow build on research with these urban neighbourhood forums carried out between 2013 and 2015. The aim of the research was to assess the impact of neighbourhood planning on civic engagement in urban areas. It was conducted through a preliminary review of online resources including the constitutions of neighbourhood forums, their applications for designation, council decision papers, minutes of meetings, consultation strategies, draft and final plans. This was followed by interviews with the chairs and secretaries of neighbourhood forums, observation at forum meetings and separate interviews with the relevant officers from the planning authority. The national sample represents only a minority of neighbourhood plans, and it is important to acknowledge that there are other stories that could and will be told about neighbourhood planning. But the findings from this research reveal clearly the differing contentions of community sustainability that emerged in neighbourhood planning. The sections that follow identify three dynamics in urban neighbourhood planning that (1) revive contentions around greater democratic engagement in sustainable development, (2) evidence a convergence between sustainability and strategies of resilience and community enterprise and (3) indicate innovation in social sustainability through the recruitment of reciprocity and neighbouring as community practices.

13.3 Sustainable Communities and Collective Action

The introduction of neighbourhood planning in England differed from previous incarnations of sustainable communities in addressing a collective public rather than an assemblage of disempowered individuals (Brownhill 2009). Under the spatial development policies that evolved from the sustainable communities plan (ODPM 2003), residents were addressed as individuals whose participation should be guided by rational choice and residents' organisations were caricatured as selfish interest groups, mocked as NIMBYs (Not in My Back Yard) and presented as unrepresentative of the public will (Bailey 2010; Lowndes and Pratchett 2012). Plan-making by communities, and especially villages and parish councils, was promoted, but these plans were granted no statutory weight. Collective participation was frowned upon as a disruption of the free exchange of market information or as a privileged voluntarism undermining the even-handed process of representative democracy (Barnes 1999; Newman 2001). The innovation of neighbourhood planning was in establishing a clear framework for integrating collective participatory democracy into the top-down plan-making of the local authority. The community was established as a political entity, or planning polity, and legal recognition was awarded to community groups as collective actors. Public participation acquired a narrow political domain in which autonomous collectives could debate questions of needs and resources, challenge the power and knowledge of managerial and professional elites and generate their own spatial development strategies. It was as a political entity rather than as an amorphous and individually imagined public that communities entered the neighbourhood plan-making process.

The opportunity to develop an autonomous vision made neighbourhood planning appealing to residents who had experienced a sense of powerlessness and marginalisation in political decision-making. A local government service with well-developed statutory mechanisms of consultation, planning can serve as a proxy for all state systems from which local people feel excluded. In some urban areas, neighbourhood planning became the focus for a generalised dissatisfaction with hierarchical decision-making and appeared to articulate a wider desire for involvement in decision-making. In the terraced streets of Preston, a town in North West England, the neighbourhood planning forum, Friends of Fishwick and St. Matthews, was set up by local residents who had formerly served as community representatives on a sustainable communities programme for their area. When neighbourhood renewal funding ended and the local regeneration team were withdrawn, the residents set up their own community group and began work on a neighbourhood plan. The secretary of Fishwick & St Matthews neighbourhood forum explained her motivation for starting a neighbourhood plan:

> I guess this was the reason I got involved, I just realised how much contempt there is, overt contempt, shown to people from deprived neighbourhoods. And I guess the planning process in that particular instance, as far as I'm concerned, confirmed everything that I thought… and I was just absolutely enraged…and I just felt that it's the general attitude of public servants towards people in deprived neighbourhoods, the way that they, they just don't count, and that's how it feels.

Her anger at the marginalisation of deprived areas, and the exclusion of communities from development decisions, motivated her engagement in neighbourhood planning. She saw it as an opportunity to begin to reverse hierarchies of power and to affirm the local community as the privileged source of knowledge and of more democratic service planning. The concept of sustainable communities became here an appeal to equality and a challenge to spatial injustice.

Development decisions were not always the main concern for community groups but spatial plan-making gave them a platform to identify and campaign for the public services they needed. The statutory powers of neighbourhood planning made it attractive to those groups looking for an institutional route to express their opposition to the impact of state welfare reductions. The decision of Manchester city council to close the local library in the suburb of Northenden was the trigger that fired the local community group to start a neighbourhood plan. Buoyed up by their sit-in protest over the closure of the library, the campaigners saw a neighbourhood plan as an opportunity to move beyond being 'informed, not consulted', as the secretary of the Northenden forum said:

> The council conflate the two ideas, they think informing is consulting and so, there was a lot of frustration on many issues all over…because I just think this community's been treated so unfairly actually, it's just not right and that's it.

Neighbourhood planning conferred a defined spatial identity on communities in their pursuit of sustainability, and it could have collectivising effects in reaffirming a sense of 'living in nearness' (Kearns and Parkinson 2001). Neighbourhood forums had to submit their proposed community boundary to the local authority for approval and boundary-setting necessitated participatory negotiation to define the spatial limits of 'people's felt sense of identity' (Stoker 2004, p. 125). Agreement over the boundary of a neighbourhood plan could have unifying effect, bringing residents' and community groups together in a shared vision for their area. The fact that this vision, once approved, would acquire statutory weight gave greater credibility to the community action of residents' groups. In the community of Holbeck, in the city of Leeds, three separate residents' groups came together to establish a neighbourhood forum that provided one focus for relations with the municipal authority. The community activists found it easier to tackle problems as a combined force. They had long experience of community action and, over a period of 7 years, had worked with the local authority on plans for a housing private finance initiative (PFI) but in neighbourhood planning they saw a unique opportunity to finally take the lead in producing their own 'people's vision for the next 15 years' as the chair of the forum explained.

> I think the thing is in the past plans and developments for lots of areas have been imposed from above, with very little consultation. Here, we have got a chance to consult at an early stage. And whether or not what we say gets implemented, at least we have got a voice and we are being listened to.

Neighbourhood planning gives legal recognition to the community as a representational space of empowered democracy and provides a statutory framework that can be used to champion claims of equality and justice in sustainable development planning. It does so, however, only ideally. The next section explores the limits of neighbourhood planning and the visions of sustainability it can inspire.

13.4 Sustainability and Community Enterprise

Neighbourhood planning was designed to 'create the conditions for communities to welcome growth' (Clark 2011) and its spatial planning powers are intended to be used to achieve sustainability without impacting on development markets. Although it enables communities to set out a 20-year vision for their neighbourhood, the neighbourhood plan has no connection to national or local investment planning. The only source of investment available to town and parish councils, and urban neighbourhoods who produce a neighbourhood plan, is a levy on any private development that takes place. The amount received from this Community Infrastructure Levy is dependent on market demand for land in the area. Market towns and rural parishes that have land sites attractive to the large volume house builders will receive a quarter of the revenues accruing from the Levy while suburban neighbourhoods on the urban fringe will also benefit once they have agreed a neighbourhood plan. Public investment in schools, community facilities and infrastructure will, to a significant extent, be dependent on the site acquisition strategies of the property market and inequalities in public spending will become increasingly linked to the uneven geography of capitalist growth (Clarke and Cochrane 2013).

In the deprived east end of Preston, the community of Fishwick and St. Matthews thought neighbourhood planning was an opportunity to improve the quality of their inner city environment. The main problems they sought to tackle were traffic pollution and the lack of green space. The mission statement of the neighbourhood forum is to be 'a better, cleaner, safer place to work and live in' (FOFS 2014). The secretary explained the vision of a sustainable community that motivates her:

> I am optimistic because I look down the road and I don't see what you see. I see a pretty town, nice firms and shops, nice pretty painted houses, people that are happy and, you know, not that sort of atmosphere of just tough life.

The opportunities for changing Inner East Preston were, however, very limited without public investment. Development sites were few, and there was little market interest, and the changes the community wanted to see required the creation of new parks and road improvements that required significant public finance. The Preston council planning officer working with Fishwick and St. Matthews explained her concerns over the limit of what the plan could achieve:

> I have this worry that it's one thing to write a plan but how do you actually put it into action? It is the delivery which is the difficult part. I mean there's no harm in having a few aspirations, but the area won't be completely transformed. It will still be the same area.

The limits of neighbourhood planning are very apparent in urban areas of deprivation. The right to agree a development plan is no right at all without a development market. The secretary of the neighbourhood forum understands this:

> You've got the power and opportunity to influence planning policy, but what does that mean to someone who lives somewhere where they can't park, where the housing stock is very poor, you know what does it mean? It means nothing.

The only source of investment for the Fishwick and St. Matthews community in Preston comes from charitable donations, and they have benefited from a lottery grant, under the Big Local programme, which will enable them to carry out some environmental works. Patronage and donations aside, the expected course of action for communities marginalised by capital growth is to become economic actors and create their own development market, an approach adopted on the outskirts of the midlands city of Northampton, where a neighbourhood plan is being led by a voluntary association, under the project name, Growing Together. This voluntary association used to manage a £10 m service contract for the municipal authority under the former neighbourhood renewal programme. When the funding was withdrawn, and in the absence of any further public investment, the association formed a new community management board, successfully applied for a Lottery grant and, with the support of the municipal authority, began work on a neighbourhood plan for the environmental works and community facilities funded by the Lottery. But the lack of any other source of public investment limits the ambitions that can be planned for, as the co-ordinator explained:

> To be honest I don't think there's any possibility of any sort of visionary vision for this area within the economic circumstances. It's a very difficult area to have a sort of bright, clear vision of the shining city in the sky in 20 to 30 years' time.

In the absence of the sort of spatial redistribution strategy envisaged under the original Sustainable Communities Plan, Growing Together will act to generate its own development market. They envisage that the community group will be constituted as a Community Interest Company or charitable trading company, and that it will bid to deliver limited local services for the Borough Council. This strategy appears to exemplify the self-reliance and resilience expected of communities as an attribute of sustainability. The neighbourhood plan has provided the community with a statutory framework through which the self-management and the acquisition of public assets can be envisaged. But, it also provides the community with a model of participation and democratic engagement that will characterise the services and bring them closer to community control. The plan co-ordinator said:

> It's a part of my vision for the 15 to 20 year future that Growing Together will be established as a legal entity which is able actually to deliver services, but to deliver them in a way that is obviously managed by the local community. So effectively you've got the old Victorian intervention of the people voting for representatives who will manage the services, but you've brought it much closer into neighbourhoods, you've effectively brought it into the 21st Century.

The Growing Together group define sustainability in terms of participation, co-production and self-management even as they embrace the self-reliance required under a policy of spatial liberalism. But the opportunities to create community as a space of care, empowerment and redistribution have diminished (Staeheli 2003). Sustainable development has been detached from accompanying concerns for inequality and political exclusion and now appears defined wholly in terms of economic self-sufficiency (Lepine and Sullivan 2007). The only option available to the Growing Together group is to develop the capacity of residents for enterprise in the hope that sustainability can be achieved by market mechanisms. But social enterprises do not represent a community and there are limits to their attachment to place and the empowerment and democratic engagement in local services that they can deliver (Amin et al. 2002). Without support from the local state, a social enterprise is unlikely to flourish in a deprived community (Trigilia 2001). The next section explores how the discourse of community enterprise and self-reliance may be shaping the local welfare state and creating innovation in the governance of public services, and it investigates the role of the local state in the resilience of sustainable communities.

13.5 Sustainability and Social Justice

In neighbourhood planning, the community was established in putative opposition to the local welfare state, as a more sustainable provider of services. The Localism Act 2011 awarded rights to communities to bid to run public services and take over public assets alongside the right to enact neighbourhood plans. The main beneficiaries were not, of course, local communities, but the multi-national companies and global finance markets involved in privatisation and outsourcing (Fyfe 2005). Many neighbourhood plans, however, included strategies for asset transfer, and community trading companies were set up to run public services. For market theorists, the capacity of community to generate social capital through unpaid reproductive labour provides a model for the re-engineering of public services around a new and more enterprising public. The sustainable community as social capital offers a moral economy with 'rights and duties about consuming and repaying existing side by side with rights and duties about giving and receiving' as the anthropologist Marcel Mauss (1954, p. 11) once observed. In anthropological studies at any rate, social capital represented an economic model of social relations that could replace the universal services of the welfare state with a responsibility on citizens to actively engage in their own social sustainability. Applied as political strategy of austerity by the coalition government in England after 2010, this vision of welfare pluralism founded on the potential of social capital promised a 'radically different form of local governance' (Lowndes and Pratchett 2012, p. 22). The quality of welfare services and content of social provision would depend on market systems of distribution, favouring those affluent communities traditionally associated with social capital and the skills, education, contacts and ability to donate higher levels of unpaid labour to voluntary service (Farnsworth 2012).

Despite the political expediency of this new vision of partial social sustainability, its suitability as public policy was questioned, particularly in relation to deprived urban neighbourhoods that appeared purposely excluded from this new localism in welfare provision. The expectation that the community, as a non-gender-specific reference to women's unpaid care work, could take over from the state was arguably unrealistic, and openly unfair, given changes in labour force participation and the impact of public spending cuts. There was a vital and continuing role for the state in providing the infrastructure and equality of service that could encourage community reciprocity (Lowndes and Pratchett 2012). Conversely, some saw the potential for a progressive localism to emerge from this return to community sustainability, recalling traditions of self-help and co-operation established by the labour movement, and pointing to the origins of the National Health Service in local initiatives of mutual aid (Featherstone et al. 2011). The fragmentation of community initiatives suggested an opportunity to rebuild social sustainability from the bottom-up by fusing the traditions of municipal socialism with the legacy of community action in a new infrastructure of state services (Clarke 2009). Interpreted through these idealistic perspectives, the sustainable community could be asserted as a challenge not only to the collectivism of the welfare state, but also to its hierarchical governance. As the urban social movements of the 1960s and 1970s showed, the incorporation of communities as managers and providers of welfare services disrupts not only the universality of the service, but also its hegemony of knowledge and power (Cockburn 1977).

In the east Pennine town of Accrington, the neighbourhood of Clayton-le-Moors and Altham is approaching this new regime of community welfare through asset transfer and neighbourhood planning. A community interest company was formed to take over a former local authority service centre called Mercer House, with a management board that included local councillors, the council chief executive and the head of a local housing association. Many of the same personnel also serve on a community development board for the neighbourhood, a council consultative committee with devolved spending powers. The community development board is producing a neighbourhood plan, while the community interest company is continuing to acquire assets from the council and bring them back into use. A former Housing Market Renewal Area, Clayton-le-Moors and Altham once benefited from regeneration funding and had its own local neighbourhood management team. Now, the one-stop shop that was the hub of the neighbourhood management project is run by local volunteers recruited by the Mercer House community interest company. Another two council buildings have been taken over by the community company to provide training and youth services. It is part of the Mercer House philosophy that public services that are run by local volunteers become more truly public and that the role of the local state should be to encourage residents to 'own' or take responsibility for achieving the service outcomes they require. The local councillor who leads the neighbourhood planning forum, and heads up the community interest company, explains:

> Council buildings need to be self-sustaining and if they aren't, I'm not saying they'll go tomorrow but they will go and these people have got to understand that. So really we're acting to bring these community centres back into full use.
>
> But it is not just the community assets. It's getting the community involved and that community involvement is vital to the success. We couldn't change the way the council worked, but we have seen the change in attitude since we've run it, because we're not council and that's something we're having to get away from. It's trying to bring it back to the community so they understand that it's theirs, it belongs to the community.

In Clayton-le-Moors and Altham, austerity is an opportunity to recast the local welfare state according to the principles of reciprocity and mutual aid. The neighbourhood development plan becomes a blueprint for the social outcomes identified by the community while asset transfer passes the responsibility for achieving these outcomes to residents themselves. As the plan co-ordinator said:

> If the community can come up with a plan that addresses all these issues, and sets out what this township is going to be like in the next 10–15 years and that is all done by the community, that'll be great because it shows the community's in the driving seat, steering this and it's not something that's being imposed by the local authority.

But neighbourhood planning does not adapt easily to the role of community service planning as the Clayton forum are aware. While their public consultation is about schools, housing, safety and dog fouling, the plan is supposed to allocate land for development, and there is little private market interest in this densely populated neighbourhood. Nor is it easy to plan for a sustainable community, when essential services like the primary and secondary schools no longer sit within the influence of the local authority, but operate, like housing, in something resembling a free market. The neighbourhood plan in Clayton-le-Moors and Altham cannot provide a vision for the sustainable development of the community, and instead becomes a catalogue of needs that can only be met by voluntary endeavour. The plan is a statement of community resilience, a testimony to the ability of deprived neighbourhoods to survive on their own resources and develop their own services. As the plan co-ordinator said:

> There was a lot of government money put into things like neighbourhood management, House Market Renewal by the last government. When that all got cut, Mercer House has shown what can be done without all that financial investment coming into an area. And if this model can be replicated to other places then yeah, let's do it because it really is the community taking hold of their assets and running them how they want to do it.

This representation of community control disguises the continuing role played by councillors and local authority officers in the management of this asset-transfer strategy. The leadership of Mercer House remains in the hands of professionals and retired councillors, and the production of the neighbourhood plan depends on the guidance and the support from officers in the planning authority. Rather than a model of community resilience, the transfer of assets to a community interest company appears to be a council strategy to reduce costs by harnessing the unpaid labour of reciprocity according to a rationale that argues that participation in the delivery of public services makes them more effective. Although presenting this as innovation in the community control of public services, the local state retains its commanding role and has outsourced delivery to volunteers. The reciprocity assumed to be available to regenerate the neighbourhood is a normative representation of community mobilised to support a dispersal of local authority into new forms of governance. Public assets are presented as embodied social capital; they are the tangible evidence of a sustainable community. Beneath this representation, we can perceive the transfer of public services across the gendered divide into the unpaid labour of domestic and neighbourly care. Rather than provide a framework for sustainable development, the neighbourhood plan becomes a design for resilience in the face of service withdrawal. Sustainability is the ability to survive without economic growth or redistribution.

13.6 Conclusion: The Future of the Sustainable Community

This chapter has traced the contrary representations of the sustainable community in English spatial development planning since 2003. Our concern has been with the chameleon-like properties of community as a place-holder for practices of sustainability. We have argued that these practices draw on the representation of community as a demarcated and gendered space of social reproduction. The value of the sustainable community has been the flexibility with which its practices can be adapted as behavioural norms to the requirements of the formal economy. They can be applied variously as reciprocity and neighbouring, resilience and self-sufficiency, responsibility and enterprise. Sustainability can be relegated to the margins as reciprocity or harnessed to economic growth as social capital and similarly, as the spatial representation of unpaid care work, the community can serve as a reservoir of precarious labour, a classification of land values or a polity of democratic governance.

The concept of the sustainable community has allowed the priorities of resource conservation and environmental protection to be relegated to the domestic and private sphere of social reproduction. Far from achieving a balance between economic, social and environmental objectives, sustainable development has jettisoned its ethical concerns behind a gendered divide so that the burden of ameliorating economic excesses is borne by those least able to affect it. This strategy has not been without challenge and the sustainable community has provided the rationale for a continuing conflict over the resources of collective consumption and the share of the surplus that should be allocated to social reproduction. Submerged in the informal economy of domestic and neighbourly care, sustainability has emerged as a model of resistance to the market expressed in the values of socialisation, mutual aid and collective action. The sustainable community represents a gendered class politics of redistributive justice. In its capacity as social capital or resilience it offers, too, an ambiguous and contradictory model for the reinvention of the local welfare state, shaping public services that are more participative and receptive to local knowledge, but that are dependent on the exploitation of unpaid labour and the unequal distribution of limited personal resources.

In tracing the trajectory of spatial development planning in England, we have evidenced a metamorphosis in representations of the sustainable community. The effective redistribution of public goods and the promotion of national standards of social and environmental protection were originally components in a planned programme of economic growth. As the balance shifted towards the successes of capital accumulation, the sustainable community was metaphorically shunted from its private sphere into the private sector where its residents were to be rendered capable of economic self-management. The community was imagined as a territory of sceptical consumers, not beneficiaries of public redistribution but generators of private wealth. The onset of neighbourhood planning in 2011 signalled the wholesale transfer of responsibility for sustainable development from state and market to the unpaid labour of community. Neighbourhood planning imagined the community as a market place in which sustainability was bought and development rights sold, while neighbourhoods were deployed as a sorting mechanism to provide a hierarchy of economic opportunities. The distribution of public goods became heavily dependent on the geographical whims of the property market and the task of communities was to attract development while seeking to mitigate its negative effects. Under neighbourhood planning, an uneven landscape unfolded where affluent areas drained their resource of social capital to ameliorate undesirable development and acquire public goods, and urban neighbourhoods experimented with welfare mutualism to withstand the severity of austerity. A plurality of sustainable communities appeared, each a precarious attempt to use a spatial planning framework to defend the values of social reproduction against the impact of uneven development.

The production of the sustainable community as a receptacle for social and environmental concerns necessitated the transfer of limited statutory authority to community organisations. Neighbourhood planning, in particular, provided statutory representation to the community as a territory of empowered participatory democracy. It gave further definition to an institutional framework of community self-management and enterprise that carried with it contentions of equality and spatial justice framed within an agenda of public sector outsourcing. While spatial inequality has radically widened, inequalities in representation have been marginally reduced through the advancement of collective participation and community control. The frustration and anger of marginalised communities has been given collective voice and provided with a ration of statutory power. Local authorities and community groups have seized on the partial powers of neighbourhood planning to maintain strategies of spatial justice, drawing on limited sources of patronage to revive models of mutualism and articulate a more participatory model of service planning and delivery. This is a patchwork politics of place, structured by the demands of capital accumulation into winners and losers. The disputed concept of the sustainable community now inspires a plethora of projects attempting to regulate an unrestrained development market or fill a vacuum in state investment planning. The future of sustainability will be etched in these precarious attempts to piece together a new umbrella of environmental and social protection.

References

Abel EK, Nelson M (1990) Circles of care: work and identity in women's lives. SUNY Press, Albany

Allen C (2008) House Market Renewal and social class. Routledge, London

Allen J, Cochrane A (2010) Assemblages of state power: topological shifts in the organisation of government and politics. Antipode 42(5):1071–1089

Amin A (2004) Regions unbound: towards a new politics of place. Geogr Ann 86 B(1):33–44

Amin A, Cameron A, Hudson R (2002) Placing the social economy. Routledge, London

Arnstein S (1969) A ladder of citizen participation. Am Inst Plan J 35(4):216–224

Bailey N (2010) Understanding community empowerment in urban regeneration and planning in England: putting policy and practice in context. Plan Pract Res 25(3):317–332

Bailey N, Pill M (2011) The continuing popularity of the neighbourhood and neighbourhood governance in the transition from the 'big state' to the 'big society' paradigm. Environ Plan C 29:927–942

Barnes M (1999) Users as citizens: collective action and the local governance of welfare. Soc Policy Adm 33(1):73–90

Bourdieu P (1986) The forms of capital. In: Richardson JE (ed) Handbook of theory of research for the sociology of education. Greenwood Press, Westport

Brownhill S (2009) The dynamics of participation: modes of governance and increasing participation in planning. Urban Policy Res 27(4):357–375

Bulmer M (1986) Neighbours: the work of Philip Abrams. Cambridge University Press, Cambridge

Burns D, Hambleton R, Hoggett P (1994) The politics of decentralisation. Macmillan, Basingstoke

Castells M (1978) City, class and power. Macmillan, Basingstoke

Clarke N (2009) In what sense 'spaces of neoliberalism'? The new localism, the new politics of scale, and town twinning. Politi Geogr 28(8):496–507

Clark G (2011) First communities to use powers to bring growth, jobs and homes to their neighbourhoods. Department of Communities & Local Government Press Release (DCLG), London

Clarke N, Cochrane A (2013) Geographies and politics of localism: the localism of the United Kingdom's coalition government. Politi Geogr 34:10–23

Cockburn C (1977) The local state: management of cities and peoples. Race Class 18:363–376

Cooke B, Kothari U (eds) Participation: the new tyranny. Zed Books, London

Cowley J, Kaye A, Mayo M, Thompson M (1977) Community or class struggle? Stage 1, London

Dalla Costa M, James S (1980) The power of women and the subversion of the community. In: Malos E (ed.) The politics of housework. (Chap. 11). Allison & Busby, London, 160–195

DCLG (2011a) Localism bill: neighbourhood plans and community right to build impact assessment. Department for Communities and Local Government, London

DCLG (2011b) Neighbourhood planning regulations. Department for Communities and Local Government, London

DCLG (2012) National planning policy framework. Department for Communities and Local Government, London

Dikec M (2012) Space as a mode of political thinking. Geoforum 43(4):669–676

Durose C, Rees J (2012) The rise and fall of neighbourhood in the new labour era. Policy Polit 40(1):39–55

Elden S (2004) Understanding Henri Lefebvre: theory and the possible. Continuum, London

Farnsworth D (2012) Who will control localism. Town Ctry Plan 81(1):20–23

Featherstone D, Ince A, Mackinnon D, Strauss K, Cumbers A (2012) Progressive localism & the construction of political alternatives. Trans Inst Br Geogr 37:177–182

FOFS (Friends of Fishwick & St Matthews) (2014) Inner east preston neighbourhood development plan 2014–2029 submission version. Preston. Friends of Fishwick & St Matthews

Forrest R, Kearns A (2001) Social cohesion, social capital and the neighbourhood. Urban Stud 38(12):2125–2143

Fukuyama F (1995) Trust: the social virtues and the creation of prosperity. Penguin Books, London

Fyfe N (2005) Making space for neo-communitarianism? The third sector, state and civil society in the UK. Antipode 37(3):536–557

Griggs S, Roberts M (2012) From neighbourhood governance to neighbourhood management: a rolled-out neoliberal design for devolved governance in the UK? Local Gov Stud 38(2):183–210

Hall S, Massey D (2010) Interpreting the crisis. Soundings 44(Spring 2010):57–71

Harvey D (2000) Spaces of hope. Edinburgh University Press, Edinburgh

Hastings A, Bailey N, Besemer K, Bramley G, Gannon M, Watkins D (2013) Coping with the cuts; local government and poorer communities. Universities of Glasgow & Heriot Watt, Edinburgh & Glasgow

Haughton G, Allmendinger P (2013) Spaces of neoliberal experimentation: soft spaces, postpolitics and neoliberal governmentality. Environ Plan A 45:217–234

Hoggart R (1957) The uses of literacy. Chatto & Windus, London

Kearns A, Parkinson M (2001) The significance of neighbourhood. Urban Stud 38(12):2103–2110

Lefebvre H (1991/1974) The production of space. (Trans. D. Nicholson-Smith). Blackwell, Oxford

Lepine E, Sullivan H (2007) More local than local government. In: Smith I, Lepine E, Taylor M (eds) Disadvantaged by where you live? The Policy Press, Bristol, pp 83–104

Lowndes V, McCaughie K (2013) Weathering the perfect storm? Austerity and institutional resilience in local government. Policy Polit 41(4):533–549
Lowndes V, Pratchett L (2012) Local governance under the coalition government: austerity, localism and the 'big society'. Local Gov Stud 38(1):21–40
Lupton R (2013) What is neighbourhood renewal policy for? People Place Policy 7(2):66–72
Macleod G, Jones M (2007) Territorial, scalar, networked, connected: in what sense a 'regional world'? Reg Stud 41(9):1177–1191
Massey D (1994) Space, place & gender. Oxford, Polity Press
Mauss M (1954) The gift: forms and functions of exchange in archiac societies. (Translated from the French by Ian Cunnison). Cohen & West, London
Mayer M (2000) Urban social movements in an era of globalisation. In: Hamel P, Lustiger-Thaler H, Mayer M (eds) Urban movements in a globalising world. Routledge, London
McCulloch A (1997) You've fucked up the estate & now you're carrying a briefcase! In: Hoggett P (ed) Contested communities: experiences, struggles, policies. Policy Press, Bristol
Mitropoulos A (2005) Precari-Us? Mute 29. <http://www.metamute.org>. Accessed 16 Feb 2012
Mollenkopf J (1981) Community and accumulation. In: Dear M, Scott A (eds) Urbanisation and urban planning in capitalist society. Methuen, London, 319–337
Newman J (2001) Modernising governance: new labour, policy and society. Sage Publications, London
Newman J (2012) Making, contesting & governing the local: women's labour and the local state. Local Econ 27(8):846–858
ODPM (2003) Sustainable communities: building for the future. Office of the Deputy Prime Minister, London
Painter J (2009) Rethinking territory. Antipode 42(5):1090–1118
Painter J, Orton A, Macleod G, Dominelli L, Pande R (2011) Connecting localism & community empowerment: research review and critical synthesis for the ARHC connected community programme. (Project Report). Durham University, Dept. of Geography & School of Applied Social Sciences, Durham
Putnam R (2000) Bowling alone: the collapse and revival of the American community. Simon & Schuster, London
Raco M (2005) Sustainable development, rolled-out neoliberalism and sustainable communities. Antipode 37(2):324–247
Reith Report (1946) 'New towns committee: final report', Cmnd. 6876. HMSO, London
Rose N (1999) Powers of freedom: reframing political thought. Cambridge University Press, Cambridge
Seabrook J (1984) The idea of neighbourhood: what local politics should be about. Pluto Press, London
Staeheli L (2003) Women and the work of community. Environ Plan A 35:815–831
Stoker G (2004) New localism, progressive politics & democracy. Polit Quart 75(Supp 1):117–129
Swyngedouw E (2004) Globalisation or 'Glocalisation'? networks, territories and rescaling. Camb Rev Intern Aff 17(1):25–48
Thame Town Council (2012) Thame neighbourhood Plan. Tibbalds Planning & Urban Design, London
Titmuss R (1970) The gift relationship: from human blood to social policy. George Allen & Unwin, London
Trigilia C (2001) Social capital and local development. Eur J Soc Theory 4(4):427–442
Upper Eden Community Interest Company (2012) Upper eden neighbourhood plan.
Wallace A (2010) New neighbourhoods, new citizens? challenging community as a framework for social and moral regeneration under new labour in the UK. Intern J Urban Reg Res 34(4):805–819
Williams R (1967) Culture and society. Chatto & Windus, London
Williams F (1993) Women & community. In: Bornat J, Pereira C, Pilgrim D, Williams F (eds) Community care: a reader. Macmillan, Basingstoke

Williams C, Windebank J (2000) Helping each other out? community exchange in deprived neighbourhoods. Commun Dev J 35(2):146–156
Williams C, Aldridge T, Tooke J (2003) Alternative exchange spaces. In: Leyshon ARL, Williams C (eds) Alternative economic spaces. Sage Publications, London, 151–168
Wilson J (2013) The devastating conquest of the lived by the conceived: the concept of abstract space in the work of Henri Lefebvre. Space and Culture 2013 16:364 (originally published online 15 May 2013) doi:10.1177/1206331213487064
Young M, Wilmot P (1962) Family and kinship in East London. Penguin, Harmondsworth

Chapter 14
Social Sustainability and the Housing Problem

Jamileh Manoochehri

14.1 Introduction

The issue of sustainability which started as an environmental concern has been defined in a number of ways, each reflecting a particular approach. While the term 'sustainability' has been in use for around three decades , the most referred to aspects are environmental, economic and social sustainability. Environmental sustainability is perhaps the most easily quantified. The impact of development on the ecology of the earth is to be kept to a minimum. The embodied energy of materials used on site, the energy consumption of the development once complete and lived in are measured or projections of such consumption are considered in order to gauge the impact of the development. Economic sustainability is probably the most easily quantifiable of the three measures of sustainability. The project simply cannot be carried out unless it functions as an economic proposition.

Social sustainability is more complex. As Bramley and Power (2009) suggest, a significant proportion of sustainable development rhetoric now refers to the importance of social equity (Bramley and Power 2009) and at policy level, we see references to social sustainability in documents published by the DETR (1997 and 2001, cited in Bramley and Power 2009).

There are, however, international agreements and documents that put social sustainability on a par with economical and environmental sustainability (UNCSD 2012).

J. Manoochehri (✉)
Leicester School of Architecture, De Montfort University, The Gateway,
Leicester LE1 9BH, UK
e-mail: jmanoochehri@dmu.ac.uk

© Springer International Publishing Switzerland 2016

M. Dastbaz et al. (eds.), *Building Sustainable Futures*,
DOI 10.1007/978-3-319-19348-9_14

14.2 Sustainability as a New Paradigm

In order to offer a more expansive definition of social sustainability, a connection is sought with the concept of the community. Brindley proposes that 'implicit in the notion of "social sustainability" is a pattern of social relations within the city that is more sustainable than current patterns' and that it amounts to a critique of the social condition of the city today and a 'reaction against the consequences of dispersal, segregation and exclusion' (Brindley 2003, p. 54). The latter three aspects of present-day cities were the focus of some attention during the first decade of the new millennium.

In 1998, the government commissioned a report by the Urban Task Force (UTF). Its report was one of the several documents published in the last throes of the twentieth century advocating a new approach to the development of sustainable cities in Britain (Egan 1998). In its report Towards an Urban Renaissance (1999), the UTF promoted the city as a dense, lively and diverse place. The report identified certain criteria as crucial in creating such a city; among these were the site and setting, character, public realm, mixing of activities, mixing tenures, sustainable buildings and environmental responsibility.

These criteria were identified as central to the creation of a sustainable urban environment. It might be said that out of the three components of sustainability, the social aspect is the most discriminating and with the most immediate impact, especially where social housing is concerned. Becker et al. suggest that there is a relevance to social sustainability of notions such as social equity and social justice—associated with social structure and related to gender equality and political participation (Becker et al. 1999).

The link between social sustainability and the notion of community is alluded to in other documents too (Bramley and Power, 2009). New formulations were offered by the UTF for new developments. The adoption of 'mixed use', 'mixed tenure' and 'high-density' developments became the main pillars of a new orthodoxy at the turn of the millennium, aimed at creating a thriving and lasting community. Mixed tenure and mixed use have had many examples in cities of Britain and Europe where thriving communities earn their living and create lively and attractive places to live, work and visit. Higher densities have had a different past.

Definitions of a number of key factors that determine the quality of housing have remained roughly the same for a number of decades although orthodoxies and ideas of 'best practice' have changed. In the last 20 years, given that shortage of social housing has been a constant within many variables, studies and policies have tried to make the most of what there is—the existing urban fabric-and have tried to find efficiencies. Density is an obvious factor in determining the number of dwellings that the city can contain, and the amendments to the Welfare Reform Bill[1] were introduced partly in relation to the issue of density.

[1] The Bill was approved in the House of Commons on 24 October 2012 by 260 votes to 206.

There has been research that links social sustainability to other aspects of social life. Littig and Griessler (2005) link social sustainability to a number of core indicators:

- Satisfaction of basic needs and the quality of life—individual income, poverty, income distribution, unemployment, education, housing conditions and health, security as well as subjective satisfaction with work, health, housing, income and environment.
- Social justice (as articulated within the sustainability discourse) and social coherence. Social justice covering justice regarding the distribution of economic goods (e.g. income); as well as the broader philosophical definition implying equal opportunity regarding quality of life and participation in society, in education and in further education.
- Aspects of social coherence, e.g. integration into social networks and involvement in activities as volunteers as well as measures for solidarity and tolerant attitudes (Littig and Griessler 2005).

Some of these indicators have clear spatial representations and some are manifested in relationships between individuals or between systems. Housing contains both characteristics. Various aspects of the physical environment and the way we relate to it have been studied through the concept of environmental psychology and its link to the people's lives (Rapoport 1969). Rapoport tries to find a link between sociocultural factors, built forms and the interactions between people and the built environment. Not only does the environment affect us physically, but it conveys a meaning too (Rapoport 1969 cited in Franklin 2001). Rapoport (1982) studies the way residential (as well as other) environments are perceived—positively or negatively. This kind of input may affect the experience or perception of inhabitants about their environment and may determine their satisfaction with it and their assessment of its quality. It appears that the 'subjective' dimension of the experience of living in a place is indisputable—we each perceive our environment according to our past experience, and our values and expectations play a part in what we think of a 'place'. Therefore, the way we perceive our environment will impact on how we interact with it, how we project our existence in it and on whether we will be a willing party to sustaining it.

The built environment affects us not only as individuals but also as communities and as social groups. The study carried out by Glaeser and Sacerdote (2000) is of interest in that they have examined the connection between housing structure and social cohesion. They studied high-rise buildings and compared them to single-family dwellings. They asked about the levels of the occupants' sense of citizenship and behaviour associated with different building types. They summarise their findings thus, 'residents of large apartment buildings are more likely to be socially connected with their neighbours, perhaps because the distance between neighbours in apartment buildings is lower than in houses', and rather importantly that 'apartment residents are less involved in local politics, presumably because they are less connected with the public infrastructure and space that surrounds them'; and 'street crime is also more common around big apartment buildings'. The authors of the study attribute this latter finding to the fact that there is less connection between the people in apartments and the streets that surround them. The conclusion of Glaeser and Sacerdote somehow explains why in an estate with high density and low-income tenure, a sense of ghettoisation and of the erosion of the sense of citizenship may result. Any process of marginalisation of a community is detrimental to the whole society and a detriment to social sustainability. These conditions are the result of state policies.

The impact of social and housing policies in the last 25 years has been among the most pronounced with regard to low-income housing, with significant consequences on both the demography of social housing and the composition of its stock. The 1980 Housing Act opened the way for Britain's local authority tenants to purchase their property at highly subsidised prices. Local authorities were removed from the housing development equation, which they had dominated since the earlier decades of the twentieth century and were turned into property managers. This role too, was challenged soon as large scale 'transfers' pressed ahead and removed social housing stock from the control of local authorities and placed them in the control of housing associations. The result of these policies was manifold. The emphasis on home-ownership as a pillar of the central government policy was accompanied by low interest rates, encouraging anyone with a regular income to try to purchase a property, through mortgages from building societies and banks. The rush-to-purchase created high demand which in turn raised the prices of properties, including the local authority stock now entering the market, as established tenants bought their council-rented housing at high discounts only to sell in due course, at market prices and at substantial profit.

The sales benefited many council tenants in the short term. However, the overall effect of the policy was to privatise social housing. Once they had entered the market, the local authority stock would command different prices according to the desirability of their location, their environment, size and general marketability. The end result was that the most desirable units, in the best areas, were successfully privatised. This led to the local authorities being left with the least desirable stock. With a number of exceptions, the marginalisation of local authority housing was complete. This marginalisation and the perception of a stigmatised community are among the factors that militate against social housing[2] today.

In the years following the UTF report in many new developments, the requirement for mixed-tenure was translated into locating social housing adjacent to private housing. By definition, units for sale need to be sufficiently desirable for purchasers to invest in them. The most advantageous units as far as access, views or spatial design quality is concerned, were assigned for sale. As social housing tenants are allocated their dwellings and do not exercise as much choice as the purchasers. As an example, The Greenwich Millennium Village was based on the Masterplan designed by the Architect Ralph Erskine and was intended as a flagship of the efforts

[2] In this text, social housing refers to state-assisted housing, which is mainly Local Authority housing or that provided by Registered Social Landlords, which constitute Housing Associations

by the government of the day towards sustainable development. The location or orientation of the units intended for Housing Association tenants betrays a weakness in the approach, as the more desirable units are secured for private market.

The spatial distribution of low-income housing is affected and determined by new policies. How do the current trends in the spatial distribution of social housing affect social sustainability? It can be argued that spatial and social-network distributions are among the most tangible aspects of social sustainability. These are related, in turn, to access, space standards and social cohesion in its broadest sense. Each of the elements needs requisite support.

As McKenzie (2004, p. 13) defines it, when present, social sustainability is 'an asset'—a 'social capital'—which 'allows for coherence and the ability to overcome hardship' and it is related to whether 'social capital is growing or diminishing' where natural, human, social and institutional projects are the sites in which this social capital is used or produced. According to McKenzie indicators that determine level of social sustainability include equity of access to services, equity between generations, cultural integration, political participation and mechanisms for expression of collective needs and strengths, for political advocacy and for community action to advance these.

The concern for the social capital and social sustainability is a continuation of similar concerns that have affected policy and practice, albeit with a different narrative. In the UK, the past hundred years have seen housing policies range from policies aimed at overcoming shortage of housing, providing housing for a country recovering from war and as a means of preparing for a brave new world made possible by new technologies and materials. This has been most visible in social housing. Over the years, social housing has been redefined and reshaped. Local Authorities and Housing Associations shared the provision of housing to those who qualified for access to them. The main difference between the two has been the fact that since the 1988 Housing Act, Local Authorities have had no power to develop, whereas SRLs still do. The housing provided by the latter often takes the form of a portion of a larger, private development with a social housing component. The 2011 Localism Act (DCLG, 2011) introduced viability as a prerequisite; therefore, if a developer can demonstrate that the provision of affordable housing is not profitable, they can request that the council's planning policy requirement is reduced or waived (Bowie 2014).

The profile of low-income housing in the social housing sector has changed over the years. The percentage of social-rented housing and private-rented housing reversed from early twentieth century to mid-century and now, they have returned roughly to where they were in the early twentieth century. In 1918, 23 % of households were in owner occupation, 1 % were social renters and 76 % were private renters. In the period after WWII and its universalist ethos, the social-rented sector rose steadily and was at 31.7 % in 1981 before the marketisation of social housing has commenced. During this time, private renting stood at 11 %. The privatisation of social housing, and various mechanisms to facilitate this, including tax exemptions for mortgages or readily available credit changed the composition of tenures, so that in 2011–2012, they stood at 65 % owner-occupiers, 17 % social renters and 17 % all private renters (English Housing Survey, Table FAI21 (S108)). It is important to note that the term 'owner' conceals the fact that out of the 65 % owner-occupiers only 32 % own their homes outright and 33 % are mortgaged.

In the intervening period, one very significant change has occurred: the unravelling of the belief that the state needs to intervene in order to provide for those whose income excludes them from the market. The fact that after the Second World War, this effort became a form of universal service meant that social housing which catered for the low income was not exclusive to the low income.

Rather importantly, in post-war UK, social housing tenants included better-off members of the working class as well as white-collar workers and professionals. Parker Morris standards, which recommended minimum space standards for housing in 1961, did so for all housing and intended the standards as minima.

The ascendancy of the neo-liberalism, its focus on the withdrawal of the state from such provision, and the accelerated privatisation of social housing in the form of subsidised sale of social housing stock to its tenants has meant that local authorities in Britain have been left with a limited supply. The result has been that in the past few years, the demographic of council tenants has changed.

Government policies such as Right to Buy and the recent Help to Buy have escalated the commodification of housing. Those who can afford a mortgage enter the market to become part owners, and those who cannot, remain in the rented sector.

The marginalisation of this tenure is compounded by their relative lack of choice. A recent report by the Office for National Statistics found that household spending matters to aspects of personal well-being. It is relevant in the study of residualisation and marginalisation of social housing to query how the well-being of the inhabitants is affected. The connection between social cohesion and social sustainability (Littig and Griessler, 2005) is clearly absent in a residualised housing sector.

It is suggested that such residualisation includes changing characteristics of social housing tenants, and that the inhabitants of social housing are likely to have rising levels of unemployment or declining incomes in comparison to the general population (Pearce and Vine 2014).

In a study carried out in 2003–2004 in five higher density wards, in London, LSE researchers recorded a number of findings related to higher densities in social housing and the dissatisfaction among the residents.

In terms of housing design, the research found that the residents ranked dwelling size third in the order of importance after security and sound insulation and before good quality open space and privacy. Out of the six successful features mentioned, personal outdoor space and generous space standards also stand out.

The key to the quality of life within high-density housing in less well-off areas is related to the amount of time an occupant spends in their home, and to how long this time is interrupted by engagements outside the home in pursuit of purposeful activity that in itself encourages a sense of well-being. Lifestyle appears to be a significant aspect of how we occupy our homes. The MORI study (cited in LSE research shows a higher rate of satisfaction among those who make more frequent trips out of London, either on holiday or weekend breaks, and that 'the time residents spend working and socialising away from their homes may also impact significantly on their attitudes to dense neighbourhoods (LSE research, p. 71). In a study of high-density college dormitories, Baum and Valins (1977) found crowding-related stress and divergent social behaviour and that the architectural design of human environments influence mood and behaviour.

In its physical manifestation, it is argued that inequity and social exclusion become 'areas of deprivation with reduced access to public services' (Brook Lyndhurst 2004; Macintyre et al. 1993; cited in Dempsey et al. 2011). The change in housing policy and provision in recent decades is believed to have increased social inequality, and the policy of targeting welfare has encouraged the better-off into home ownership and has led to an increasingly residual-rented sector (Murie 1997).

Moreover, studies have also shown that the social-rented sector's social base has narrowed to exclude middle or high earners (Burrows 1997; Forrest and Murie 1983, 1990). Lee and Murie (1999) consider the spatial and social divisions in the city and suggest that the debate on residualisation could go beyond the consideration of council and social-rented housing. They suggest that it is wrong to assume that the concentrations of deprived housholds are purely in the council sector (Lee and Murie 1997). What happens to social cohesion, when residual housing is the only form of housing accessible to the low-income? The concept of social cohesion, has been connected to solidarity and order (Lee and Murie 1999) and to functioning communities. Social cohesion is seen as a by-product of family and kinship, work and work place and is the 'glue' that holds people together who work in local places and meet and have relationships through work (ibid). Lee and Murie (1997) consider transient populations and lack of commitment and investment in these areas and suggest that the functioning of council housing neighbourhoods is not about social mix and balance, but about the economic marginalisation of inhabitants that coincides with isolation and marginalisation in terms of other resources and 'unstable populations' (p. 636).

The broader definitions of social sustainability and social cohesion[3] appear to be more useful and broad enough to cover the most significant aspects of margin-

[3] Returning to the concept of social cohesion, the definition needs expanding and refining. In 2003, a cohesive community was defined by the ODPM-appointed committee as one where: 'there is a common vision and a sense of belonging for all communities; the diversity of people's different backgrounds and circumstances are appreciated and positively valued; those from different backgrounds have similar life opportunities; and strong and positive relationships are being developed between people from different backgrounds in the workplace, in schools and within neighbourhoods'. The definition of cohesion was influenced by the fact that the report followed disturbances in Oldham, Bradford and Burnley in the summer of 2001 and it was commissioned in order to advise on long-term integrated programmes in order to achieve social cohesion.

Ethnicity and immigrant communities feature in many such reports or research. A study funded by Joseph Rowntree Foundation found that any racialised tensions between residents of multi-ethnic neighbourhoods had 'material underpinnings around struggles for resources such as employment, housing and the physical infrastructure of the neighbourhood' (Policy Studies Institute 2007)

alised housing. By definition, marginalised housing is one that does not enjoy all the diverse constituents that make a thriving community. This marginalisation is to a large extent defined by the economic marginalisation of its inhabitants. Social cohesion is bound to be related to commonalities and a sense of common identity. These, it may be argued, have been produced in the past through common employment, schools, religious practice, local community activity and the like. The identity of communities is formed by the work that the members of the community engage in. Their professions and their trades define them. It is perhaps wrong to try to find the answer in everything except meaningful employment opportunities that are linked to the locality in some way, in order to enhance local identity and social cohesion.

14.3 A chronology of Ideas and policies on Social Housing till 2009

The study of housing policy in the last hundred years reveals an intermittent preoccupation with the concepts that are identified as central to social sustainability, among these were access to housing of socially acceptable standards, space standards and social cohesion. The level at which each was set was determined by the dynamics and the requirements of the socio-economic system reproducing itself. I propose that the housing policies of the last hundred years or so were responding to acute concerns about issues that affect and to a large degree define social sustainability. As such they are bound to be important to study as lessons for the future. Key milestones in housing policy from the Tudor Walters Report onwards each reflects the social values that have a bearing on the way social capital is seen and the society itself is understood.

There have been a number of significant milestones in housing policy in Britain: Tudor Walters Report (1918), Dudley Report (1944), Parker Morris Report (1961), and the Housing Act of 1988. The later versions of these reports and housing Acts have not had quite the same impact as the earlier ones. The Urban Task Force report of 1999 commissioned by New Labour and the National Planning Policy Framework of 2012, which was brought in by the coalition government of Conservative and Liberal Democrats, shows the clearest shift from their context, with a new approach. Each of these affected the methods, forms and standards of housing provision and the demography of its inhabitants to various degrees.

The first housing policies of the twentieth century emerged out of the sanitation policies of the earlier century but only after regulation seemed to have become absolutely necessary for the maintenance of order and averting the risk of widespread disease and social disintegration (Swenarton 1981; Merrett 1979).

When it became clear to policymakers that something had to be done, the play between quantity and quality of provision kept resurfacing. In deciding on a priority, generally, two factors appear to be at work, factors that have remained evident in the provision of social housing ever since. These related to the assessment of who should benefit from state-funded provision of housing and what is deemed to be an acceptable standard for housing the low income.

The statements made at the time reflect the value judgments that were at play in selecting the recipients of state assistance. The picture that Charles Booth (1902) paints of the period with the poorest inhabitants mapped out at the heart of London, mirrors of problems of access to housing of a socially acceptable standard nearly a century on. The main difference may be in the spatial distribution of the problem. As the living conditions of the poor worsened at the turn of the twentieth century and congestion increased, those who could afford to leave central areas of the city left for new lower density suburbs and the poor crowded into the vacated dwellings. A century on, the spatial distribution of accommodation exclusive to the wealthy appears to have been reversed, at least in London. Studies of the spatial distribution of social-rented, private-rented and owner-occupied properties in 2014 by Office for National Statistics show the social-rented sector is well-represented in the geographical centre of London and the owner-occupied properties dominate the band around this core. The core is occupied mainly by the rented sector, whether private or social.

At the turn of the twentieth century, the suburbs around London and the Metropolitan cities were targeted for providing better low-cost housing.

The Housing (no.2) Act of 1914 was passed less than a week after the declaration of the World War I. Swenarton attributes the passing of this act to the fear that the outbreak of the war would lead to a suspension of building and therefore to widespread unemployment in the building trades (Swenarton 1981).

The outbreak of World War I was a landmark in redefining social and political expectations. Under David Lloyd George as the Prime Minister of the Wartime Coalition Government (1916–1922), Christopher Addison became the Minister of Health. The 1919 Act, known as the Addison Act, was a significant step in housing provision as it made housing provision the responsibility of the state and gave local authorities the task of developing new housing and rented accommodation for working people [Available from: http://www.parliament.uk/about/living-heritage/transformingsociety/towncountry/towns/overview/councilhousing/; Accessed: 28.02.2015]. Addison became a key figure influencing housing standards for at least another decade. He was aware of the potential for centralised planning in order to meet the massive demand for housing and considered that the Ministry of Munitions was in a unique position as it functioned both as a manufacturing and an employment agency. He declared that never before had 'the State assumed such extensive responsibilities for directing and originating production… It provided a new kind of opportunity' and suggested that '…use ought to be made of the exceptional situation to secure a better and more humane standard of working conditions' (Addison cited in Swenarton 1981, p. 50).

Prior to the post-war promises being made, the early efforts to respond to the shortage of housing were influenced by the Garden City movement. However, in 1915, the Treasury curbed any avoidable expenditure stating that in order to curb expenditure, 'the erection of houses … on Garden City lines must be abandoned' (PRO T132, 1915, cited in Swenarton 1981, p. 51). The Garden City movement however had become influential. The architect Raymond Unwin was included in the committee that drew up the Tudor Walters report—published in 1919. The report advocated garden-city standards of house design, layout, circulation and the environment for working-class housing. These aspects had been limited previously to a wealthy minority. In internal layout and equipment and density, the recommendations were an improvement on the best municipal housing of the previous years (Whitham 1982). Against this background, on the day following the Armistice, the Prime Minister Lloyd George announced a general election and undertook to secure 'habitations fit for the heroes who have won the war', (The Times, 13 Nov. 1918, cited in Gilbert 1970). At this time, through the Tudor Walters Report, the Addison Housing and Town Planning Etc Act of 1919 came into being.

In 1917, mindful of the enormity of the problem, the 'Royal Commission on the Housing of the Industrial Population of Scotland, Rural and Urban' Cd 8731 (HMSO 1917) wrote the state alone was in a position to assume responsibility and to undertake the necessary building schemes with assistance from the state, if necessary (cited in Whitham 1982).

Would there be building of minimal standards in order to alleviate the pressure of the housing shortage? The Workmen's National Housing Council (WNHC) argued against building down to the 'poverty standard of the ill-paid and unorganised classes' (cited in Swenarton 1981). They called for housing estates that would be aesthetically pleasing, equipped with communal facilities, private gardens, a bath and a supply of hot and cold water (ibid).

The narrative of working people's entitlement to quality housing was lost after the war. The press played their part in this process by characterising it as extravagance (Swenarton 1981). For a period after WWI, the needs and expectations of the people with low income were intense and the state's policies appeared to be responsive to these. The Addison Act of 1919 represented this attitude.

The earlier Garden City Movement left an indelible mark on the need to improve design quality in social housing.

The general election in November 1922 brought in the Conservative Party, whose government in turn revived the private sector in housing. The Housing Act of 1923, known as the Chamberlain Act, obliged local authorities to pass on the subsidy to private builders. The local authorities were able to build only if they could demonstrate that private builders could not meet local needs.

At this time, there were no safeguards against lowering of standards by local authorities. The reduction in space standards in 1923 Housing Act was accompanied with lower standards in layout and environmental standards. By the middle of the 1920s, the architects' panels were discontinued and late in the decade and in the early 1930s, the design of social housing dropped in design quality. Whitham (1982) argues that the council houses of this time were still better built and better laid out than those by private builders.

Slum clearance remained a prominent aspect of the housing problem at this time and affected the consideration of typology and density, which moved away from the Garden City idea and towards blocks of flats. This typology came to lack the environmental standards of earlier council housing and local authorities offered improved domestic equipment in order to compensate for this (Whitham 1982).

In 1929, the second Labour Government came to office. The Minister of Health, Arthur Greenwood, adopted a general needs housing policy, as stipulated in the Wheatley Act. Greenwood's Housing Act of 1930 required larger local authorities to submit 5-year plans for dealing with their slums. Whitham (1982) attributes the move away from general needs housing in 1931 to an economic crisis. In 1933, the Housing (Financial Provisions) Act charged local authorities with mainly slum clearance. Bowley (1945) suggests that by this time the ethos of good-quality housing provision was replaced with a return to the sanitary policy of the nineteenth century.

A number of features stand out in social housing in the first third of the twentieth century. The most important is the recognition that the state needs to take action in solving the housing problem. The needs and expectations of people on low income were intense and the state had to respond to them. The Garden City movement created an understanding of the need to improve design quality in housing for the low-income. General needs housing became distinct from targeted housing, with the Labour Party favouring the former and the Conservative Party the latter.

As war broke out in 1939, house building in the UK ceased. In 1944, the Minister of Health appointed Lord Dudley to head a Sub-Committee of the Central Housing Advisory Committee and to 'make recommendations as to the design, planning, layout, standards of construction and equipment of dwellings for the people throughout the country' (HMSO 1944).

The Manual considered Site Planning and Housing: 'Where big authorities must build a very large number of new houses, it may be possible to plan a new self-contained community based on a new centre of employment. In such a community due regard must be paid to industrial, social, educational, and recreational centres and their relations to the new development as well as to accommodation for the different classes of people who make up a well-balanced residential neighbourhood' (ibid, para. 16–19). Much attention was paid to the quality of the environment, and there was a recommendation for 'improving thermal and acoustic insulation within buildings and the across party walls and floors—measures that affect the quality of the inhabitants' lives significantly (ibid). The authors tried to pre-empt any financial arguments against their recommendations.

The idea of council housing being at the service of satisfying general needs was the starting point of the Housing and Town Planning Act of 1919 and shared by Liberal-dominated coalition, and Labour governments alike. This approach that was continued in the Housing Acts of 1924, 1946 and 1949 and the Housing Rent Subsidies act of 1975 all of which aimed to achieve this objective. Council housing in the view of Labour at this point was not intended solely for the poor, the underprivileged or the population of traditional working-class areas.

14.4 Social Consensus

The earlier restrictions by law on local authorities, to provide housing for working class people, were replaced in 1949 by 'a more comprehensive, universalist policy' and subsequent legislation 'placed a duty on local authorities to give reasonable preference to persons occupying insanitary or overcrowded houses, or who had large families, or who were living under generally unsatisfactory housing conditions—those people, in fact, who could prove housing need' (Burke 1981, p. 17).

In his 1949 Act, as Minister responsible for housing, Aneurin Bevan removed the restrictions on the local authorities to provide houses for the working class only. Instead, they could introduce 'the tapestry of a mixed community', where the doctor, the grocer, the butcher and the labourer lived in the same street (Foot 1973).

Similarly, William Beveridge, the Liberal politician, in his foreword to The Rehousing of Britain by Madge (1945), reiterated his commitment to raising housing standards as a means of redressing social inequalities. In doing so, he was redressing the 'greatest inequalities between different sections of the community' there was no question that 'good housing—far better housing than we have at present—is the indispensable foundation for health, efficiency and education'… This was a far-sighted vision, as it was 'a waste of money to build hospitals to cure disease if families are forced to live in houses that breed disease. It is a waste of money to build schools for children who must return every night to squalid, crowded, unhealthy homes' (Madge 1945).

The whole society and the wholesome view of the life for the citizen drove part of the agenda at this time. Beveridge's argument was for a new and better society where human and financial resources do not go to waste through lack of forward planning and vision. In this, Beveridge was not alone. He represented a left of centre stance of his Party, and it has been argued that his position reflected the left-ward shift of the electorate. (Knight 2006). This shift was not confined to the electorate but some of the political establishment as well.

In 1951, the Conservative government sanctioned a lowering of standards in social housing. A different ideology was trying to achieve ascendancy. The decline in standards started with Circular 38/51 of April 1951. The Circular prompted reductions in circulations space while maintaining living space standards. Within 3 years, the average area of a 5-bedroom space house had fallen by 11 sq. m. (Merrett 1979). By 1959, the average floor areas for a new 3-bedroom house had fallen to 897 sq. ft. (83.3 sq. m.) (Malpass and Murie, 1994, p. 75). In key periods, in the 1950s, the government presented the market as the major mechanism by which to solve the housing problem. In this period, the principal measure of success became the number of dwellings to be built and politicians were acutely sensitive to the rate of new house building (Bullock 2005).

Under Macmillan the Conservative housing minister, the government built 319,000 houses in 1953 and 348,000 in 1954, against the promised 300,000 houses per annum to which they had committed the Conservative government. This was achieved as the result of Macmillan's decision to relax the standards set by the Labour housing minister, Dalton, in 1951. During his tenure, the rigorous standards set by Bevan were also lowered (Sullivan 1996). In January 1954, having reached the total target a year ahead of time, Conservatives hit a great electoral success, one as effective as that of 1980s when they championed the sale of council houses.

14.5 Universalism and social policy

The rise of the post-war social democratic consensus coincided with social democratic governments in office in much of the Western world. This in turn signalled a shift in power balance, in the social attitudes and the political climate of the time. Whether there was consensus and to what degree, is debated. The term Butskellism was coined pejoratively in 1954 to emphasise the difficulty of distinguishing between the economic policies of the Conservative Chancellor of the Exchequer R.A. (Rab) Butler and Hugh Gaitskell his opposite number in the Labour Party. Two significant aspects in this consensus are cited as 'a commitment to equality' and 'an attempt to mitigate the worst aspects of inequality' (Seldon 1991, cited in Gladstone 1995). Kavanagh and Morris (1989) described the consensus as the finding of parameters of what was practicable, affordable and politically acceptable.

The ideological influence on policies came partly from the intelligentsia and economists of varying views. In the aftermath of WWII, the ideas of John Maynard Keynes became dominant. Keynes promoted the government's use of fiscal techniques to control the economy, especially on the demand side, and thus full employment. During this time, there was also William Beveridge whose influence gave rise to a commitment to the provision of social welfare to the country's citizens as well as control of essential sectors of the economy

Opposed to this stance were those who pressed for the freedom of the market and tried to change the focus of policy from the state providing welfare and equal opportunity for its citizens, to directing resources to critical problems, such as replacing the worst housing, and halting the effective albeit partial decommodification of housing. The 1957 Act, which decontrolled new tenancies encouraged privatisation, was an instrument in this direction. The government argued that if a free market in housing was desirable and long-term, private landlords had to be able to charge rents that would give them an adequate return. Rent rises followed and the incentive to free-rented property led to unscrupulous landlords abusing tenants.

The ideological or theoretical input of academics like Mannheim was crucial in the tone and direction of the policies. Karl Mannheim of the London School of Economics was one of the ideologues who were asked to work on Conservative reconstruction policy (Harris, 1986). At the time, a study commissioned by the government envisaged the state engendering a powerful moral order on the society, centred on Christianity. On the practical side, Harris (1986) suggests that it proposed compulsory day release for young workers to acquire proper attitudes to citizenship, and the public schools would be taken over by the Board of Education to reduce the class division in society.

14.6 Homes for Today and Tomorrow

The argument for sustainability has the projection of the future at its core. In this respect the reports such as Parker Morris' mirror the concern for building a sound (sustainable) the future, and its ethos is relevant to the arguments for socially sustainable developments. The 1960s was an important period in the development of new ideas, the most significant indication being the Parker Morris Report. Health, education and social security were the high-profile aspects of the welfare state. At specific periods, housing joined them as an aspect worth prioritising. Overall, the welfare state was an ambitious project that relied on massive state intervention and with its introduction Britain came to have 'one of the most comprehensive and powerful planning systems in the world' (Hall 1973, p. 40).

The marked decline in standards in the 1950s had produced a counter-reaction. In 1959, a new design committee was set up under the chairmanship of Sir Parker Morris. The report of the committee was published in 1961 under the title of Homes for Today and Tomorrow. It cited the increase in living standards since 1944 as the main reason for revising the specification of new housing. The Report's main proposals dealt with space, including storage space, and heating within the home. The committee criticised current practice 'Homes are being built at the present time which not only are too small to provide adequately for family life but also are too small to hold the possessions in which so much of the new affluence is expressed' (cited in Malpass and Murie 1994, p. 75).

The Parker Morris committee took account of the new living patterns and expectations and it considered its report to be 'applicable to private enterprise and public authority housing alike' (HMSO 1961, p. 1). The Parker Morris report had the future firmly in its sights: 'Homes are being built at the present time which not only are too small to provide adequately for family life but also are too small to hold the possessions in which so much of the new affluence is expressed. Moreover, many of them have steep stairs and mean halls and landings. Such places cannot be expected to meet the needs of their occupiers today, still less to hold their value in the long term' (ibid, p. 4). The report improved on the Dudley Committee (1944) recommendations. It suggested, 'And where for reasons of economic stability there may have to be a choice between standards and numbers built, we think that in future it should not be the standards that are sacrificed' and 'the homes of the country are one of its most important assets—assets which must be built to a standard at which they are likely to give reasonable satisfaction and therefore hold their value over the years' (ibid, p. 6). The Parker Morris report came to reflect the values of a period in British social history when universalist ideas were in ascendancy.

14.7 Selectivism and Social Policy

When in power the Conservative Party embarked on the process of the marginalisation of council housing through its promotion of the private market, and used taxation—or its removal—to do this. In 1963, the tax on owner-occupiers was abolished but they continued to get tax relief on the money they borrowed. The capital gains to be made by a landlord selling his property to an owner-occupier became irresistible. In the 1950s and 1960s alone, around 2 million private landlord properties were sold to owner-occupiers (Holmans 1987).

In practice, due to IMF demands in late 1976, major spending cuts had been undertaken under the Labour Government and the idea of a smaller state was gaining ground before 1979. Various instruments started to be used to promote the market and (purportedly) to reduce public expenditure. In the 1970s, the Labour Party was also taking seriously the idea of the promotion of home-ownership. This shift in attitudes was also becoming visible in the electorate. Later in the decade, with the adoption of monetarism by the Callaghan administration and in response to the Green Paper, Housing Policy: A Consultative Document (Department of the Environment 1977a), Labour abandoned the policy.

Furthermore, council housing became a target of public expenditure cuts, and public sector house building fell from 173,800 in 1975 to 80,100 in 1979, its lowest level since the 1930s and insufficient to satisfy even welfare needs (Pearce and Stewart 1996). Improvements also fell—from a peak of 188,000 in 1973 (under the Heath Conservative government) to 111,000 in 1979, and in addition, many councils reduced their repairs and maintenance expenditure causing delays in getting repairs done, and frustration to tenants (ibid).

In 1979, the Conservative Party led by Margaret Thatcher came to power with a massive majority. One of the Party's promises was that it would legislate for greater owner-occupation and the retreat of the state from main arenas of economy and social provision. If every decade marked the inception of what was to come in the following one, the 1970s may be said to have been the years when the idea of the withdrawal of the state and the rise of the market as a mechanism of control started gaining momentum. The ideas of free-marketers began to be seen in the policies promoted by politicians and most significantly embraced by the Prime Minister, Margaret Thatcher.

At this time, deregulation and selectivism were deemed to be the most appropriate approaches in social policy. The argument for selectivism was applied to most arenas of the social and civic life. As indicated in the Conservative Party's publication, The Next Move Forward (1987), the government in the eighties was adamant: 'Municipal monopoly must be replaced by choice in renting'. A process of privatisation of council housing started by which the tenants could elect to transfer the management of their estate from the local authority to another body. This was coupled with the policy of giving generous incentives to tenants to purchase their dwelling.

Under the 1980 Housing Act, the government gave council tenants with a 3-year history of tenancy, a right to buy their homes at generous discounts. These ranged from a third to half the value of the property and in some cases in later years (1989) there were discounts of 70 % given. With such incentives, there was a massive surge in the process of the sale of council houses, reaching close to a quarter of a million a year by 1982. Between 1979 and 1987, more than a sixth of the total stock of council houses was sold and between 1979 and 1994 the share of the total housing stock owned by local councils fell from nearly a third to little more than a fifth. Sales began to slow down as most of those who had the capacity to buy, and some that really did not, had bought their houses. This left behind in local authority and housing association tenure mostly the poorest groups in society. In 1974, just under half the tenants in social housing were in the bottom 40 per cent of the income distribution; by 1991, the proportion was three-quarters with only 40 per cent of them in employment (Glennerster 1995). The government policies had started a process of complete change in the demographic make-up of those in social-rented housing.

In the 1980s, central government removed the councils' subsidies, which were used in order to keep down the level of council house rents. This led to the councils having to raise their rents and bringing them closer to the 'fair rent levels', which applied to private and housing associations tenants. Those who could not afford the rent and proved they had low income were entitled to claim housing benefit from the Department of Social Security (Glennerster 1995). In the late 1980s, the government housing policy had to change again in order to take account of effect of their earlier policies on the housing stock. Glennerster points out that by 1987 there was little to be gained from the sale of council houses to individuals as a method of further reducing the size of the council sector, especially as by now the remaining tenants were too poor to purchase their dwellings. The Government's solution to overcoming this slowdown in the privatisation of council housing was to give the tenants, who were unhappy with their councils, the right to transfer the ownership of their house to private landlords or housing associations. In order to facilitate this, a central government agency would take over failing estates, improve them and then sell them off to the private sector. Moreover, if the enterprise was to be attractive to the private sector, it was important to give landlords the right to let at free market rents.

As the Conservative Party manifesto had promised, the 1988 Housing Act removed rent control in the private sector. Central government could designate an area of public housing and create a Housing Action Trust to manage the houses, eventually disposing of them. The notion that individual tenants could get a landlord to take over their house was dropped as impractical and replaced by an arrangement, which put the onus on private landlords or housing associations. They were given the right to bid to take over an estate or part of it and the tenants had a right to vote on the proposal and veto it, under some complex voting rules, and in addition, under the 1989 Local Government and Housing Act, local councils lost their right to subsidize council rents out of local revenue (Glennerster 1995). The philosophy behind these measures was that of giving more 'market power to tenants'. As a result of the 1988 Act, gradually housing associations of varied kinds and sizes came to be the new main providers of social housing. The legacy of the 1980s and much of the 1990s remained the worsening social problems on the most deprived estates and a lowering of standards in social housing. At times of shortage of housing, the effort to raise quality becomes an uphill one. The provision of greater numbers of units becomes such a priority that the issue of quality comes to be viewed as a luxury. There is a pattern in the 1930s, 1950s, 1970s and post 1980s.

Consecutive governments have repeatedly acknowledged that there is still a shortage of housing in Britain. Reportedly, "taking all the concealments into account, in the United Kingdom there was a substantial shortage of housing by 1994, approaching 3 million' (Balchin 1994). Also in 2003, dealing with the issue of the cost of housing for those who wish to purchase their homes, the Bank of England reported that the 'chronic shortage' of new homes will keep prices in London and the south-east high, predicting a cumulative shortage of half a million homes in the region by 2021 unless the decline in new house building is reversed" (Denny and Collinson 2003).

The social policies that set the scene for housing policies during the various decades in the twentieth century had determined the standards of housing available to those who needed it. This had varied from the idea of entitlement of everyone in need to a state-provided home of socially acceptable standard, to one where the worst housing and the poorest tenants (to-be) would be targeted and the others were to take their chances in the market. In this respect, the notion of acceptable standards is an important, although variable one. The social values of the time may be assessed by what constitutes acceptable housing standards at any one time.

As consumerism becomes more embedded in economic and social existence of the society, and the living patterns change, the space requirements need to be updated to suit, although this is not always the case. These considerations have been applicable in almost every decade when the major changes of attitude and policy have been visible, most notably in the aftermath of the two world wars as well as the 1960s and 1980s.

The 1960s and 1970s are pivotal periods in time as in the 1960s Britain 'a measure of prosperity reached more and more families, the instinct to accumulate more material objects also spread down the social scale, and houses which were formerly required merely to provide minimum shelter took on new roles for the storage and display of possessions' (Burnett 1986, p. 284). Burnett (1986) raises the important point that 'the desire for increased space also followed as a result of the greater time spent in the home and the greater interest taken in it'. This issue is as relevant today as it was in the 1960s as it relates the social values that define a socially acceptable or desirable lifestyle to the material living conditions of the people.

Norton and Novy (1991) observed that the post 1980s policies involved 'increasing inequalities in housing and a wider divergence between minimum standards, or housing rights, and those standards experienced by the average household, or those at the top end of the market' and found that the housing market reflected inequality of income and of economic position; and generally reflected social inequality.

14.8 Social Sustainability and Housing

Nearly, three decades of neoliberal consensus in Britain have led to the conflict between the dominance of the state or the market in determining the provision of the basic needs of housing being resolved in the interest of the latter. The main aspect of the current condition is the serious lack of access to affordable housing, the reduction in the number of units available to the low income within the social housing stock provided by local authorities or the Registered Social Landlords (Housing Associations) and the upward drive of cost of accommodation due to the supply and demand mechanism and the increase in the number of investors who purchase property in cities like London, in order to make a gain when the prices rise inevitably, while leaving them empty.

The provision of housing for the low income has changed a great deal over time and the main characteristic of this change has been the commodification of housing provision. Housing is seen as another product that should be subject to the law of supply and demand. The government website on improving the rented housing sector puts the number of households on waiting lists for social housing at 1.8 million [Available at https://www.gov.uk/government/policies/improving-the-rented-housing-sector—2; accessed 21.02.2015]. This may be translated into 4.5 million people on the waiting list, assuming an average of 2.5 members per household.

The housing stock in the UK changes from homes into assets, which were a part of a volatile market. The policies, which started in the 1980s, took the UK into a property-owning society (Lowe 2011), which was seen as a desirable development by some (King 2010), as it allowed people to have control over what they do.

The shortage of housing for those in need is a serious problem both from the point of view of those who are affected by it and also as a social phenomenon. Homelessness and inadequate housing affect the social landscape of a city and a country.

In recent years, the social aspect of housing has been redefined by new legislation. The Localism Act of 2010 introduced changes to how local authorities manage and allocate their stock. In a Guide to understanding Local Government Act, the Minister of State for Decentralisation, Greg Clark MP, explained how tenancy in Local Authority housing had moved from one of general provisions to a targeted one. Previously, almost anyone could apply to live in social housing, whether they need it or not. As social housing is in great demand and priority is rightly given to those most in need, many applicants have no realistic prospect of ever receiving a social home. The previous arrangements encouraged false expectations and large waiting lists.

In an effort to get more people into social housing, the government has launched the Affordable Rent Scheme. Under the new scheme, social landlords will be able to offer fixed-term tenancies at up to 80 % of local market rent. At present, however, tenancies tend to be for life and are at a lower rate. The scheme was first announced in the 2010 Comprehensive Spending Review in October and will be implemented in April 2011.The government predicts that by increasing the rate at which a landlord can charge rent it will mean social landlords, generally run by housing associations can build 150,000 extra homes over the next 4 years, with a £ 4.5bn capital injection from the government.

The number of housing units available for rent at affordable rates has diminished due to the numerous schemes that are designed to assist the privatisation of social housing. In 2015, sales of Local Authority stock occur under the Right to Buy, Preserved Right to Buy, Right to Acquire, Social HomeBuy, Voluntary Purchase Grant and other outright or shared equity sales to tenants. The existing Local Authority stock may be disposed of to the private sector, too. Central Government has facilitated, through legislation, the privatisation of the existing social housing (Housing Statistical Release, 16 October 2014, Social Housing Sales, England, 2012–2013, Department of Communities and Local Government).

It will also give greater flexibility to housing associations and allow them to build more homes. If you are an existing council or social housing tenant, you will not be affected by the change to lifetime tenancies and if you are already on one of these you will continue to do so. But for those entering into the scheme, although it will mean more people qualify it could mean they will need to pay more rent than they would under the old scheme. But ultimately it should result in a wider proportion of people paying less rent. So, while it benefits more people and gives them a better chance of being able to afford rental accommodation, there will be those few people that it disadvantages.

The pros and cons of government policies that are based on encouraging sale of properties have been well rehearsed. The promotion and increase in private ownership is the key aim of the policies adopted by the coalition government that was formed after the 2009 General Elections in UK. The central government's role has also been increased while the role of the local government has been reduced to a peripheral one.

There has been a rise in the proportion of people in employment who are homeworkers. According to the Office of National Statistics, in January to March 2014, 4.2 million people, amounting to 13.9 % of those in work, were homeworkers. This was the highest rate of homeworking since comparable records began in 1998. [Available from: http://www.ons.gov.uk/ons/rel/lmac/characteristics-of-home-workers/2014/sty-home-workers.html; Accessed 18.02.2015]. ONS states that of these homeworkers, around 1.5 million people worked within their home or its ground, while the remaining 2.7 million people used their home as a base but worked in different places. Homeworking has increased at a rate of 13.9 % in January to March 2014. The jobs of these homeworkers tend to be of higher skilled roles, ranging from managers or senior officials, professionals or associate professionals; and those within skilled trades. This amounts to 73.4 % of homeworkers being in high-skill roles in the economy, with a median hourly pay of £ 13.23 in comparison with £ 10.50 an hour for other workers. The most common roles for male homeworkers were roles in the construction industry, while for women the top roles included childminding and care work (ibid).

According to NSO, 63 % of homeworkers are self-employed; they are more likely to be older individual, with 38.3 % being aged 65 or over. This varies geographically. Out of the top five local authorities in UK, in this respect, West Somerset has the highest percentage of workers who use their home for work at 25.7 %. The lowest percentage belongs to Kingston Upon Hull (at 5.2 %), where the prevailing industries are in retail, heavy industry, education, transport and port services, and therefore, working from home is difficult (ibid).

The comparison of the time that homeworkers spend working is significant. According to NSO, 40.2 % work 30 h or less, 34.5 % work between 30 h and 45 h and 25.3 % work for more than 45 h. This compares to 26.4 %, 42.1 % and 31.4 % respectively, in the case of employees. The most common job role for homeworking employees is identified as Sales Accounts and Business Development Managers, and the most common job role for homeworkers who are self-employed is identified as Farmers (ibid). Homeworkers are also more likely to work more extreme hours of 45 per week or more. These statistics demonstrate the trend that leads to a higher demand on the home and the importance that it plays in an individual's life or that of the society and even the economy.

Fewer affordable homes are becoming available. Developers build in order to maximise their profit and the new planning laws are removing the requirements that imposed quotas of social housing on new developments.

The percentage of affordable housing is diminished[4]. The danger of the current situation, as has been argued in the Parliament, is that people who have lived their whole lives in an area are forced to scatter to the periphery, due to property prices and rent being so high in more central areas of London. In addition, the debate is also linked to the dominance of the market mechanism in the provision of housing which in turn leads to a reduction by local authorities of the quota of affordable housing in any new development. This process leads to a 'hollowing out of engagement in the community' (Thornberry 2015) and a hollowing out of the city. This is connected to the purchase of property by wealthy investors who have no need or intention of living in the property, resulting in a tangible effect on the built environment and the locality. The spatial consequence of this market-driven, speculative process is that whole areas of the city are left unoccupied while many individuals and families who are in need of housing have to look for accommodation in more distant areas, where employment opportunities might be lower in comparison.

[4] A particular case raised in the Parliament relates to a development in Islington. According to the Member of Parliament for Islington South and Finsbury, the Local Authority 'sold a large site formerly owned by the public, where the developers are proposing to build 88 % luxury flats' [Emily Thornberry, the Labour MP for Islington South and Finsbury led the debate on the housing crisis in the Parliament on 6th February 2015, Available at: www.emilythornberry.com/news/99/22/Housing-Crisis-Debate; Accessed: 14.02.2015]

14.9 Conclusion

The experience of 20th century housing and social policy in the UK, in its most progressive periods, shares the concerns about social coherence, quality of life, and social justice, that are integral to the definition of social sustainability today. The instruments for achieving these have varied from strengthening a sense of community, social connections and networks, to the issues of continuity and change.

The effort to build new housing and to solve the problem of slums and the shortage of acceptable housing was the cornerstone of the legislation WWI. The Welfare state after WWII was part of a regeneration of the society and the system, following the destruction resulting from the War. The changes in the 1950s to the attitude towards meeting the housing shortage, that of quantity over quality, were clearly seen to be counter-productive and led to the reaction that emerged in the spirit of optimism that was embodied in the Parker Morris report of 1961. This was in large part based on an interest in social capital and of the idea of the universal and the social. The highest space standards were set in the 1960s when the state used resources at its disposal to build for a cohesive society with a future. Many of the criteria we associate with social sustainability today, were present in the formation of the universalist policies of the period. During periods when market relations have dominated, the focus has shifted from the social to the individual, from regulation to de-regulation and from the state to the market as the regulator of provision of housing for the low-income. Since the 1980s and the dominance of the neo-liberal 'consensus', the social aspect of social policy has been under-represented in the UK.

The dominance of market relations has been shown to be inadequate in solving the pressing housing needs. The housing problem can be solved only with complete regard for the long-term issues, as appear to be articulated in the concept of social sustainability. This requires attention to the impact of poor-quality housing, low-space standards, and of dispersal on the inhabitants.

The conflict between the interests of the market and the provision of social services has been apparent in the history of social housing in the past century. Time and again when the needs of those with the least purchasing power has become prominent in the political agenda, the state has assumed the position of the arbiter in favour of the citizen, whereas when the market mechanism has been paramount, the state has acted as the arbiter in favour of capital. A sustainable future has to consider investing in its social capital in order to secure both producers and consumers. Housing and the need for socially acceptable homes is paramount in meeting the needs of the individual and the collective. The narrative around the issue of social sustainability offers a relevant frame of reference towards finding a solution. The argument for a cohesive society with universal values and broad access to housing of socially acceptable standards continues to be a relevant one.

References

A Plain English Guide to Localism Act (2011) Department for Communities and Local Government, November 2011, DCLG Publications. https://www.gov.uk/government/uploads/system/uploads/attachment_data/file/5959/1896534.pdf. Accessed 21 Feb 2015

Balchin P (1994) Housing policy—an introduction, 3rd edn. Routledge, London

Baum A, Valins S (1977) Architecture and social behaviour: psychological studies of social density. Lawrence Erlbaum Associates, Publishers, Hillsdale

Becker E, Jahn T, Stiess I (1999) Exploring uncommon ground: sustainability and the social sciences. In: Becker E, Jahn T (eds) Sustainability and the social sciences, a cross-disciplinary approach integrating environmental considerations into theoretical reorientation. Zed Books, London

Booth C (1902) Life and labour of the people in London, 1902 edn. Third series: Religious Influences, vol. II, pp 68–71

Bowie D (2014) Planning reform, housing reform and social polarisation: options for the future development of London, ENHR

Bowley M (1945) Housing and the State, Allen and Unwin

Bramley G, Power S (2009) Urban form and social sustainability: the role of density and housing type. Environ Plan B: Plan Des 36(1):30–48

Brindley T (2003) The social dimension of the urban village: a comparison of models for sustainable urban development. Urban Design International (2003) 8, 53–65, Palgarve, Macmillan Ltd

Bullock N (2005) Re-assessing the post-war housing achievement: the Impact of war-damage repairs on the new housing programme in London, twentieth century British history, Vol. 16, No. 3. Oxford University Press, Oxford, pp 256–282

Burke G (1981) Housing and social justice: the role of policy in British housing. Longman, p 17

Burnett J (1986) A social history of housing—1815–1985, 2nd edn. Routledge, London

Burrows R (1997) Contemporary patterns of residential mobility in relation to social rented housing in Englan. University of York, New York

CityForm (2006) The sustainable urban form consortium, funded by EPSCRC

Dempsey N, Bramley G, Power S, Brown C (2011) The social dimension of sustainable development: defining urban social sustainability. Sustain Dev. doi:10.1002/sd.41719 289300 (Wileyonlinelibrarycom)

Denny C, Collinson P (2003) Economy bounces back—house price rises confound fears of property slump—Guradian 6th August 2003

Document (Department of the Environment (1977a)) Conservative party's publication The Next Move Forward (1987) https://www.gov.uk/government/policies/improving-the-rented-housing-sector—2. Accessed 21 Feb 2015

Dudley Report 1944 - The Design of Dwellings, London, HMSO, MHLG

Egan J (1998) Rethinking construction: report of the construction task force. HMSO, London

Foot M (1973) Aneurin Bevan. A bibliography. Vol. 1: 1879–1945, Vol. 2 1945–1960, Four square

Forrest R, Murie A (1983) Residualisation and council housing: aspects of the changing social relations of housing tenure. J Soc Policy 12(4):453–468

Forrest R, Murie A (1990) Residualisation and council housing: a statistical update. University of Bristol, Bristol

Franklin BJ (2001) Discourses of design: perspectives on the meaning of housing quality and 'good' housing design, 2001. Housing. Theory Soc 18:79–92

Gauldie E (1974) Cruel habitations: a history of working-class housing 1780–1918. George Allen and Unwin, London

Gilbert BB (1970) British social policy 1914–193. Cornell University Press, Ithaca

Gladstone, et al, 1995, The Welfare State and the State of Welfare, in British Welfare, Past, Present and Future, UCL Press

Glaeser EL, Sacerdote B (2000) The social consequences of housing. HIER Harvard Institute of Economic Research, Discussion Paper Number 1915, December 2000. Harvard University, Cambridge. http://papers.ssrn.com/paper.taf?abstract_id=263369

Glennerster H (1995) British social policy since 1945. Blackwell, Oxford and Cambridge
Hall P (1973) The containment of Urban England, Vol 1. Allen and Unwin, London
HMSO (1917) Command paper no. Cd 8731. Her Majesty's Stationery Office
HMSO (1944) Ministry of Health and Ministry of works, housing manual (See also, Dudley Report), published by Her Majesty's Stationery Office
HMSO 1961, Parker Morris Report, Homes for Today and Tomorrow, Department of the Environment, London, Her Majesty's Stationery Office
Holmans AE (1987) Housing policy in Britain: a history. Croom Helm, Beckenham
Housing Statistical Release (2014) 16 Oct 2014, Social housing sales, England, 2012–13, Department of Communities and Local Government. http://www.ons.gov.uk/ons/rel/lmac/characteristics-of-home-workers/2014/sty-home-workers.html. Accessed 18 Feb 2015
Kavanagh D, Morris P (1989) Consensus politics from Attlee to Thatcher, Blackwell Oxford Publishers, Oxford
King P (2010) Housing policy transformed—the right to buy and the desire to own. The Policy Press, University of Bristol
Knight N (2006) Governing Britain since 1945. Politico's Publishing Ltd
Lee P, Murie A (1997) Poverty housing tenure and social exclusion. The Policy Press, Bristol
Lee P, Murie A (1999) Spatial and social divisions within British cities: beyond residualisation. Hous Stud 14(5):625–640
Littig B, Griessler E (2005) Social sustainability: a catchword between political pragmatism and social theory Int J Sustain Dev 8(½):2005
Lowe S (2011) The housing debate. Policy Press, University of Bristol
Madge J (1945) The rehousing of Britain (Target for tomorrow). The Pilot press ltd (1945)
Malpass P, Murie A (1994) Housing policy and practice, 4th edn. Macmillan Press Ltd
McKenzie S (2004) Social sustainability: towards some definitions, working paper series No.27. Hawke Research Institute, University of South Australia, Magill, South Australia
Merrett S (1979) State housing in Britain. Routledge Kegan Paul, London
Murie A (1997) The social rented sector, housing and the welfare state in the UK Hous Stud 12(4):437–462
Norton A, Novy K (1991) Low income housing in Britain and Germany, Edited by Alan Norton and Klaus Novy, A project of the Anglo-German Foundation for the Study of Industrial Society, 1991
Pearce M, Stewart J (1996) British political history, 1867–1995: democracy and decline, 2Rev Ed edn (22 Aug 1996) Routledge
Pearce J, Vine J (2014) Quantifying residualisation: the changing nature of social housing in the UK. J Hous Built Environ 29(4):657–675
Rapoport A (1982) The meaning of the built environment: a non-verbal communication approach. Sage, Beverly Hills
Seldon, A (1991) The conservative party since 1945, Macmillan, London
Sullivan M (1996) The development of the British Welfare State. Harvester Wheatsheaf, Prentice-Hall
Swenarton M (1981) Homes fit for heroes—the politics of architecture on early state in Britain. Heinemann Educational Books
Thornberry E (2015) Labour MP led the debate on the housing crisis in the Parliament on 6th February 2015. www.emilythornberry.com/news/99/22/Housing-Crisis-Debate. Accessed 14 Feb 2015
UNCSD (2012) Report of the United Nations Conference on Sustainable Development, Rio de Janeiro, Brazil, 20-22 June 2012 [accessible from: http://www.uncsd2012.org/content/documents/814UNCSD%20REPORT%20final%20revs.pdf, Accessed: 26.02.2015]
Urban Task Force (1999) Towards an urban renaissance, Department of the environment, transport, local government and the regions, HMSO Government Papers and websites. http://www.parliament.uk/about/living-heritage/transformingsociety/towncountry/towns/overview/councilhousing. Accessed 28 Feb 2015
Whitham D (1982) The first sixty years of council housing, in the future of council housing, ed. John English, Croom Helm

Index

© Springer International Publishing Switzerland 2016
M. Dastbaz et al. (eds.), *Building Sustainable Futures*,
DOI 10.1007/978-3-319-19348-9

MIX
Papier aus verantwortungsvollen Quellen
Paper from responsible sources
FSC® C105338

If you have any concerns about our products,
you can contact us on
ProductSafety@springernature.com

In case Publisher is established outside the EU,
the EU authorized representative is:
Springer Nature Customer Service Center GmbH
Europaplatz 3, 69115 Heidelberg, Germany

Printed by Libri Plureos GmbH
in Hamburg, Germany